AF738043

BIOCHEMISTRY RESEARCH TRENDS

ISOFLAVONES: BIOSYNTHESIS, OCCURRENCE AND HEALTH EFFECTS

BIOCHEMISTRY RESEARCH TRENDS

Additional books in this series can be found on Nova's website
under the Series tab.

Additional E-books in this series can be found on Nova's website
under the E-book tab.

BIOCHEMISTRY RESEARCH TRENDS

ISOFLAVONES: BIOSYNTHESIS, OCCURRENCE AND HEALTH EFFECTS

MELANIE J. THOMPSON
EDITOR

Nova Biomedical Books
New York

Copyright © 2010 by Nova Science Publishers, Inc.

All rights reserved. No part of this book may be reproduced, stored in a retrieval system or transmitted in any form or by any means: electronic, electrostatic, magnetic, tape, mechanical photocopying, recording or otherwise without the written permission of the Publisher.

For permission to use material from this book please contact us:
Telephone 631-231-7269; Fax 631-231-8175
Web Site: http://www.novapublishers.com

NOTICE TO THE READER

The Publisher has taken reasonable care in the preparation of this book, but makes no expressed or implied warranty of any kind and assumes no responsibility for any errors or omissions. No liability is assumed for incidental or consequential damages in connection with or arising out of information contained in this book. The Publisher shall not be liable for any special, consequential, or exemplary damages resulting, in whole or in part, from the readers' use of, or reliance upon, this material.

Independent verification should be sought for any data, advice or recommendations contained in this book. In addition, no responsibility is assumed by the publisher for any injury and/or damage to persons or property arising from any methods, products, instructions, ideas or otherwise contained in this publication.

This publication is designed to provide accurate and authoritative information with regard to the subject matter covered herein. It is sold with the clear understanding that the Publisher is not engaged in rendering legal or any other professional services. If legal or any other expert assistance is required, the services of a competent person should be sought. FROM A DECLARATION OF PARTICIPANTS JOINTLY ADOPTED BY A COMMITTEE OF THE AMERICAN BAR ASSOCIATION AND A COMMITTEE OF PUBLISHERS.

Additional color graphics may be available in the e-book version of this book.

Library of Congress Cataloging-in-Publication Data

Isoflavones : biosynthesis, occurrence, and health effects / editor, Melanie J. Thompson.
p. ; cm.
Includes bibliographical references and index.
ISBN 978-1-61728-113-6 (hardcover)
1. Isoflavones. I. Thompson, Melanie J.
[DNLM: 1. Isoflavones--biosynthesis. 2. Isoflavones--therapeutic use.
QU 220 I85 2010]
QP671.F52I86 2010
615'.321--dc22

2010015609

Published by Nova Science Publishers, Inc. † New York

Contents

Preface

Isoflavonoids are a valuable group of natural products and are primarily synthesized in leguminous plants, where they play important roles in plant defense. They also have significant health benefits for animals and humans and have been ascribed various estrogenic, antiangiogenic, antioxidant, and anticancer activities, leading to their popularity as dietary supplements. This book gathers and presents topical research data on isoflavones including bioactive isoflavones and their implications in cancer treatment and prevention, soy protein and isoflavones in the prevention of cardiovascular disease and the processing techniques on isoflavone profiles.

Chapter I - Isoflavones are naturally occurring molecules present in significant amount in many leguminous, some of which are edible plants. Isoflavones exhibit similarities in their chemical structure with steroidal estrogens. This allows them to trigger some of the endogenous physiocellular signalization processes of estrogens that are known to occur naturally within living organisms. In fact, isoflavones mediate numerous health effects mainly related to their estrogenic and/or anti-estrogenic actions. Some of these effects are estrogen receptor-mediated physiological phenomenona such as those linked to the hypothalamo-hypophyso-gonadal axis. Moreover, because the expression of estrogen receptors is highly enriched in neural tissues, it was not surprising to found out that isoflavones were potent neurotrophic and neuroprotective compounds as it is the case for estrogens. Functional estrogen receptor proteins are also expressed in some central nervous system territories, including the hippocampus and the cerebral cortex, that are well thought to be devoted to higher cognitive brain functions such as the memory function. More particularly, the hippocampus expresses both nuclear and membrane estrogen receptors known to mediate local physiocellular alterations of the neuronal phenotype, namely dendritic spine development, synaptic plasticity and modulation of the Ca^{++} trafficking rate between different compartments of the cell. These physiocellular effects are thought to be part of the adaptative chemical and architectural changes sustaining some processes of the mnemonic information storage. This review provides *in vitro*, animal, and clinical data intending to highlight the impact of isoflavones on the physiocellular features of the hippocampal formation plasticity and, therefore, on the hippocampus-dependent cognitive brain function. In summary, isoflavones not always exhibit all the mechanistic effects of estrogens but do, however, display some own specific mechanisms of signalization. Nevertheless, the ongoing efforts for the search of neuroprotective selective estrogen receptor-modulating molecules that do not trigger the adverse side effects of estrogens still deserve continuous investigation. More specifically, the putative neuronal survival-promoting and promnesic features of isoflavones, need further exploration to get full acceptance as beneficial natural substances against age-triggered alterations of general health including hippocampal-dependent memory decline.

Chapter II - The use of functional foods in the diet is one of the latest trends in a strategy of intervention to reduce risk factors for the development of cardiovascular diseases. Soybeans and derived foods has attracted much attention of health professionals since its consumption is associated with a reduction of cholesterol levels in blood and other risk factors. In spite of the mechanisms and components responsible for the potential effect of soy foods for the prevention of cardiovascular diseases are not yet completely identified, the most recent studies suggest that several factors are inter-related, such as the functional composition of soybeans, the synergism between components in food, individual metabolism and genetics as well as other risk factors

Chapter III - Isoflavones, together with lignans, and coumestans represent the three classes of phytoestrogens, compounds that have an estrogen-like and antioxidant activity and are structurally similar to estradiol. Isoflavones are widespread in a variety of leguminous plants such as red clover, kudzu root and soy and are subjected to increasing attention due the possibility of use as potential therapeutic agents. In fact, several studies suggest that consumption of isoflavone-containing foods is associated with the lowest level of cholesterol in blood, and prevention of cardiovascular disease, prostate, colon and breast cancers, osteoporosis, and menopausal symptoms in people whose soy-rich diet allow to increase of 100 times the concentration of endogenous estrogens. Isoflavones are associated with the protein fraction of soybeans and are present under several chemical forms: the β-glucosides genistin, glycitin, daidzin, their 6”-*O*-malonyl-glucosides and a 6”-*O*-acetyl-glucosides and the aglycone forms genistein, daidzein and glycitein. Processing greatly affect the content of total isoflavones and the concentrations of each of their chemical forms. Among soy foods, the highest concentration of isoflavones is detectable in the whole soybeans whereas soy milk, tofu, tempeh, miso, fermented bean curd, bean paste and the so called “second-generation soyfoods” (soy hot dog, tempeh burger, soy noodles) show isoflavone contents ranging from 2 to 46% with respect to the whole soybeans. These reductions are mainly due to dilution of soy protein with other ingredients and heat treatments that also result in changes of the isoflavone profiles. In fact, the predominant isoflavones in soybeans and non-fermented soy products (soy protein, soymilk derivates, etc.) are malonyl glucosides, whereas fermented soy foods (miso, tempeh) contain mainly aglycones. These changes affect the metabolic activity of isoflavone-containing foods since the glycoside forms are biologically inactive and need to be cleaved by the β-glucosidases from intestine in the glucosyl residue and the biologically active aglycones.

This review gives an insight of the effects of processing and processing parameters (extraction, fermentation, pasta-making, addition of coagulants, addition of protease, temperature, pH) on isoflavone content, profile and antioxidant activity of soy foods.

Chapter IV - Edible plants provide the human with hundreds of non-nutritional phytochemicals which are recognized as beneficial, such as isoflavones. Likewise other polyphenols, isoflavones may undergo extensive transformations during passage through human digestive tract, especially in the colon, where members of the complex commensal microbiota are capable to carry out synergistically a broad range of metabolic transformations affecting the fate and the biological activity of phytochemicals. Diverse bacterial species occurring in the large intestine hydrolyze the glucose conjugated forms of isoflavones, releasing the corresponding aglycones, which may undergo further microbial conversions (especially reductions) giving rise to a wide spectrum of isoflavone-derived compounds. The end-products of microbial transformations are subjected to substantial person-to-person variation, reflecting the impact of the colonic microbiota, since intestinal bacteria may greatly increase or compromise the biological activity of dietary isoflavones. In fact, certain reduction products (e.g. S-equol) possess superior antioxidant and estrogen-like activities, while others are precursors for C-ring cleavage and lead to isoflavone degradation. Extensive research has been performed to characterize the intestinal bacteria which are responsible for isoflavones transformations. In particular, many efforts are being carried out to identify single bacteria that convert the major soy isoflavone daidzein into S-equol, in the perspective to enrich soy products with such a valuable isoflavone-derived metabolite. Besides, diverse bacteria which are capable of deglycosylation, demethylation, and various reduction reactions are increasingly being identified. Nonetheless, only bifidobacteria and lactobacilli, which produce β-glucosidase that hydrolyze isoflavone glycosides into the corresponding aglycones, are currently exploited to improve the biological activity of soymilk.

Chapter V - Flavonoids are bioactive agents present in several plants that are used for food. These foods include legumes, soy, fruits, vegetables and different parts of plants. Vast amount of literature exits implicating the role of flavonoids in influencing changes in biological processes in cancer cells. The ability of flavonoids to induce changes in receptor activation, anti-estrogenic properties, tyrosine phosphorylation and cell signaling makes these agents important in cancer prevention and treatment and development of anticancer drugs. The antioxidant properties of flavonoids have also been extensively investigated. The above-mentioned cellular mechanisms are known to play critical roles in carcinogenesis, including metastasis and angiogenesis. Several isoflavones are in clinical trials and there has been epidemiological evidence for their role in cancer prevention and in decreasing incidence of several cancer types. Development of derivatives of isoflavones and associating these natural compounds with nanomaterials to give better efficacy and pharmacokinetic properties is some of the future areas of research which will have impact on incidence and treatment of an array of cancer types. This review will summarize some of the critical findings in the field and discuss the future directions in use of isoflavones in treatment and prevention of several selected cancer types.

Over half a million people in US die of cancer every year. Lung cancer is the major contributor of deaths due to smoking. The progression of cancer is a multi-step process and can be interrupted at several steps. The impact of consumption of isoflavones has been a topic of research and discussion for decades. To understand how isoflavones may block progression of cancer, it is necessary to discuss the process of carcinogenesis.

Chapter VI - Flavonoids are bioactive agents present in several plants that are used for food. These foods include legumes, soy, fruits, vegetables and different parts of plants. Vast amount of literature exits implicating the role of flavonoids in influencing changes in biological processes in cancer cells. The ability of flavonoids to induce changes in receptor activation, anti-estrogenic properties, tyrosine phosphorylation and cell signaling makes these agents important in cancer prevention and treatment and development of anticancer drugs. The antioxidant properties of flavonoids have also been extensively investigated. The above-mentioned cellular mechanisms are known to play critical roles in carcinogenesis, including metastasis and angiogenesis. Several isoflavones are in clinical trials and there has been epidemiological evidence for their role in cancer prevention and in decreasing incidence of several cancer types. Development of derivatives of isoflavones and associating these natural compounds with nanomaterials to give better efficacy and pharmacokinetic properties is some of the future areas of research which will have impact on incidence and treatment of an array of cancer types. This review will summarize some of the critical findings in the field and discuss the future directions in use of isoflavones in treatment and prevention of several selected cancer types.

Over half a million people in US die of cancer every year. Lung cancer is the major contributor of deaths due to smoking. The progression of cancer is a multi-step process and can be interrupted at several steps. The impact of consumption of isoflavones has been a topic of research and discussion for decades. To understand how isoflavones may block progression of cancer, it is necessary to discuss the process of carcinogenesis.

Chapter VII - *Dioscorea* spp. (yam) is a rich source of carbohydrate crop in the diet and also resource of different phytohormones. Among *Dioscorea* spp, especially, *Dioscorea bulbifera* of aerial-potato has widely been used as their traditional folklore medicines for their treatments such as thyroid disease, tumor, diretic disturbance and inflamation in the tropical and subtropical areas. The diverse compounds in *Dioscorea bulbifera* have been found and their functionality has been examined in various researches such as their antibacterial and antitumor activity. The purpose of this review paper is to focus and describe their compositions found and functional activities.

Chapter VIII - There is evidence that soy consumption benefits health by contributing to the prevention of chronic diseases due several factors, including isoflavones present in this legume. The isoflavones from soybean are known to have several biological properties such as estrogenic, antimicrobial and antitumoral (breast and prostate) activities, besides inhibiting the activity of enzymes linked to the cellular division. These properties are shown to be more accentuated in the aglycone than glycoside forms. Their presence in soy products have been reported with a number of beneficial health effects. In addition, both USA and Brazil are great soybean producer in the World. The analysis of isoflavone form contents into the soy products take an important role in the understanding of consumer of these products. In this way, with the germination process, the soybean has a positive effect in improving nutritional status including isoflavones. However, the isoflavone content varies with the period of germination. The process of germination showed no significant difference in amount of glycitin, genistein, malonyl daidzin, malonyl genistin, daidzein, glycitein and genistein, although theirs numerical reductions after germination. There were only differences in the amount of daidzein and malonyl glycitin between the samples. Some authors have been reported that the total concentration of isoflavones increased after 1 day of germination and then slowly decrease these concentrations. This increase and decrease observed in the content of isoflavones may be associated with the conversion of isoflavones to other flavonoids. The germination time of 48 hours not improved the amount of isoflavones. Different cultivars present several amount of isoflavones and forms of isoflavones. When the grains were cooked to be soft the malonyl form decreased, and aglycone and glycoside forms increased. It was observed that the heating treatment transformed the malonylglucosides into glucoside isoflavones. After heat treatment at 121°C for 30 min, nearly all malonyl isoflavones were converted into glucoside, but acetylisoflavones were not detected via RP-HPLC analysis. On the other hand, gamma radiation also promoted some reduction on isoflavone profiles, but this effect depends of humidity into soy products. Irradiation from ^{60}Co interfere at amount of isoflavones forms at doses up 2 kGy, glycosides and malonyl forms have tendency to down, but aglycones increase with radiation. For example, when soybean flour was studied, the gamma-radiation had a little effect (loss of 9.9%, w/w) in total isoflavone. However, soy isoflavones are likely to have the beneficial effects when found in aglycone forms in many processes will be proposed to obtain those aglycone isoflavones.

Chapter IX - Isoflavonoids play important roles in plant defense, exhibit various health-promoting activities, and have attracted great interest from the pharmaceutical and nutraceutical industries as drug targets and dietary supplements. The isoflavonoids are identified primarily in leguminous plants. They are mainly synthesized through the central phenylpropanoid pathway and the specific isoflavonoid branch pathway. Many key enzymes in the central pathway have been elucidated structurally, providing essential insights into early stages of isoflavonoid biosynthesis. Significant progress has been made on structural studies of enzymes in the isoflavonoid branch pathway. Structures of two isoflavonoid-specific NADPH-dependent reductases revealed how enzymes specifically recognize isoflavonoids and catalyze stereo-specific reductions. Several isoflavonoid methyltransferases and glycosyltransferases were also characterized structurally. Bioactivities and functions of isoflavonoids are largely dictated by their methylation and glycosylation patterns. The structural analysis of these enzymes provides a basis for understanding the complex biosynthetic process and the selective substrate- and regio-specificities. Structure-based enzyme engineering is becoming a new strategy for synthesis of bioactive isoflavonoids.

Chapter X - Genistein and Biochanin (A) isoflavones are shown to exist in two different conformations which are the 90-degrees completely twisted geometry and the 50 degrees less twisted one, with respect to the C_3—$C_{1'}$ single bond. Specific interactions with the solvent cage as well as self association processes seem shifting the isoflavone from the perpendicular conformation towards the less twisted one. From the UV-visible photophysics investigations it's revealed, in both isoflavones, that monomeric species can't exist alone even at very low solute concentration (~10^{-6}M), the self association process occurs already in this concentration range. The theoretical simulation, using analytical atom-atom pair potential, predicts a self dimer in a slipped non sandwich, face to river, perpendicular structure.

Chapter XI - Soy isoflavones have been identified as dietary components having an important role in reducing the incidence of breast and prostate cancers in Asian countries. The isoflavone polyphenol genistein in soybean is considered to be a potent chemopreventive agent against cancer. Several studies, both *in vitro* and *in vivo*, have clearly shown that genistein is a promising therapeutic candidate for cancer chemoprevention as it is able to cause inhibition of human cancer growth via pathways that lead to apoptotic cell death. Genistein has pleiotropic molecular mechanisms of action affecting estrogen mediated signaling pathways, cell cycle regulation, apoptosis and cell growth signal cascades. Of particular interest is the observation that genistein has been found to induce apoptosis in cancer cell lines but not in normal cells. We have proposed a novel mechanism for the cytotoxic action of plant polyphenols including genistein, which has implications for its chemopreventive properties against cancer. We believe that such a mechanism better explains the preferential cytotoxicity of genistein towards cancer cells. In this review we discuss various mechanistic approaches that have been proposed to account for the chemopreventive properties of genistein.

Chapter XII - Isoflavones are a class of plant secondary metabolites, related to flavonoids which are most commonly known for their antioxidant activity. Flavonoids are a type of polyphenol [Arranz *et al.*, 2009] derived from the shikimate pathway, responsible for plant defense and flower color. Plants use isoflavones and their derivatives as phytoalexin compounds to ward off disease-causing pathogenic fungi and other microbes. Isoflavones are produced via a branch of the general phenylpropanoid pathway that produces flavonoid compounds in higher plants.

In: Isoflavones Biosynthesis, Occurence and Health Effects ISBN: 978-1-61728-113-6
Editor: Melanie J. Thompson, pp. 1-81 © 2010 Nova Science Publishers, Inc.

Chapter I

Isoflavones: a Race after the Rescue of the Ageing Hippocampus

Khalid Jamali[1,2], Aline Marighetto[3] and Catherine Bennetau-Pelissero[1,2*]

[1]Université de Bordeaux, UMRS, ENITA de Bordeaux, 33175 Gradignan France
[2]Université de Bordeaux, Inserm U862, Neurocentre Magendie, Université Bordeaux 2, 33077 Bordeaux, France
[3]Université de Bordeaux, CNIC, UMR 5228 CNRS, Université Bordeaux 1, 33405 Talence France

Abstract

Isoflavones are naturally occurring molecules present in significant amount in many leguminous, some of which are edible plants. Isoflavones exhibit similarities in their chemical structure with steroidal estrogens. This allows them to trigger some of the endogenous physiocellular signalization processes of estrogens that are known to occur naturally within living organisms. In fact, isoflavones mediate numerous health effects mainly related to their estrogenic and/or anti-estrogenic actions. Some of these effects are estrogen receptor-mediated physiological phenomenona such as those linked to the hypothalamo-hypophyso-gonadal axis. Moreover, because the expression of estrogen receptors is highly enriched in neural tissues, it was not surprising to found out that isoflavones were potent neurotrophic and neuroprotective compounds as it is the case for estrogens. Functional estrogen receptor proteins are also expressed in some central nervous system territories, including the hippocampus and the cerebral cortex, that are well thought to be devoted to higher cognitive brain functions such as the memory function. More particularly, the hippocampus expresses both nuclear and membrane estrogen receptors known to mediate local physiocellular alterations of the neuronal

* Corresponding author: Unité Micronutriments Reproduction Santé, ENITA de Bordeaux, 1, cours du Général de Gaulle, 33 175 Gradignan FRANCE, Tel : 33+(0)5 57 35 07 52, Fax : 33+(0)5 57 35 07 39, e-mail : c-bennetau@enitab.fr

phenotype, namely dendritic spine development, synaptic plasticity and modulation of the Ca^{++} trafficking rate between different compartments of the cell. These physiocellular effects are thought to be part of the adaptative chemical and architectural changes sustaining some processes of the mnemonic information storage. This review provides *in vitro*, animal, and clinical data intending to highlight the impact of isoflavones on the physiocellular features of the hippocampal formation plasticity and, therefore, on the hippocampus-dependent cognitive brain function. In summary, isoflavones not always exhibit all the mechanistic effects of estrogens but do, however, display some own specific mechanisms of signalization. Nevertheless, the ongoing efforts for the search of neuroprotective selective estrogen receptor-modulating molecules that do not trigger the adverse side effects of estrogens still deserve continuous investigation. More specifically, the putative neuronal survival-promoting and promnesic features of isoflavones, need further exploration to get full acceptance as beneficial natural substances against age-triggered alterations of general health including hippocampal-dependent memory decline.

Keywords: Estrogens, isoflavones, hippocampus, memory.

I. Introduction

Estrogens belong to a family of signalling molecules, called sex steroid hormones that are endogenously produced in animals. Multiple sites of production of endogenous estrogens have been identified in mammals, including the gonads (Mendelson *et al.* 2005), the adrenal glands (Peltier *et al.* 1992), the adipose tissue (Thijssen *et al.* 1993) and the central nervous system (Garcia-Segura, 2008), to name a few. It is well established that estradiol (E_2) derives from cholesterol through several metabolic steps where the rate-limiting, ultimate step is represented by the enzymatic aromatization of testosterone to E_2 (Ohlsson *et al.* 2009). This aromatization occurs in all the tissues mentioned above, and more specifically, in both cerebral neuronal cells and glial cells (Garcia-Segura 2008). In the mammalian females, the ovaries represent the major site of synthesis and endocrine release of estrogens. Through the blood circulation, the ovaries are capable of supplying the rest of the organism with estrogens, during most of life ages. Within the family of estrogens, E_2 represents one of the most biologically active molecules. Besides, phytoestrogens are plant derived molecules (Albertazzi and Purdie 2008, Cornwell *et al.* 2004, Fitzpatrick 2003) with structural characteristics resembling those of E_2. They exhibit two hydroxyl groups on the opposite sides of the molecule at a distance of 10 Å forming an angle similar to the one that exists between the two hydroxyl groups of E_2 (Katzenellenbogen *et al.* 1980). Five major classes of phytoestrogens have been identified: isoflavonoids, flavonoids, stilbenes, lignans and coumestans. Isoflavones are a subclass of isoflavonoids that includes genistein, daidzein, glycitein, biochanine-A and formononetin (Figure 1). Isoflavones are present in significant amount in some food plants such as soy, kudzu or clover (Setchell 2001). Under their natural form, unprocessed isoflavones occur mainly as glucoside (Figure 1), malonyl or acetyl molecules (for example, in soybean hypocotyls) (Anderson & Wolf, 1995) that are chemically bound to a protein matrix. When ingested, these isoflavone-glycosides are subjected to enzymatic deconjugation in the gut tract. This leads to the release of the aglycone moiety being the active molecule. Other enzymatic transformations end up with the production of even more powerful metabolites. It is the case, for example, for equol (Figure 1), which can arise from formononetin or daidzein as a result of the action of gut tract-harboured bacterial flora (Setchell *et al.* 2002). Further enzymatic transformations are intended to facilitate the transport and/or the excretion of assimilated isoflavones. They take place in the enterocytes, hepatocytes or kidney cells and lead to the formation of glucuronids and sulfate conjugates (Rowland *et al.* 2003).

II. Dietary Isoflavones

Due to their structural resemblance to the active steroid, E_2 (Figure 1), which allows them to display a relatively high affinity for E_2 receptors (ERs) (Belcher and Zsarnovsky 2001, Kuiper *et al.* 1997, 1998), isoflavones are able to trigger directly some of the endogenous physiocellular signalization processes of estrogens that are known to occur naturally *in vivo*. Indeed, a large set of data did establish that isoflavones were capable of influencing welfare and of relieving some age-triggered health alterations in humans, such as menopausal symptoms which include mild memory dysfunctions, advancing age-related mood state dysregulations (depression), bone metabolic defects (osteoporosis) and vasomotor-originating thermogenic discomforts (hot flashes) (Setchell and Cassidy 1999, Cornwell *et al.* 2004, Ferrari 2004, Kreijkamp-Kaspers *et al.* 2004, Geller and Studee 2006, Zhao[2] and Brinton 2007, Hooper *et al.* 2009). In this context, isoflavones were also shown to affect numerous physiological parameters in animal models (Clarkson *et al.* 1995, Lephart *et al.* 2004, Usui *et al.* 2006, Xiao 2008) some of which concern neurochemical and hormonal interactions within the hypothalamo-pituitary-gonadal axis (Lephart *et al.* 2005, Setchell and Cassidy 1999). Empirically, the estrogenic impact of isoflavones came to light as adverse anti-reproductive and pro-abortive effects in sheep grazing on red or subterranean clover pastures (Bennetts *et al.* 1946, Braden et al., 1967). Subsequently, the resulting studies have established that the incriminated components were isoflavones from clover, also present in soy and kudzu, and that these molecules were able to distort the normal neurochemical/hormonal signalization within the hypothalamo-hypophyso-gonadal axis (Findlay *et al.* 1973). These occurrences were made possible, most likely, because isoflavones, to some extent, are able to replace faded E_2 within the molecular interactions governing estrogen signalization (Kuiper *et al.* 1997, 1998). Isoflavones are also capable of dislodging existing E_2 from the signalizing molecular tangle (Thompson *et al.* 1984; Kuiper *et al.* 1997, 1998) and, thereof, functionally sliding into specific intracellular signalization pathways taking place in various cell/tissue/organ systems (Watson *et al.* 2007). Isoflavone-triggered, E_2-mimicking signalization does also occur in the higher central nervous system (Belcher and Zsarnovsky 2001). Another interesting aspect of isoflavones is reflected by their ability to modulate the endogenous estrogen metabolism itself. Although all the used doses may not fall into a physiological concentration range, in cell culture models, isoflavone molecules were shown to increase the levels of enzymatic activity and/or of mRNA expression of the aromatase P450-*CYP19* (Edmunds *et al.* 2005, Ye *et al.* 2009) through the activation of the intracellular PKCα/P38/ERK1/2/CREB signalling pathways (Ye *et al.* 2009). In contrast, isoflavones were also shown to downregulate aromatase P450 and/or 3β-hydroxysteroid dehydrogenase (HSD) activities in other cell culture systems (Pelissero *et al.* 1996, Blomquist *et al.* 2005, Rice *et al.* 2006, Wang[1] *et al.* 2008). Besides all their powerful signalization properties (see below), this special characteristic of isoflavones to have leverage with the steroidogenic enzyme activities in peripheral tissues might also be of interest in the putative neurosteroid-metabolizing enzyme-related dysfunctions during normal and/or physiopathological (neurodegenerative) aging of the hippocampal formation in humain subjects.

Following the ingestion of a soy-derived product, the main isoflavones found in the circulating plasma are genistein and daidzein (Lephart *et al.* 2004, Bennetau-Pelissero *et al.* 2003, Vergne *et al.* 2007). Equol has also been detected as circulating isoflavane in equine (Marrian and Haslewood, 1932), rodent (Lephart *et al.* 2004, Mathey *et al.* 2007) and ruminant (Woclawek-Potocka *et al.* 2008) species. In some human subjects (approximately, between 40 to 60% of people consuming soy products, so referred to as equol producers) equol has been detected in plasma and urine (Bennetau-Pelissero *et al.* 2003, Mathey et al., 2006). Because of their lipophilic structure, circulating isoflavones in their aglycone form can cross freely all biological membranes including the blood-brain barrier. As a consequence, these molecules are able to reach all the physiological compartments and organs within the body, including the central nervous system, endocrine glands, cardiovascular system and squeleton (Fitzpatrick 2003, Wuttke *et al.* 2003, Cornwell *et al.* 2004, Ferrari 2004, Lee *et al.* 2005, Lephart *et al.* 2005, Usui 2006, Moutsatsou, 2007, Watson *et al.* 2007, Xiao 2008, Carlson *et al.* 2008). These compartments and organs are vital sites for fine-tuned regulations of bodily homeostasis, of which regulations are deeply impacted by an intense signalization originating from imported (McEwen 2002, McEwen and Milner 2007, Spencer *et al.* 2008) and/or locally (Kretz *et al.* 2004, Rune and Fortscher 2005, Agis-Balboa *et al.* 2006, Ish *et al.* 2007, Hojo *et al.* 2008) synthesized E_2. This occurrence was made possible because the upper mentioned compartments and organs are highly enriched in sites of specific and active interaction with E_2. Within the large class of E_2 effect-mediating molecules, new members of the three families of ERs (Maggiolini and Picard, 2010), ER-interacting proteins [PELP1, a praline-, glutamic acid-, and leucine-rich protein-1, (Brann *et al.* 2008)] and non-ER/E_2-binding proteins (Valverde *et al.* 1999, Curtis *et al.* 2002, Sarkar *et al.* 2008) are continuously still coming into light.

III. Estrogen Receptors

After decades of active research, three subtypes of ERs, arising from 3 different genes, have been identified: estrogen receptor α (ERα), estrogen receptor β (ERβ) (Kuiper *et al.* 1997,1998) and G protein-coupled estrogen-receptor (GPER, also designated as GPR30) (Maggiolini *et al.* 2004, Prossnitz *et al.* 2007, Maggiolini and Picard, 2010). In humans, the ERα gene structure has been revealed as a deoxyribonucleic acid (DNA) sequence containing 8 exons interrupted by 7 long introns (Zhao *et al.* 2005). The resulting protein (human ERα, hERα) displays 5 functional domains referred to as A/B, C, D, E and F (Zhao *et al.* 2005). The human ERβ (hERβ) gene and protein structures have been found to be constructed in a similar manner to those of the hERα (Zhao *et al.* 2005). However, the hERα protein (67kD) was shown to contain longer amino-acid sequence than that of the hERβ (59kD), although various splice variant proteins could be detected in different cell systems including the brain cells (Zhao *et al.* 2005).

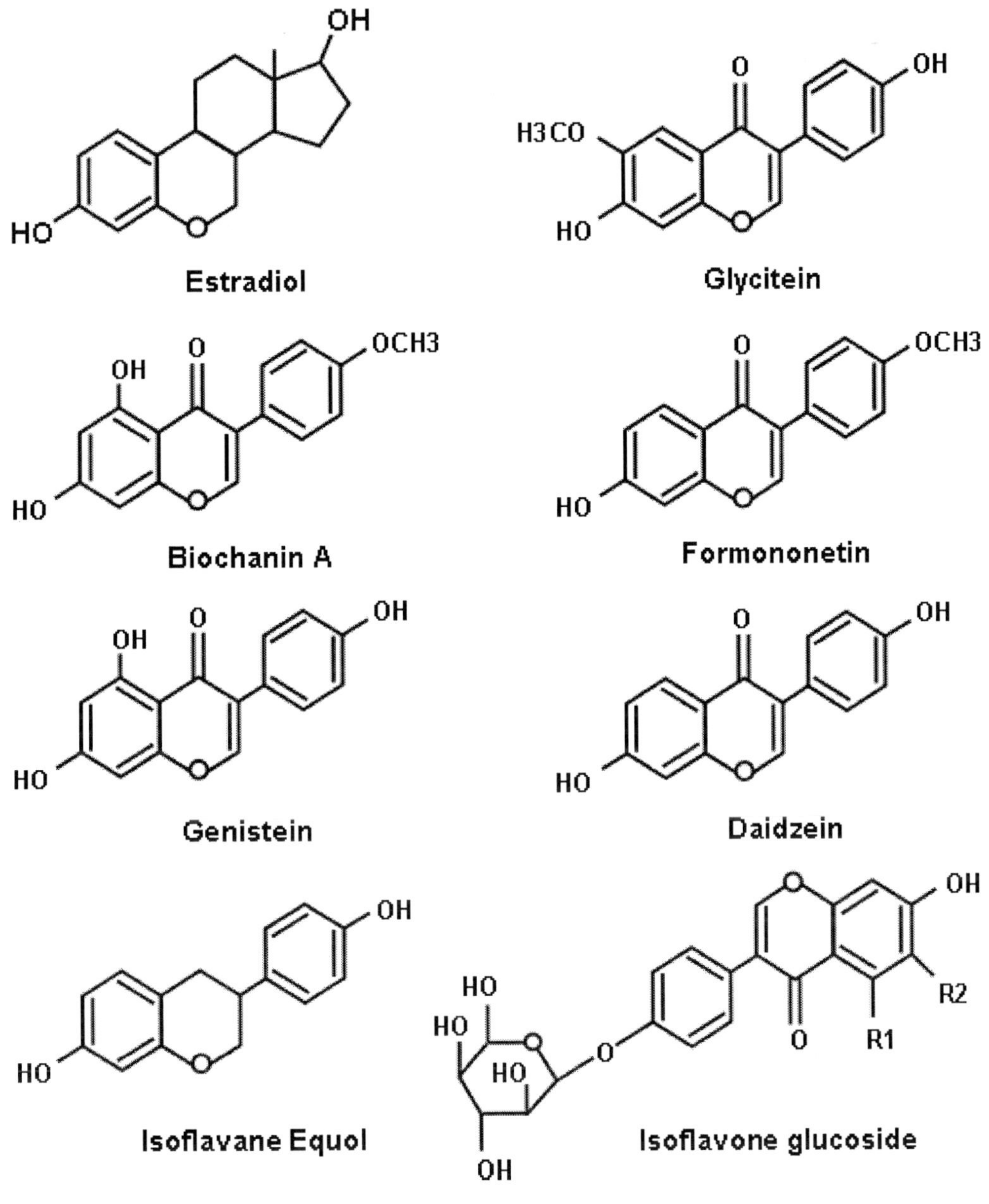

Figure 1. Molecular structures of estradiol, isoflavone, isoflavane and isoflavone-glucoside species. Notice the close resemblance of isoflavones (glycitein, biochanin A, formononetin, genistein, daidzein) and the isoflavone metabolite, isoflavane, equol to estradiol. Dietary isoflavones occur in vegetables mainly as glucoside molecules: daidzin ($R_1 = R_2 = H$), genistin ($R_1 = OH$; $R_2 = H$) and glycitin ($R_1 = H$; $R_2 = OCH_3$).

III.1. Molecular Features of Estrogen Receptors

Both of ERα and ERβ are members of the superfamily of steroid receptors that are also specifically liganded by (1) other sex steroid-hormones, such as androgen (for androgen receptors) and progesterone (for progesterone receptors) or (2) adrenocorticoid hormones such aldosterone (for mineralocorticoid receptors) and cortisol/corticosterone (for glucocortcoid receptors). In order to get full transcriptional activity, ERs have to dimerize either as homo-dimers or as hetero-dimers. Dimerized ERs are able to bind, directly, a unique and specific DNA consensus sequence called ERE for estrogen responsive element and, then, to modulate the transcriptional rate of down-stream genes (Bennetau-Pelissero *et al.* 2000, Rune *et al.* 2002, Gruber *et al.* 2004, Schreihofer 2005). E_2 bound ERs can also regulate gene expression indirectly, through the estrogenic tethering pathway, by interacting with DNA-bound transcription factors, such as specificity protein-1 (SP-1) or activator protein-1 (AP-1) (Gruber *et al.* 2004). The mechanism of ER-mediated modulation of gene expression is referred to as E_2-elicited genomic effects (or nuclear-initiated steroid signaling) that have been widely documented in various cell systems. Because it requires mRNA expression, subsequent protein synthesis and post-translational maturation of proteins, this mechanism of action of E_2 has been labelled as "slow" or "long-term" cell response. Recently ERα and ERβ were shown to be present also in cell membranes mediating rapid effects (Mhyre and Dorsa 2006) through MAPK-dependent pathways (Zao and Brinton 2007).

The G protein-coupled estrogen-receptor (GPR30) is member of the 7 transmembrane spanning segments, ligand-activated, plasma membrane-bound receptors that are capable of interacting with a plasma membrane-integrated G-protein. This is the reason why these molecules are referred to as "G" protein-coupled receptors, GPCRs. When liganded by an extra-cellular signalling molecule, GPCRs change their tri-dimensional conformation to trans-activate a guanosine-di-phophate (GDP)-bound, multi-subunit ($\alpha\beta\gamma$) protein, leading to the exchange of GDP against a guanosine-tri-phosphate (GTP) and, therefore, to the individualization of α_{GTP} complex. The latter is able to activate an adenylyl cyclase which hydrolyses adenosine-tri-phosphate (ATP) to cyclic adenosine-mono-phosphate (cAMP) and adenosine-di-phosphate (ADP). Consecutively, the abrupt intracellular rise in the cAMP concentration levels, acts as a transduction signal for the interaction between the upper-mentioned extra-cellular signalling molecule and the ligand-activated GPCR. The duration of that interaction, does not exceed few seconds-to-few minutes time. As opposed to the classical ER-mediated transduction of the E_2 signal, GPCR-mediated transduction is referred to as "rapid" or "fast" cell response (Belcher and Zsarnovsky 2001, Watson *et al.* 2007, Levin 2009).

III.2. Brain Distribution of Estrogen Receptors

In mammals, the brain is the conductor of most vital bodily functions that include metabolic and reproductive neuroendocrine regulations, as well as motor and cognitive behaviours. These brain functions are deeply impacted by E_2. Thus, documenting the neuroanatomical localization of ERs, precisely, is the first requisite to a broader deciphering of E_2 signalization throughout the brain. Therefore, the cerebral distribution of ERs has been documented using 6 major ensembles of morphological and molecular techniques:

(1) Protein-binding probing. For example, radio-labelled agonist or antagonist,
(2) Immunological detection of tissue protein antigens. For example, immunohistochemistry,
(3) Immunological detection of electrophoretically-separated extracted protein antigens. For example, western blotting,

(4) Complementary hybridization of tissular messenger ribonucleic acid (mRNA). For example, *in situ* hybridization,
(5) Complementary hybridization of extracted mRNAs and subsequent electrophoretic separation of specific nucleotides. For example, ribonuclease protection assay,
(6) Enzymatic *in vitro* amplification of extracted mRNAs. For example, reverse transcription-coupled polymerase chain reaction (RT/PCR) with qualitative, semi-quantitative, quantitative or real-time measurement variants.

It is noteworthy, that some alternative procedures could consist of a mixed protocol between 2 or more techniques from the aforementioned list of technologies. Nonetheless, the majority of resulting findings, but not all of them, have agreed to draw, somehow, a more or less coherent and unified picture between the relative patterns of distribution of ER-binding sites, ER mRNAs and ER proteins. In the same context, the physical isolation of living cells arising from specific brain territories, along with the use of pharmacological tools, allowed neuroscientists around the world to gain invaluable insights into the physiological signature of E_2-mediated signalization.

III.2.1. Estrogen receptor α

By mean of immunohistochemistry, the distribution of ERα protein was addressed in the mammalian brain. The occurrence of ERα-immunoreactivity (ERα-IR) was shown to be clustered in brain areas such as the cerebral, prefrontal and piriform cortices, as well as the hippocampus, bed nucleus of the stria terminalis, medial nucleus of the preoptic area, medial nucleus of the amygdala, arcuate nucleus/periventricular zone of the hypothalamus, ventromedial hypothalamic nucleus, periacqueductal gray area, dorsal raphe nuclei, pons and locus coeruleus (Weiland *et al.* 1997; Rune *et al.* 2002; Mitra *et al.* 2003; Rune and Frotscher 2005; Kalita *et al.* 2005; Mehra *et al.* 2005; Ishunina *et al.* 2007; Montague *et al.* 2008; VanderHorst *et al.* 2005; Hazell *et al.* 2009, VanderHorst *et al.* 2009). Within the hippocampal formation, the most heavily stained subdivisions were the cornu ammonis (CA) field 1-4 (CA1-4), dentate gyrus (DG) hilus and subiculum. The cerebral ERα immuno-staining was mainly visible in the nuclear compartment. In the hippocampus, ERα-IR was localized in cytoplasm, neuronal processes and nuclei as shown by both ERα-immunohistochemistry (Milner *et al.* 2001, Rune *et al.* 2002, Rune and Frotscher 2005) and ERα-immunoblotting following cell fractionation (Kalita *et al.* 2005).

In the hippocampal formation, the investigation of the mRNA expression confirmed the presence of ERα mRNA transcripts as shown by ribonuclease protection assay (Kalita *et al.* 2005), by *in situ* hybridization (Rune *et al.* 2002) and by RT/PCR (Ishunina *et al.* 2007, Schreihofer and Redmond 2009).

III.2.2. Estrogen receptor β

Immunohistochemical studies have documented the distribution of ERβ throughout the mammalian brain. The expression of ERβ-immunoreactivity (ERβ-IR) was shown to be less strong, but more extended throughout the brain, than the ERα-IR. Indeed, ERβ-IR was detectable in the cerebral cortex, piriform cortex, medial septum, vertical limb of diagonal band, bed nucleus of the stria terminalis, medial nucleus of the preoptic area, medial nucleus of the amygdala, arcuate nucleus/paraventricular nucleus/periventricular zone of the hypothalamus, supraoptic nucleus, ventromedial hypothalamic nucleus, hippocampus, reticular substantia nigra, ventral tegmental area, periacqudectal gray area, dorsal raphe nuclei, superior olive nucleus, locus coereleus, spinal trigeminal nucleus and cerebellar Purkinje cells (Milner *et al.* 2005, Nomura *et al.* 2005, VanderHorst *et al.* 2005, Hazell *et al.* 2009). More specifically, within the hippocampal formation, the most evidently stained subdivisions were the CA1-3, DG, subiculum and entorhinal cortex (Rune *et al.* 2002, Mitra *et al.* 2003, Kalita *et al.* 2005, Mehra *et al.* 2005, Milner *et al.* 2005, Rune and Frotscher 2005, Chung *et al.* 2007, Hazell *et al.* 2009). In addition, ultrastructural investigation of the ERβ immuno-staining revealed the presence of ERβ-IR in the plasma membrane, nuclear and cytoplasmic compartments, as well as in neuronal processes including dendrites and axon terminals (Milner *et al.* 2005). These findings were confirmed by ERβ-immunoblotting following cell fractionation (Kalita *et al.* 2005).

In the hippocampal formation, the investigation of the mRNA expression confirmed the presence of ERβ mRNA transcripts as shown by ribonuclease protection assay (Kalita *et al.* 2005), by *in situ* hybridization (Rune *et al.* 2002) and by RT-PCR (Kuiper *et al.* 1997, Chung *et al.* 2007, Schreihofer and Redmond 2009).

III.2.3. G protein-coupled estrogen-receptor (GPR30):

The presence of GPR30-immunoreactivity (GPR30-IR) was demonstrated within all the major neuro-anatomical subdivisions of the mammalian brain including the piriform cortex, medial nucleus of the preoptic area, paraventricular nucleus, supraoptic nucleus, arcuate nucleus/peri-ventricular zone of the hypothalamus, striatum, medial nucleus of the amygdala, hippocampal formation, midbrain/pons area, locus coereleus, spinal trigeminal nucleus and cerebellar Purkinje cells (Brailoiu *et al.* 2007, Sakamoto *et al.* 2007, Matsuda *et al.* 2008, Hazell *et al.* 2009). Within the hippocampal formation, the most heavily stained subdivisions were the DG hilus, subiculum, ventral part of the CA1-CA3 and entorhinal cortex. The GPR30 immuno-staining was mainly seen in cytoplasms and fibres in hippocampal neurons. The immunoblotting of GPR30 protein also confirmed its presence in the hippocampal formation of both male and female rats. The expression of GPR30 mRNAs was also detected in the CA1, CA2, CA3 and DG neurons (Matsuda *et al.* 2008, Schreihofer and Redmond 2009).

III.2.4. Conclusion

The documentation of [^{125}I]estrogen binding sites did corroborate some of the major findings obtained in immunohistochemical studies, specifically, in the hippocampus (Shughrue *et al.* 2000, Milner *et al.* 2008). When comparing the relative distribution of each ER, we can observe that the co-existence of ER subtypes occur in many regions throughout the brain. Indeed, all the three ER subtypes were found to be expressed in numerous neuro-anatomical areas such as piriform cortex, medial nucleus of preoptic area, bed nucleus of stria terminalis, paraventricular nucleus, arcuate nucleus/periventricular zone, ventrolateral part of the ventromedial hypothalamus nucleus, periacqueductal gray area, dorsal raphe nuclei, locus coereleus, the spinal trigeminal nucleus and the hippocampal formation. Coexpression of both ERα and ERβ in the same hippocampal pyramidal cell layers, namely CA1 and CA3, was reported in both male and female rats (Rune *et al.* 2002). Furthermore, neuronal colocalization of both ERα and ERβ was also reported for the hippocampus (Mehra *et al.*

2005). The ultrastructural localization of ERα and ERβ revealed that these ERs could be found on the plasma membrane, as well as in nuclear, cytoplasmic and synaptic compartments of hippocampal neurons. The GPR30 protein was also linked to plasma membrane-localized functions. All together, these findings may imply that ERs play some important role in the mechanisms of rapid signalization in the hippocampal formation. Besides, ERs are, undeniably, robust transcription factors, regulating gene expression in countless estrogen-targeted tissues. When bound by E_2, GPR30 is also able to elicit a rapid production of cAMP (Thomas and Dong 2006, Filardo *et al.* 2007) that may, ultimately, affect gene expression. Taken into account all the transduction features of E_2 signal through ERα, ERβ and GPR30, it is conceivable that immediate, short-term and long-term signalling mechanisms may cooperate in concert in order to shape adequately the architecture of synapses animating the hippocampal neuronal networks. This would explain some of the well known and largely documented effects of E_2 on cognition and, more specifically, on hippocampal-dependent learning and memory functions. In addition, because ERα and ERβ can colocalize within the same hippocampal neurons (Mehra *et al.* 2005), it is reasonable to speculate that multiple ER-mediated signalizations do coexist in the same cells of the hippocampal formation.

IV. Brain Signalization by Estrogens and Estrogen Receptors

A large body of data made it evident that, beyond the reproductive function, ERs are both active players and molecular targets in a number of physiological regulations involving estrogen signalization in the mammalian higher central nervous system. Bioelectrical, neurochemical, structural and behavioural aspects of E_2 signalization, have been identified in the hippocampal formation of all mammalian species examined so far. In the present section, we are going to recall briefly some of the main aspects of estrogen signalization interesting the hippocampal formation physiology.

IV.1. Estrogen Signalization through Estrogen Receptors

The mechanism of E_2-elicited genomic modulation of gene expression that has been extensively and repeatedly documented in various peripheral cell systems was also identified in the mammalian brain. Accordingly, at a number of brain sites that may include the hippocampal formation neurons, estrogens have been shown to regulate various genes such as low voltage-activated (T-type) Ca^{++} channel α_1 Cav3.1-isoform, tryptophan hydroxylase, serotoninergic 5HT1A/2A receptors, progesterone receptors, glutamic acid decarboxylase, anti-apoptotic bcl-2 proteins, brain-derived neurotrophic factor (BDNF), glial fibrillary acidic protein (GFAP), laminin, (McEwen 2002, McEwen *et al.* 1997, Qiu et al. 2006, Scharfman and MacLusky 2006, Foy *et al.* 2008). Such a regulation certainly require cell trafficking of ERs from the cytoplasmic compartment to the nuclear location. Indeed, for example, E_2 induced nuclear translocation of ERα, but not of ERβ, to CA1 pyramidal neuronal nuclei (Rune *et al.* 2002).

In the context of rapid signalization by E_2, ERα and ERβ were localized in cell membranes where they initiate rapid effects through MAPK-dependent pathways (Zhao[1] and Brinton 2007). Also, the expression levels of hippocampal GPR30 were shown to be similar in both genders in adult animals, but higher levels could be measured in male rat neonates (post-natal day 2, PND2) as compared to female rat neonates (Matsuda *et al.* 2008). This suggests that E_2 signalization through GPR30 may play some dimorphic functional role in the developing hippocampus. In the female rat dorsal root ganglion (DRG), ovariectomy (OVX) resulted in decreased number of GPR30-IR neurons and estrogen replacement reversed this effect (Takanami *et al.* 2010) implying that circulating ovarian steroids may regulate the transcriptional rate of the GPR30 gene. Elsewhere, in GPR30-transfected HEK293 cells (a cell model devoted of ERα, ERβ and GPR30) E_2 was shown to bind isolated membrane preparation with high specificity and to elicit the production of cAMP within less than 20 minutes (Thomas and Dong, 2006). This confirmed the rapid-signal transducing function of membrane-associated GPR30 protein. In keeping with this, E_2-treatment of dissociated hippocampal cell cultures resulted in increased cAMP response element binding protein (CREB) phosphorylation (phosphorylated CREB, pCREB) within 1h, and this increased phosphorylaion was still detectable at 6h, 12h and 24h post-treatment before it was downregulated below the control level at 48h (Lee[1] *et al.* 2004). This activated intracellular signallization seamed to be triggered through both Ca^{++}-calmodulin-dependent kinase II (CaMKII) and mitogen-activated protein kinase (MAPK) pathways, but to a much lesser extent, through Akt pathway (Lee[1] *et al.* 2004).

IV.2. Estrogen Signalization through Neuronal Bioelectrical Activity

In neuronal cells, estrogen signalization can take place at the plasma membrane as a direct interaction with ion channel proteins or neurotransmitter receptors. Indeed, E_2 has been shown to bind directly to hippocampal L-type voltage-gated Ca^{++} channels (VGCC) and to potentiate VGCC-sustained currents (Sarkar *et al.* 2008). Non diffusible extra-cellular E_2, also, binds directly the large-conductance, voltage-gated, Ca^{++}-activated, potassium channel (MaxiK), regulatory β1-subunit resulting in an increased sensitivity of the pore-forming, α-subunit (*Slo*), to intracellular Ca^{++} concentrations ($[Ca^{++}]_i$) (Valverde *et al.* 1999). In addition, the expression of *Slo* in the hippocampal formation was consistently demonstrated by *in situ* hybridization, immunohistochemistry, radioligand-binding, RT/PCR and western blotting, and the related protein was also localized to presynaptic elements including the mossy fibers and perforant path terminals (Knaus *et al.* 1996, Pacheco *et al.* 2008) implying that MaxiK channels may translate some E_2 signalizing effects to a rapid presynaptic neurotransmission at hippocampal principal neurons. In an epileptic model of rat (displaying high levels of recurrent spontaneous seizures) MaxiK signalization was dramatically downregulated in the hippocampus (Pacheco *et al.* 2008). Altogether, these findings suggest that MaxiK-mediated, E_2-elicited modulation of glutamatergic neurotransmission might play some significant role in the bioelectrical signalization of hippocampal formation neurons. However, this issue remains unresolved because the hippocampus expresses MaxiK β4 subunit, but not the β1 or β2 subunits, at least in the guinea pig (Jamali *et al.* 2003). Elsewhere, E_2 has been demonstrated to bind directly neuronal ACh receptor protein resulting in an allosteric change-mediated potentiation of the ACh-evoked currents under specific configurations of the α/β subunit-related composition of the channel (α_4/β_2 and α_4/β_4, but not α_3/β_2) (Curtis *et al.* 2002). Respective mRNAs to these ACh receptor subunits have been detected in a large population of both human and rodent hippocampal neurons (Hellström *et al.* 1999, Son *et al.* 2008), including glutamic acid decaboxylase-67 (GAD67)-expressing [most likely, γ-aminobutyric acid (GABA)-synthesizing] interneurons (Son *et al.* 2008). In addition, E_2 modulation of bioelectrical activity involving GPCRs does also occurs in brain neurons (Kelly and Wagner 1999, Kelly and Levin 2001, Kelly and Rønnekleiv 2008). In a different context, E_2 did consolidate LTP induction through MAPK-dependent and -independent mechanisms in young (4-5 week-old) rats (Kim *et al.* 2002).

IV.3. Estrogen Signalization through Cerebral Neurotransmitter Systems

The brain distribution of ERs was shown to be closely related to serotoninergic, catecholaminergic and cholinergic neuronal systems (Nomura *et al.* 2005, VanderHorst *et al.* 2005, Aydin *et al.* 2008, Yanagihara *et al.* 2008, VanderHorst *et al.* 2009) that, along side with GABAergic and glutamatergic neuronal systems, are well known to impact deeply the cognitive abilities (McEwen *et al.* 1997, McEwen 2002, McEwen and Milner 2007, Maki and Dumas 2009). These observations, and many others, have led to the presently widely spread notion of the existence of a robust E_2-mediated control of the activity of brain neuronal neurotransmitter systems (Spencer *et al.* 2008). Indeed, for example, intracerebro-tissular administrations of E_2 into the cholinergic neuronal population-containing, medial septal-diagonal band of Broca area (Lam[1] and Leranth 2003) or into the serotoninergic neuronal population-containing, median raphe nucleus (Prange-Kiel *et al.* 2004, Prange-Kiel and Rune 2006) result in a dramatic increase of spine density in the CA1 stratum radiatum. E_2-treatment also downregulate hippocampal norepinephrine-elicited inositol phospholipid hydrolysis in OVX rats (Favit *et al.* 1991) and serotonin-IR (Prange-Kiel and Rune 2006). In the same context, administration of a high dose (by daily oral gavage with 1 mg/kg b.w., for 6 weeks) of letrozole, an aromatase inhibitor, to normal intact rats resulted in increased levels of hippocampal monoamine (dopamine and noradrenaline) contents (Aydin *et al.* 2008). Finally, local infusion of E_2 into the hippocampus did decrease the occurrence of serotoninergic fibers in both the stratum lacunosum moleculare and the stratum radiatum (Prange-Kiel and Rune 2006) suggesting that E_2 may increase the synaptic release and/or decrease the enzymatic synthesis/reuptake of serotonin at the raphe nuclei-arising serotoninergic axonal terminals.

IV.4. Estrogen Signalization through Neuronal Morphological Adaptations

Morphological changes do occur in hippocampal formation neurons in response to E_2. Accordingly, E_2 is capable of shaping the architectural features of hippocampal afferent/efferent neuronal networks and both embed presynaptic and postsynaptic elements, alike. For example, E_2-treated OVX young rats displayed higher levels of axospinous synaptic density in the stratum radiatum as compared to vehicle-treated OVX rats of the same age (Adams *et al.* 2001). Also, E_2 increases the spine density levels in the stratum radiatum molecular layer (Rune *et al.* 2002). Additionally, local infusion of E_2 increases the number of spine synapses in the stratum radiatum of the CA1 region (Prange-Kiel and Rune 2006).

Finally, E_2 did upregulate, whereas the ER antagonist, ICI 182,780, did downregulate axonal outgrowth (vonSchassen *et al.* 2006).

IV.5. Estrogen Signalization through Synaptic Proteins

Biochemical adaptations do occur in hippocampal formation neurons in response to E_2. As a critical part of the E_2-mediated cellular adaptations, E_2 also modulates synaptic proteins, namely synaptophysin, spinophilin, postsynaptic density-95 protein (PSD95) and growth cone-associated protein-43 (GAP43). Indeed, E_2 increased the abundance of immunoreactive presynaptic protein, synaptophysin, in the stratum radiatum molecular layer (Rune *et al.* 2002). Administration of E_2 to gonadectomized rats resulted in a dimorphic effect on stratum radiatum spinophilin-immunopositive fibers in the sense that E_2 increased spinophilin-IR in OVX female rats whereas it decreased spinophilin-IR in orchidectomized male rats, although E_2 did not affect hippocampal levels of spinophilin mRNA expression in both genders (Lee[1] *et al.* 2004). Fourty-eight hours after E_2 treatment synaptophysin, syntaxin and spinophilin were upregulated in the stratum radiatum and the stratum oriens of the CA1 region, and spinophilin also was upregulated in the DG hilar area and in the stratum lucidum of the CA3 (Brake *et al.* 2001). Similar results were obtained for the hippocampal CA1 region of the female Rhesus Macaques *(Macaca mulatta)*, after a menstrual cycle-long E_2-treatment, whereas no significant effect could be found in the molecular fields of the DG (Choi *et al.* 2003). The modulatory effects of E_2 upon hippocampal synaptic plasticity could be ER subtype-specific in the sense that specific agonists to ERα or ERβ triggered different profiles of expression of synaptic molecules such as presynaptic proteins, synaptophysin, vesicular GABA transporter (VGAT), vesicular glutamate transporter-1 (VGlut1) or postsynaptic proteins, PSD95, spinophilin and α-amino-3-hydroxyl-5-methyl-4-isoxazolepropionate receptor (AMPA)-type glutamate receptors (AMPARs), GluR1/2/3 (Waters *et al.* 2009).

IV.6. Estrogen Signalization through Neurotrophins

In the context of the nurturing effects of a given brain structure toward its corresponding neuronal terminal inputs, the discovery of neurotrophins and their effects, did throw an invaluable light upon the intimate mechanisms of reciprocal signalization exchanges between the targeted brain area and the one that send axonal fiber afferents to this brain area. Hence, the hippocampal formation was identified as one of the major brain sites for the production of signalizing neurotrophins that are known to be deeply impacted by E_2. For example, it has been reported that E_2 increases hippocampal nerve-growth factor (NGF) mRNA levels (Pan *et al.* 1999) and the resulting protein increases tropomyosin-receptor kinase-A (TrkA) and ERK/MAPK activity within the hippocampus-projecting medial septo-basal forebrain neurons (Williams *et al.* 2006). These hippocampal physio-molecular modifications in response to E_2 administration were associated with improved spatial working memory (Pan *et al.* 2000) through a better recruitment of cholinergic neurons (Pan *et al.* 1999). In the same context of E_2-triggered production of neurotrophins, the hippocampus was shown to produce BDNF which, in turn, affects local synthesis of neuropeptide Y (NPY) (Scharfman and MacLusky 2006). Also, BDNF-mediated E_2-elicited signalizations exert, locally, deep impacts on neurotransmitter neuronal systems arising from the hippocampal formation or from other brain areas.

IV.7. Estrogen Signalization through Hippocampal Estrogens

Another interesting aspect of E_2 signalization resides in the fact that this molecule can also be considered as a neurosteroid. Indeed, E_2 is synthesized locally within neuronal and glial cells of the hippocampal formation (Hojo *et al.* 2004, Kretz *et al.* 2004, Rune and Fortscher 2005, Agis-Balboa *et al.* 2006, Ishii *et al.* 2007, Garcia-Segura 2008) where the testosterone-aromatazing enzyme, aromatase (cytochrome P450 depending enzyme *CYP19*), and the dehydro-epiandrosterone synthase (cytochrome P450 depending enzyme *CYP17α*), were detected preferentially in pre- and postsynaptic ultrastructural elements of the CA1-3/DG principal neurons (Hojo *et al.* 2004). In the Rhesus monkey hippocampus, aromatase-IR is present in CA1-3/DG within NeuN (a neuronal marker)-, but not GFAP-coexpressing cells, implying that aromatase is preferentially expressed in neurons (Yague *et al.* 2008). In addition, the hippocampal aromatase-IR, that was highly enriched in dendritic processes suggesting that the aromatase enzyme may play some key role in the mechanisms of synaptic plasticity, did also display interesting coexpression profiles within hippocampal calretinin-, calbindin- and parvalbumin-immunopositive neurons (Yague *et al.* 2008). The pharmacological blockade of hippocampal endogenous E_2 synthesis using the aromatase inhibitor, letrozole, or an RNA silencer (siRNA) directed against steroidogenic acute regulatory (StAR) protein, resulted in a significant downregulation of spine formation as well as of synaptic proteins, namely synaptophysin, spinophilin and GAP43 (Rune *et al.* 2002, Rune and Fortscher 2005). In keeping with this, letrozole administration to normal intact rats resulted in increased expression levels of hippocampal neural cell-adhesion molecules (N-CAM), the 140 and 180 isoforms (Aydin *et al.* 2008). Aromatase knockout (ArKO) female mice, that are estrogen deficient animals, displayed significantly higher mRNA expression levels of hippocampal N-methyl-D-aspartate (NMDA) receptor (NMDR) subunit NR2B, and to a lesser extent, that of NR1 and NR2A, as compared to wild-type animals (Boon *et al.* 2005). Treatment of hippocampal slices with letrozole, resulted in a downregulation of spinophilin-IR and both densities of spine synapses and buttons in the stratum radiatum (Prange-Kiel and Rune 2006). Letrozole administration did also result in a dose-dependent, downregulation of hippocampal neuronal synaptophysin-IR in dispersed cell cultures (Prange-Kiel[1] *et al.* 2006, Prange-Kiel *et al.* 2006). Finally, letrozole and the ER antagonist, ICI182780, did downregulate the expression levels of GAP43 (VonSchassen *et al.* 2006).

IV.8. Estrogen Signalization through Estrogen Receptor-Interacting Proteins

It is noteworthy to evoke some other specific mechanisms of ER-mediated signalization occurring in the hippocampal formation and involving a different type of signalling molecules, such as PELP1. This protein is a steroid receptor-interacting molecule which is implicated in the transcriptional regulation of gene expression in steroid targeted cells. Indeed, PELP1 was shown (1) To interact with ERα, probably with ERβ, too, (2) To be expressed in a number of E_2-targeted, ER-expressing brain regions including the hippocampus, (3) To be coexpressed within hippocampal ERα-immunopositive neurons, (4) To be recruited to the promoters of ERα-targeted genes such as progesterone receptor, pS2 and insulin-like growth factor (IGF), (5) To modulate cyclin D1 mRNA expression, and (6) To be associated with an increased phosphorylation of STAT3 (at least, when PELP1 is over-expressed in MCF7 cells, and given the fact that estrogen enhances brain phosphorylation of STAT3) through Src-MAPK-dependent pathway (for an extensive review on PELP1, see Brann *et al.* 2008).

IV.9. Estrogen Signalization Modulating Mnemonic Behaviours

In mammals, the animal behaviours are deeply impacted by estrogen signalizations. Specifically, the mnemonic characteristics constitute, most likely, the behaviourally expressed translation of a large part of the biological effects we reported above and that are triggered, directly or indirectly, by estrogens at the cognitively competent brain sites. For example, E_2-replacement in OVX rats improved the performance of spatial working memory monitored in an 8-arms radial maze when E_2-treatment was administered soon after the surgical removal of the ovaries but not after 5 months of E_2 withdrawal (Daniel *et al.* 2006). Similar results were obtained in the Morris water maze (Talboom *et al.* 2008). This implies that circulating estrogens are essential for the maintenance of efficient functioning in the cognitively competent brain areas at least in females. It does also suggest that the ability of the female hippocampus to synthesize its own E_2 is not sufficient for a normal behavioural expression of mnemonic acquisition, or that this ability is impaired under longterm E_2 withdrawal. Moreover, peripheral administration (0.2 mg/Kg b.w.) or intra-tissular dorso-hippocampal infusion (5.0 μg/side) of E_2 resulted in enhanced object recognition memory consolidation through a membrane-initiated ERK activation (Fernandez *et al.* 2008). Contrasting results were reported throughout the oestrous cycle where increased levels of plasma E_2 were associated with diminished performance of spatial memory as assessed in the Morris water maze (Frye 1995). This occurrence led to the suggestion that E_2-mediated increase of newly formed synapses in the hippocampal area could be detrimental to the spatial reference memory. However, agonist-specific activation of ERβ, but not of ERα, was associated with increased expression of hippocampal synaptic proteins, increased LTP potentiation and improved performance in nonmatch-to-sample tasks in hippocampus-dependent spatial memory (Liu *et al.* 2008). Quite surprisingly, though, letrozole administration to normal rats, also, did improve the consolidation of spatial memory (Aydin *et al.* 2008). Therefore, the question of the dose and/or of the activated ER-subtype, seams to be essential. However, ArKO female mice, did acquire the spatial reference memory task in the Morris water maze as well as wilde-type mice did (Boon *et al.* 2005). Altoghether, these findings show that there is no simple equation between the E_2 level and the cognitive memory performance. The improvement of memory under low levels of E_2, might suggest that neurophysiological adaptations are still taking place in the E_2-deprived brain in order to rescue hippocampus-dependent memory functions. Discovering some key mechanisms within these neurophysiological adaptations may benefit greatly for the rescue of the ageing hippocampus and, more particularly, in the neurodegenerating and/or postmenopausal brains.

V. Natures of Aging

Nature has sentenced every single matter that has succeeded in even a slightest occupation of time, to an irrevocable aging process. Accordingly, both mineral and organic matters do experience more or less severe deteriorations with increasing age. In living organisms, these deteriorations inevitably concern the species as a whole entity, the population as an ensemble of individuals and the living individual subject throughout different stages of its lifespan, as well as both the physiological function within every single organism and the behavioural expression of each single living.

V.1. Aging Species

Phylogenetic evolutions of living organisms within both the animal and the vegetable kingdoms have thought us that species do appear in a given stage of earth ages. Species do also develop and strengthen over times. Sometimes they do successfully thrive and multiply to constitute large populations. Evolutionary successful species usually expand their occupation of space beyond their original inhabited ecosystem. This occurrence often throws the emerging species into the arena of the struggle for survival. After a more or less longer period of time, the formerly flourishing species do decline and then after disappear for multiple reasons. These reasons might be attributable to suddenly or gradually adverse environmental changes (physical or biological changes) and/or to an inherent age-related accumulation of crippling biological defects. The laters could be of different natures such as a genomic nature (life expectancy-reducing, immunogenic deficiency-inducing or reproductive age-shortening gene mutations) or a behavioural nature (excessive food-related specialization, for example). These phenomena and many plausible others may play a central role in the aging processes of a given species on the pace of breaking dawn of extinction.

V.2. Aging Individuals and Aging Populations (Humans)

In different parts of the world, human populations are increasingly growing and aging (Christensen *et al.* 2009, Lutz and Qiang 2002, Shrestha 2000). For example, an estimated 420 million persons were aged 65 years-or-more in the year 2000 (CDC, 2003). This number was the result of a 9.5 million increase that was recorded within one year time laps (CDC, 2003). The worldwide number of individuals aged 65 years-or-more is projected to increase by approximately 550 million to 973 million during the period 2000-2030 (CDC, 2003). These changes of demographic features are likely due, at least in the most industrialized countries, to decreased fertility and increased life expectancy that have been obviously seen in a number of human societies worldwide during the 20th century. Consequently, the ranking within the list of leading causes of deaths is changing in favour of neurodegenerative diseases, such as Alzheimer's disease (AD) (Ferrucci *et al.* 2008). Old age-related disability, morbidity and co-morbidity are also major concerns for both public health service planners and providers alike, around the world (Ferrucci *et al.* 2008). Hence, the aging-triggered physiological dysfunctions result in an increasingly worsening social burden upon worldwide human societies. Because nutrition is able to impact deeply the functioning of bodily organs, it has been, along side with the improved hygiene and pharmaceutical/medical technologies, a leading cause of increased life expectancy. Ironically, though, nutrition remains the ultimate and highly needed gateway we are hoping for in order to partially alleviate age-associated physiological dysfunctions where the memory decline may be bringing along with the most costly and challenging social burden that humanity must face in the presently began century. However, carefully examined and well documented dietary/supplemental isoflavone-related recommendations are still highly missing.

V.3. Aging Function (Mammalian Reproductive Neuroendocrine Axis)

In the mammalian brain, gonadotrophin-releasing hormone (GnRH) neurons, which lay in the preoptic/mediobasal hypothalamic areas, are well known to orchestrate the activity of the hypothalamo-pituitary-gonadal axis. By providing the endocrine pituitary cells with phasic pulses of the signalizing GnRH peptide, these neurons are able to entrain the pituitary secretion of both luteinizing hormone (LH) and follicle-stimulating hormone (FSH) in an oscillatory fashion. Down-stream of the blood circulation, LH and FSH are addressed to the ovaries which in turn, as a response to the pituitary hormonal signal, synthesize and release estrogens in a cyclic manner. Back into the brain, the laters then convey a retro-control message to GnRH neurons directly or via numerous monoaminergic/peptidergic neuronal systems. Other molecules emanating from the ovary, such as for example inhibin A/B or gonadotrophin surge-attenuating factor (GnSAF), do participate in the establishment of positive/negative sensing retro-controls exerted by the ovary upon the pituitary FSH/LH secretion. The major players within this functional loop are grouped under the name of reproductive neuroendocrine axis/system (for reviews on this axis, see Smith and Jennes 2001, McEwen 2002, Messinis 2006, Spencer *et al.* 2008). Therefore, the reproductive neuroendocrine system is responsible of cyclic features in the basal free-running physiology of most, if not all, the organs within the body. Among the organs affected by the fluctuating E_2 plasma levels, the brain is one of the most sensitive organs to circulating E_2. Within the brain, the hippocampal formation is a site of active integration of the ovary-emanating hormonal signals. Consequently, overall neuronal plasticity adaptations, namely bioelectrical, physio-molecular, neurochemical and morphogenic adaptations do occur throughout the oestrous cycle in the hippocampal formation. For example, in the mouse hippocampus, the activity markers of Akt, LIM kinase, tropomyosin-receptor kinase (TrkB), synaptophysin and PSD95 all undergo oestrous cycle-shaped fluctuations, although the performance in an object placement task remained unaffected by the oscillating endogenous estrogens (Spencer *et al.* 2008). However, the reproductive neuroendocrine system is subjected to age-over time alteration processes resulting in major accumulating dysfunctions (Downs and Urbanski 2006, Hall 2007, Downs and Wise 2009). Three major aspects in the functioning of the aged reproductive neuroendocrine axis are the loss of hypothalamic GnRH pulse secretion, the dramatic rise in LH/FSH secretion and the overwhelming downregulation of ovarian estrogen synthesis/secretion (Downs and Urbanski 2006, Messinis 2006, Hall 2007, Downs and Wise 2009).

V.4. Aging Cognition (Mammalian Hippocampal Formation Sub-Serving Memory Function)

Similarly to other vital organs within the body, the brain is deeply impacted by the passage of time (Crews 2007). As a result, the aging of overt cognitive abilities can be markedly and increasingly seen in age-advanced populations. More specifically, the hippocampal formation is no exception in the age-impacted physiologically altered cerebral structures. Thereafter, age-related hippocampus-dependent learning and memory dysfunctions do occur in elderly people. This aspect of cognitive aging has been intensively studied and its highly multifaceted and complex feature has been documented and extensively reviewed elsewhere (Halbreich *et al.* 1995, Bellino and Wise 2003, Billard 2006, Dumas *et al.* 2008, Barrett-Connor and Laughlin 2009, Frick 2009). Because the cell firing of hippocampal neurons was correlated to the position of the animal in its environment, this discovery and many others allowed the hatching of the ever alive concept of the hippocampus being the brain site where the spatial map is built-up using the integrated surrounding visual cues (Hill 1978, Olton *et al.* 1978, McNaughton *et al.* 1983). This functional characteristic of the hippocampus let it show as the mastermind of the highly developed spatial navigation art seen in mammals. Unfortunately, bioelectrical, structural and neurochemical alterations, more or less severely, do occur in the aging hippocampi of rodents, non-human primates and humans and these alterations are believed to sustain behaviourally expressed, hippocampus-related cognitive defects. Indeed, aging has been associated with altered inductions of both longterm potentiation (LTP) and longterm depression (LTD) in hippocampal neurons and with hippocampus-dependent learning deficits that was seen in rodents, non-human primates and humans (Foster, 1999, Foy *et al.* 2008, Frick, 2009, Lacreuse, 2006, Genazzani *et al.* 2007). In addition, induced LTP in CA1 pyramidal neurons arise from the fusion of two different conductances, an NMDAR-dependent component and a voltage-dependent calcium channel (VDCC)-related component (Shankar *et al.* 1998). Quite interestingly, and compared to young rats (6-9 week-old), aging rats (24 month-old) displayed a reduction in the magnitude of the NMDAR-dependent component and an increase in the magnitude of the VDCC-dependent component (Shankar *et al.* 1998) implying that compensatory mechanisms against the age-triggered alterations of neuronal excitability, are shifting the bioelectrical activity of hippocampal neurons to much higher Ca^{++} concentrations in old subjects as compared to young ones.

Age-related alterations in the hippocampal formation physiology have also been associated with impaired E_2 signalization and, more particularly, ER-related signalization (for an extensive review, see Frick, 2009) and, thereof, aging women's health is more at stake (Sherwin, 2006, 2007). For example, the number of both ERα and ERβ immunopositive neurons was shown to be affected in the hippocampus during the normal aging process. Indeed, the number of ERα immunoreactive neurons decreases by 78% and 56% in, respectively, the CA3 and the CA1 of aged female rats as compared to normal adult female rats (Mehra *et al.* 2005). Also, the number of ERβ immunoreactive neurons drops by 88% and 41% in, respectively, the CA3 and the CA1 of aged female rats as compared to normal adult female rats (Mehra *et al.* 2005). In the same study, western blotting did confirm that the amounts of ERα and ERβ proteins were highly decreased in aged female hippocampi. In the same context, ERα-IR and aromatase (P450arom)-IR were upregulated in postmenopausal women hippocampi whereas ERα and P450arom mRNA transcripts were downregulated in Alzheimer disease (AD)-deceased women hippocampi (Ishunina *et al.* 2007).

In the context of age-triggered alterations of human hippocampal nicotinic ACh receptor (nAChR)-mediated neurotransmission, it was shown by mean of RT/PCR that the mRNA expression levels of the β_2 subunit, but not the α_4 subunit, is downregulated in post-mortem hippocampi of elderly people (Tohgi *et al.* 1998). Similarly, the mRNA expression levels of the α_7 subunit, but not the α_3 or α_4 subunits, were upregulated, whereas the $[^{125}I]$-α-bungarotoxin binding levels were downregulated in post-mortem hippocampi of Alzheimer disease patients (Hellström *et al.* 1999). Furthermore, an altered signalization occurs, in aged rats, in the cholinergic basal forebrain-hippocampal pathway as revealed by a downregulation of hippocampal NGF-induced TrkA expression and ERK phosphorylation (Williams *et al.* 2006) which is believed to be crucial for memory encoding (Giovannini 2005, 2006). Recently E_2 has also been shown to interact at the hippocampal level with the cholinergic receptors as a protective agent of the cerebrovascular system that is altered during aging (Kitamura *et al.* 2009). In human hippocampi, structural and neurochemical changes have been demonstrated by mean of magnetic resonance imaging and proton magnetic resonance spectroscopy in two hippocampus-dependent memory tasks, namely virtual Morris water maze and transverse patterning discrimination tasks (Driscoll *et al.* 2003).

Under the paradigm of hormone replacement therapy, in both aged female and aged male mice (24 months of age), E_2-treatment improved memory performance in inhibitory avoidance and water maze tasks (Frye *et al.* 2005). E_2 benzoate (EB)-treatment of aged mice (27-28 month old) also improved spatial reference memory performance in high EB-dose (5 μg/day) injected but not in low EB-dose (1 μg/day) injected animals (Frick *et al.* 2002). Aged mice treated daily with 5 μg EB displayed increased expression levels of hippocampal synaptophysin-immunoreactivity, whereas the activity levels of hippocampal ChAT and GAD remained unaffected by the treatment (Frick *et al.* 2002). Elsewhere, E_2-treatment did increase axospinous synaptic density in the stratum radiatum of OVX young (4 months) rats but not in that of OVX aged (24 months) rats (Adams *et al.* 2001). On the contrary, E_2-treatment did increase the number of NMDAR, NR1 subunit, in individual synapses of the CA1stratum radiatum of OVX aged rats but not that of OVX young rats (Adams *et al.* 2001) suggesting that some lower energy-requiring cellular adaptations are taking place in the aging hippocampal neurons as an ultimate attempt to counteract the increasing age-linked defects of hippocampal neuronal excitability and/or of E_2 signal transduction.

VI. Pharmacological Effects of Isoflavones on the Hippocampal Formation

In order to decipher the impact of isoflavones on the memory function, we did collect two main ensembles of data that differ by their delivery routes of active molecules. The first ensemble of data consisted of using isoflavone molecules as pharmacological tools. This was mainly carried out in simplified hippocampus-derived cell systems including organotypic hippocampal slice cultures, dissociated hippocampal cell cultures (neuronal + glial cells) and hippocampus-derived synaptosomal membrane preparations. The corresponding data are summarized in Tables 1.1-6 where the studies are classified in the decreasing order of the isoflavone concentrations (from, nutritionally, the less plausible to the highly probable). The second ensemble of data consisted of using isoflavone molecules as dietary compounds (for example, those naturally isoflavone-rich products) or as orally-administered, supplemental compounds (for example, those with a molecularly controlled composition). The corresponding data are summarized in Tables 2.1-3, for studies in rodents; in Table 3, for a study in a non-human primate (Rhesus monkey); in Tables 4.1-5, for a sample of cognitive testing in human subjects; in Tables 5.1-8 for dietary/supplemental interventions in human subjects.

Pharmacological effects of isoflavones on hippocampal neurons have been examined in different mammalian species including humans (control and AD-deceased subjects), rats (Wistar and Sprague-Dawley strains), mouse and guinea pig (see Table 1. for a complete summary). For most of them, these studies consisted of *in vitro* investigations that used hippocampus-derived cell systems arising from embryos, neonates, young or adult subjects. However, few studies have explored the effects of isoflavones in whole animals. The mostly investigated isoflavone molecule, by far, has been genistein. This is most likely due to its inherent ability to act as a powerful competitor at the ATP-binding site on protein tyrosine kinases (PTKs), and thereof, as a potent inhibitor of intracellular PTK activity (Levitzki, 2006). Besides, PTKs have been involved in a broad range of intracellular signalizations taking place ubiquitously in different neuronal systems, not to mention non-neuronal ones, which allow genistein to affect count-less intracellular signalization pathways. Within the scale of data availability, daidzein ranks as the second mostly studied isoflavone molecule in hippocampus-derived cell systems. Much fewer data did report effects of other isoflavones, namely formononetin or of the isoflavone metabolite, the isoflavane, equol (Setchell *et al.* 2002).

VI.1. Description of Pharmacological Studies

Pharmacological studies carried out using hippocampal neurons did address isoflavone signalization under 4 main paradigms leading to 4 main modulations of the neuronal activity:

(1) bioelectrical modulation, consisting of measurable changes in the pattern of electrical activity of electro-physiologically recorded hippocampal neurons,
(2) physio-molecular modulation, consisting of quantitative and qualitative changes in the molecular phenotype of hippocampal neurons in response to a given signalling substance,
(3) neuroprotective modulation, consisting of measurable pro-survival and anti-apoptotic impacts on a given hippocampal neuronal population that has been challenged by a cell stressor,
(4) neurotrophic modulation, consisting of visually detectable transformations, under microscope, of the cellular structure of hippocampal neurons that have been exposed to a given neurotrophic molecule.

These 4 features of isoflavone signalization have been widely documented for estrogen signalling molecules in which E_2 has been the leading and the reference molecule. In the pharmacological studies we are summerazing below, genistein was mainly used as a specific tyrosine kinase inhibitor and daidzein was sometimes used in parallel as a biologically inactive structural analogue of genistein with no tyrosine kinase inhibitory effect.

VI.1.1. Bioelectrical features

The Isoflavone effects on bioelectrical properties of hippocampal neurons have been addressed by mean of electrophysiological recordings in a number of cell configurations. O'Dell *et al.* (1991) used guinea pig hippocampal slices to investigate the effect of genistein (110 μM) on the Schaffer collateral/commissural fibre stimulation-induced LTP in the CA1 pyramidal neurons. The results showed that, when applied before the titanic stimulation, genistein was able to block both the induction of LTP and the activation of a hippocamus-enriched PTK (namely, the pp60c-src$^{(+)}$: a cellular counterpart of Rous sarcoma virus tyrosine kinase) in CA1 pyramidal neurons. Similarly, Shankar *et al.* (1998) have used hippocampal slice preparations taken from young (aged 6-9 weeks) or old (aged 24 months) male, virgin Fisher 344 rats, to stimulate the Schaffer collateral/commissural fibres and to examine the effect of pharmacological genistein on electrical stimulation-induced LTP in the CA1 pyramidal neurons. They have demonstrated that the normal aging process was associated with a reduced magnitude of the NMDAR-dependent component and an increased magnitude of the VDCC-dependent component as of the relative contributions of these current components to the LTP induction (Shankar *et al.* 1998). Quite nicely, these authors have also demonstrated that 20 μM genistein reduces LTP induction in young rats and almost block it completely in aged rats. In the same context, Casey *et al.* (2002) have demonstrated that intracerebro-ventricular injection of 250 μM genistein prior to the electrical stimulation of the perforant path leads to the abolition of the expected induction of LTP in the DG granule cells. This abolition was also associated with a complete blockade of Glutamate release and Ca^{++} influx normally observed in potassium chloride (KCl)-stimulated hippocampal synaptosomes, subsequently prepared from electrically-stimulated rat brains.

Ramsey *et al.* (2005) used hippocampal slices derived from young Sprague-Dawley rats to demonstrate that acute application of the insulin-like growth factor-1 (IGF-1) results in a dramatic increase of AMPAR-mediated excitatory synaptic transmission in the CA1 region, and that previous bathing of hippocampal slices with genistein (220 μM) completely blocks this increase. Genistein (100 μM), but not daidzein (100 μM), was also able to alleviate a metabotropic glutamate receptor (mGluR)-mediated inhibition of Ca^{++}-activated potassium (K^+) currents [also called after-hyper-polarization K^+ currents (I_{AHP})] in DG neurons recorded on Wistar adult male rat-derived hippocampal slices (Abdul-Ghani *et al.* 1996). Similarly, genistein(50 μM) , but not daidzein (50 μM), was able to block 4β-phorbol ester myristate acetate (4β-PMA)-mediated potentiation of NMDA-activated currents in isolated hippocampal CA1 pyramidal neurons (Lu *et al.* 1999). In keeping with this, genistein (100 μM), but not daidzein (100 μM), was shown to reduce a depolarization-coupled induction of high voltage-activated Ca^{++} currents in CA1 pyramidal neurons (Potier *et al.* 1999). Genistein (100 and 250 μM) was capable of potentiating ACh-evoked currents in CA1 interneurons which were, presumably, GABA-ergic neurons (Charpantier *et al.* 2005). The same result was obtained by Cho *et al.* (2005) with Genistein (10 μM). However, daidzein (10 μM) was unable to potentiate the ACh-evoked currents in the same neurons (Cho *et al.* 2005).

VI.1.2. Physio-molecular features

Due to their crucial role in the neuronal signalization, the activity of ion channels was documented in the presence of isoflavone molecules in a number of hippocampus-derived cell models. This time again, genistein was mainly used as potent tyrosine kinase inhibitor. Considering the hormonal regulation of ion channel availability on the cell membrane surface, genistein (1 μM) did block efficiently the insulin-elicited, but not the AMPA- or NMDA-elicited, internalization of AMPARs seen in dissociated DG neurons taken from PND0 neonate rat pups (Beattie *et al.* 2000). The effects of isoflavones were also examined under the paradigm of neurotransmitter-mediated regulations of ion channel trafficking. Charpantier *et al.* (2005) investigated the ACh-induced internalization of ACh receptors (AChRs) in cultured hippocampal slice neurons from 2-5 weeks aged rats. They demonstrated that genistein (100 and 250 μM) was (1) capable of blocking the tyrosine phophorylation of the ACh nicotinic receptor α_7 (α_7nAChR) subunit but (2) not capable of inducing the translocation of this subunit to the plasma membrane. Similarly, Mielke *et al.* (2009) have shown that the maximal pool of α_7nAChRs was localized in the intracellular compartment and that genistein (100 μM), insulin and K^+-triggered depolarization were all unable to elicit a significant translocation of the intracellular α_7nAChR subunits. On the contrary, Cho *et al.* (2005) have reported a significant increase in the number of cell surface-localized α_7nAChR subunits under the conditions of genistein (10 μM) treatment. Additionally, site-directed mutation of each, or all of the 3 cytoplasmic tyrosine residues, did not prevent the genistein-mediated facilitation of ACh-evoked currents in transfected *Xenopus* oocytes implying that a direct phosphorylation/dephosphorylation mechanism is unlikely. Elsewhere, in hippocampal membrane synaptosomes prepared from AD-deceased or control human subjects, genistein dose-dependently (from 1 to 10 μM) stimulated the binding of sulphate 35-radio-labeled GTP-gamma stimulating subunit ([^{35}S]-GTPγ-S) to synaptosomal preparations. This effect of genistein (1 to 10 μM) was similar in both AD and control hippocampi and could also be mimicked by daidzein from 1 to 10 μM (Jefremov *et al.* 2008) indicating a non genomic estrogenic effect rather than an anti-tyrosine kinase effect.

VI.1.3. Neuroprotective features

Estrogens are well known for their robust actions as molecular shields against cell damages, and more particularly against neuronal damages that are caused by cell stressors. For this, they have, deservedly, been labelled as neuroprotective molecules. Indeed, the neuroprotective features of estrogens have been largely reviewed in previous well-documented literature (Garcia-Segura *et al.* 2001, Wise *et al.* 2001, Brann *et al.* 2007). Because isoflavones could mimic different physiological regulations normally devoted to E_2, the putative capacity of these molecules to protect brain neurons, have been challenged in experimental conditions of cell stress.

VI.1.3.a. In vitro approaches

The study of the effect of genistein on intra-endoplasmic reticulum Ca^{++} stores was undertaken *in vitro* using thapsigargin which is believed to inhibit the sarco/endoplasmic Ca^{++} ATPase (SERCA) pump. This results in a dramatic depletion of intra-endoplasmic reticulum Ca^{++} stores and leads, subsequently, to the activation of store-operated Ca^{++} channels (SOCCs) which are known to be responsible of massive Ca^{++} influx into the thapsigargin-

Table 1.1. Pharmacological effects of isoflavones on hippocampal neurons.

Molecule	Subject	Target / paradigm	Mean	Variable parameter	Effect / Conclusion	Reference
Genistein 250 μM	Male Wistar rats (250-350 g b.w.) Anesthetized animals	Electrical stimulation of the perforant path Electrophysiological recording on the DG granule cells LTP induction paradigm	Electrophysiology *In vitro* biochemical assays Pharmacological tools (icv injection of genistein) Immunoblotting DG synaptosomes	Ca^{++} influx Tyrosine kinase activity, ErK phosphorylation Phosphorylation of the α_1 subunit of the voltage-gated Ca^{++} channels	Genistein pre-treatment blocks the induction of LTP, inhibits the LTP-associated increases of Glu release and Ca^{++} influx in KCl-stimulated synaptosomes Genistein also inhibits the LTP-associated increases in α_1 and ERK phosphorylation	Casey *et al.* 2002
Genistein 220 μM	Sprague-Dawley male rats Age: 20-40 days	HPC slices Neurotrophic signalization	Electrophysiology	IGF-1-induced fEPSP	IGF-1 increases the fEPSP slope by 40% through AMPA receptors genistein completely blocks this increase	Ramsey *et al.* 2005
Genistein 110 μM	Guinea Pig	HPC slices LTP induction	Electrophysiology *In vitro* kinase assays	LTP recording kinase activity	Genistein blocks LTP induction and inhibits pp60c-src^{+} (a HPC- enriched tyrosine kinase) activity	O'Dell *et al.* 1991
Genistein 100 μM	Wistar adult male rats	HPC slices, DG neurons, Receptor-channel cross-talk	Electrophysiology	mGluR-mediated inhibition of I_{AHP}	Genistein, but not daidzein, alleviates this inhibition	Abdul-Ghani *et al.* 1996
Genistein 100 μM	Sprague-Dawley rats Age: 15-25 days	HPC slices, CA1 pyramidal neurons Channel activation	Electrophysiology	depolarization-coupled HVA Ca^{++} currents	Genistein, but not genistin or daidzein, reduces HVA Ca^{++} currents	Potier *et al.* 1999

Abbreviations: AMPA: α-amino-3-hydroxyl-5-methyl-4-isoxazolepropionate, b.w.: body weight, CA1: cornu ammonis, Ca^{++}: calcium, DG: dentate gyrus, ERK: extra-cellular signal regulated kinase, fEPSP: field excitatory postsynaptic potential, Glu: glutamate, HPC: hippocampus, HVA: high-threshold voltage-activated, I_{AHP}: transient calcium-activated (after hyperpolarization) potassium currents, icv: intracerebro-ventricular, IGF-1: insuline-like growth factor-1, KCl: potassium chloride, LTP: longterm potentiation, mGluR: metabotropic glutamate receptors, pp60c-src(+): cellular contrepart of Rous sarcoma virus.

Table 1.2. Pharmacological effects of isoflavones on hippocampal neurons

Molecule	Subject	Target / paradigm	Mean	Variable parameter	Effect / Conclusion	Reference
Genistein 100 μM	Sprague-Dawley rats Age: 2-5 weeks	HPC slices CA1 interneurons α_7nAChR-mediated signalization	Electrophysiology Binding assays	ACh-evoked α_7nAChR currents Tyrosine phosphorylation of α_7nAChR	Genistein blocks the tyrosine phosphorylation of α_7nAChR It also potentiates the ACh-evoked currents Genistein did not increase the number of cell surface-localized α_7nAChR subunits	Charpantier *et al.* 2005
Genistein 100 μM	Sprague-Dawley PND3-5 rat pups	Dissociated HPC neuronal cultures Thapsigargin-induced Ca^{++} depletion	Fura-2 Ca^{++} imaging Pharmacological tools	Thapsigargin-induced influx of extracellular Ca++ through SOCCs	Genistein remarkably reduces both the amplitude and the latency of the peak response of SOCC-mediated thapsigargin-elicited Ca^{++} influx	Koss *et al.* 2009
Genistein 100 μM	Sprague-Dawley PND8 rat pups	Organotypic HPC slice cultures α_7nAChR-mediated signalization	αBGT-binding assay, *Western* Blotting, IHC Cell fractionation, Cell-surface protein biotinylation, Pharmacological tools	Cellular distribution of α_7nAChRs	Maximal pool of α_7nAChR subunit was localized in the intracellular compartment Genistein, insulin and K^+-triggered depolarization were unable to elicite a significant translocation of the internal pool of α_7nAChR subunits to the plasma membrane	Mielke *et al.* 2009
Genistein 50 μM	Wistar rats Age: 10-24 days. Mice B6129F2/J	Isolated HPC CA1 pyramidal neurons Channel regulation	Electrophysiology	4βPMA-mediated potentiation of NMDA-activated currents	Genistein, but not daidzein, blocks this potentiation	Lu *et al.* 1999

Abbreviations: 4β-PMA: 4β-phorbol 12-myristate 14-acetate, α_7nAChR: α_7 nicotinic acetylcholine receptor, αBGT: α-bungarotoxin, ACh: acetylcholine, IHC: immunohistochemistry, NMDA: N-methyl-D-aspartate, K^+: potassium, PND : post-natal day, SOCCs: store-operated Ca^{++} channels. (see Table 1.1. for further abbreviations).

Table 1.3. Pharmacological effects of isoflavones on hippocampal neurons:

Molecule	Subject	Target / paradigm	Mean	Variable parameter	Effect / Conclusion	Reference
Genistein 0.1 μM to 40 μM	Sprague-Dawley PND1 rat pups	Primary HPC neuronal cultures βamyloid-induced neurotoxicity	Spectrofluo-metry Biochemical assays Cell imaging	Neuronal apoptotic features $[Ca^{++}]_i$ assays ROS production Caspase-3 activity	Genistein is anti-apoptotic: It decreses DNA fragmenta-tion It decreses ROS production It blocks increased $[Ca^{++}]_i$ It blocks caspase-3 activation These effects of genistein were blocked by ICI 182,780	Zeng *et al.* 2004
Genistein 10 ng/mL to 10 μg/mL	ED18 rat foetuses	Primary HPC neuronal cultures Glu- or βamyloid-induced neurotoxicity	*In vitro* biochemical assays Cell imaging	Neuronal survival plasma membrane peroxidation (LDH index) mitochondrial activity (MTT reduction)	< 10 μg/mL, genistein rescues plasma membrane integrity but not mitochondrial meta-bolism in both models Similar results were obtained with genistin At 10 μg/mL, genistein is neurotoxic for cultured HPC neurons	Zhao *et al.* 2002
Daidzein 10 ng/mL to 10 μg/mL	ED18 rat foetuses	Primary HPC neuronal cultures Glu- or βamyloid-induced neurotoxicity	*In vitro* biochemical assays Cell imaging	Neuronal survival plasma membrane peroxidation (LDH index) mitochondrial activity (MTT reduction)	Daidzein rescues plasma membrane integrity but not the mitochondrial function Similar results were obtained with daidzin Unlike genistein, daidzein at 10 μg/mL was not found to be neurotoxic for cultured HPC neurons	Zhao *et al.* 2002

Abbreviations: DNA: deoxyribonucleic acid, ED: embryonic day, LDH: lactate dehydrogenase, MTT: 3-[4,5-di-methylt-thiazol-2-yl]-2,5-diphenyl tetrazolium bromide, ROS: reactive oxygen species. (see Tables 1.1-2. for further abbreviations).

Table 1.4. Pharmacological effects of isoflavones on hippocampal neurons.

Molecule	Subject	Target / paradigm	Mean	Variable parameter	Effect / Conclusion	Reference
Genistein 10 μM	10-25 days old rats	HPC slices Stratum Radiatum interneurons α_7nAChR-mediated signalization	Electrophysiology *In vitro* bio-chemical assays	ACh-evoked α_7nAChR currents Cell surface expression of α_7nAChRs	Genistein, but not daidzein, increases ACh-evoked currents Genistein also increases the number of cell surface-localized α_7nAChR subunits	Cho *et al.* 2005
Genistein 10-10,000 nM	ED15-ED17 Swiss mouse embryos	Dissociated HPC neurons Cell cultures Glu-induced neurotoxicity	Cell imaging *In vitro* biochemical assay Pharmacological inhibitors	Neuronal apoptosis Caspase-3 activity assays Calcein visualization LDH measurement	Genistein exerted a robust dose- dependent inhibition over Glu- induced caspase-3 activation and cellular LDH depletion When applied alone, genistein did not affect these parameters Genistein effects were completely abrogated by ICI 182,780 and greatly potentiated by a GSK-3β inhibitor	Kajta *et al.* 2007
Genistein 1-10 μM	AD-deceased men and woman *vs* control subjects Age: 68-96 years old N = 28	HPC tissues from postmortem brains HPC Membrane preparations Isoflavone signalization	[35S]-GTPγ-S-binding assay *In vitro* bichemical assay	[35S]-GTPγ-S-binding to HPC plasma membrane preparations	Genistein dose-dependently stimulates [35S]-GTPγ-S-binding to the HPC plasma membrane preparations from both control and AD brains Genistein effect was similar in both AD and control hippocampi	Jefremov *et al.* 2008

Abbreviations: [35S]-GTPγ-S: sulphate 35-radio-labeled guanosine-triphosphate, AD: Alzheimer's disease, GSK-3β: glycogen synthase kinase-3β. (see Tables 1.1-3. for further abbreviations).

Table 1.5. Pharmacological effects of isoflavones on hippocampal neurons.

Molecule	Subject	Target / paradigm	Mean	Variable parameter	Effect / Conclusion	Reference
Daidzein 1-10 μM	AD-deceased male and females subjects *vs* control Subjects Age: 68-96 years old N = 28	HPC tissues from postmortem brains HPC membrane preparations Intracellular Isoflavone signalization	[35S]-GTPγ-S-binding assays *In vitro* bichemical assays	G-protein-mediated [35S]-GTPγ-S-binding to HPC plasma membrane preparations	Daidzein dose-dependently stimulates [35S]-GTPγ-S-binding to the HPC plasma membrane preparations from both control and AD brains Daidzein effect was similar in both AD and control hippocampi	Jefremov *et al.* 2008
Formononetin 10 ng/mL 100 ng/mL 1000 ng/mL	ED18 rat embryos	Primary HPC neuronal cultures Glu- or βamyloid-induced neurotoxicity	*In vitro* biochemical assays Cell imaging	Neuronal survival Plasma membrane pero-xidation (LDH index) Mitochondrial activity (MTT reduction)	Formononetin protects HPC neurons from the LDH depletion but not from the impaired mitochondrial function in both neurotoxicity models	Zhao *et al.* 2002
Equol 10 ng/mL 100 ng/mL 1000 ng/mL	ED18 rat embryos	Primary HPC neuronal cultures Glu- or βamyloid-induced neurotoxicity	*In vitro* biochemical assays Cell imaging	Neuronal survival Plasma membrane peroxidation (LDH index) Mitochondrial activity (MTT reduction)	Equol rescues plasma membrane integrity but not mitochondrial function in both neurotoxicity models	Zhao *et al.* 2002
Genistein 1 μM	PND0 (neonates) rat pups	Dissociated HPC DG neurons Hormonal regulation of channel activity	ICC Receptor Imaging	Insulin-elicited AMPA-receptor internalization	Genistein blocks insulin-elicited AMPA receptor internalization but not that induced by AMPA or NMDA	Beattie *et al.* 2000

Abbreviations: ICC: immunocytochemistry, (see Tables 1-4, for further abbreviations).

Table 1.6. Pharmacological effects of isoflavones on hippocampal neurons.

Molecule	Subject	Target / paradigm	Mean	Variable parameter	Effect / Conclusion	Reference
Daidzein 5-100 nM	ED17 Wistar rat embryos	Dissociated HPC neurons Cell cultures Neurotrophic signalization of isoflavone	Cell imaging *Western* blotting ICC Pharmacological tools Cell fractionation	Neurotrophism Immuno-visualization of Tau, GAP-43, β-actin, ERα and ERβ DiI labelling	Up to 100 μM concentration, daidzein did not impair the cell viability of cultured HPC neurons Daidzein increased axon formation and elongation in a dose-dependent manner 30 μM daidzein treatment resulted in: increased number of filopodia, increased translocation of GAP-43 and ERβ, but not ERα, to the cell membrane, increased phosphorylation of GAP-43 and PKCα The effects of daidzein were abolished by pretreatments with antagonist, ICI 182,780	Wang *et al.* 2008
Genistein 0.1, 1 or 10 mg/Kg b.w.	Wistar albino female rats 7-days OVX	DG hilus area of the HPC KA-induced neurotoxicity	Histological analysis Cell counting	Hilar neuronal degeneration (numerical quantification of neuronal loss within the DG hilus)	At 10 mg/Kg b.w., genistein showed a strong neuroprotective potency against KA-induced neuronal loss in the hilus area of the DG	Azocoitia *et al.* 2006
Genistein 250 μg/day for 7 or 12 days	Sprague-Dawley adult rats OVX females	HPC CA1 neurons from postmortem brains HPC membrane preparations Intracellular signalization of isoflavones	Transmission Electron Microscopy (TEM) HPLC	Ultrastructural features of HPC CA1 Mitochondrial ATP content in the HPC CA1 region	OVX resulted in a morphological alteration of the CA1 neuronal mitochondria without affecting their numerical density OVX decreased the HPC ATP contents Genistein treatment efficiently reversed the two impaired parameters within the HPC CA1 area	Xu *et al.* 2008

Abbreviations ATP: adenosine triphosphate, ERα: estrogen receptor α, ERβ: estrogen receptor β, GAP43: growth cone-associated protein 43, OVX: ovariectomized, KA: kainic acid, PKCα: protein kinase Cα. (see Tables 1-5, for further abbreviations).

activated cell. Using thapsigargin along side with the Fura-2 Ca^{++} imaging, Koss *et al.* (2009) were able to show that genistein (100 or 10 μM depending on the time of treatment) remarkably reduces both the amplitude and the latency of the peak response of SOCC-mediated thapsigargin-elicited Ca^{++} influx into dissociated hippocampal neurons isolated from Sprague-Dawley PND3-5 rat pups.

In the configuration of neuroprotection against chemical insult, primary hippocampal neuronal cultures derived from PND1 rat pups were used to investigate the β-amyloid-induced neurotoxicity. The results, interestingly, highlighted the anti-apoptotic mechanisms of the isoflavones within hippocampal neurons. Indeed, Zeng *et al.* (2004) have shown that, in the β-amyloid-induced neurotoxicity model, genistein (40 μM) was efficiently able (1) to decrease both the resulting DNA fragmentation and reactive oxygen species (ROS) production and (2) to block the associated increases in $[Ca^{++}]_i$ and pro-apoptotic caspase-3 activity. This study did also demonstrate that the anti-apoptotic neuroprotective effects of genistein were completely abrogated by the addition to the neuronal culture medium of the ER antagonist, ICI 182,780, evidencing that the isoflavone may act through an ER-binding, gene expression-regulating mechanism to protect hippocampal neurons against the β-amyloid neurotoxic insult. Elsewhere, ED18-derived primary neuronal cultures were used to investigate the neuroprotective effects of genistein in glutamate- and β-amyloid-induced neurotoxicity models. The results showed that, in both models and compared to control, low concentrations of genistein (<10 μg/mL) were efficiently able to rescue the plasma membrane integrity (as evidenced by the decreased lactate dehydrogenase (LDH) depletion index) but not the mitochondrial metabolism (as evidenced by the unaltered levels of 3-[4,5-di-Methyl-Thiazol-2-yl]-2,5-diphenyl Tetrazolium bromide (MTT) reduction index) (Zhao *et al.* 2002). In this study, daidzein, formononetin and equol all have been reported to have similar neuroprotective effects to genistein demonstrating, though, an estrogenic effect rather than an anti-tyrosine kinase effect. However, up to 100 ng/mL (i.e. 0.37 μM) concentration, genistein did not display any noticeable neurotrophic effect upon ED18-derived cultured hippocampal neurons (0 parameter affected, out of 6 morphological parameters that were visibly enhanced by 10 ng/mL E_2). In addition, at 10 μg/mL (*i.e.* 37 μM) concentration, genistein, but not daidzein, was rather neurotoxic for the studied neurons (Zhao *et al.* 2002) probably because of anti-protein kinase (PK) effects. On the contrary, glutamate-mediated neuronal apoptosis of dissociated hippocampal neurons isolated from ED15-17 Swiss mouse embryos was efficiently and dose-dependently opposed by increasing concentrations of genistein (0.01 to 10 μM) in the culture medium (Kajta *et al.* 2007). Again, the effects of genistein were completely abrogated by the ER antagonist, ICI 182,780. The effect is stronger in hippocampal neurons than in the other types of neurons tested, and genistein already exerts a protective effect at 0.01 μM. This low dose elicited effect would advocate for an estrogenic membrane effect. Inversely, these effects were greatly potentiated by a glycogen synthase kinase-3β (GSK-3β) inhibitor. Finally, the authors of this study have also shown that, up to a 10 μM concentration, genistein did not trigger any neurotoxic effect in hippocampal neurons (Kajta *et al.* 2007).

VI.1.3.b. In vivo approaches

Azocoitia and colleagues (2006) studied the neuroprotective effects of a single intra-peritoneal (i.p.) injection of a soy extract preparation (containing 40% of isoflavones, 0.2, 1, 2, 20 mg/Kg body weight, b.w.) or increasing concentrations of genistein (0.1, 1, 10 mg/Kg b.w.) upon the kainic acid (KA)-induced degeneration of hilar DG neurons in 7-day-ovariectomized (OVX) Wistar albino female rats. Both of the highly dosed injections of soy extracts (1, 2, 20 mg/Kg b.w.) or genistein (10 mg/Kg b.w.) were efficiently proactive in counteracting the neuronal loss within the hilus area of the DG. Interestingly, when injected alone at the highest dose, each one of the tested isoflavone preparations could not initiate any evident adverse effect (namely, neuronal loss) in the hippocampal formation. Xu *et al.* (2008) examined the effects of genistein, given as a short-term treatment (7-12 days, 250 μg/day) to OVX Sprague-Dawley female rats, upon the OVX-induced impairment of hippocampal function. Two parameters were addressed in this study: the ultrastructural feature of mitochondria and the ATP contents in the hippocampal CA1 region. It was found that genistein treatment opposed both OVX-induced impairments of hippocampal CA1 neurons, namely morphological abnormalities of mitochondria (but not their numerical density) and decreased ATP contents.

VI.1.4. Neurotrophic features

Under the paradigm of neurotrophic signalization by isoflavones, Wang *et al.* (2008) have demonstrated that, even used at a 100 μM concentration in the culture medium, daidzein does not impair the viability of dissociated hippocampal neurons isolated from ED17 Wistar rat embryos. More interestingly, daidzein dose-dependently increased both axon formation and elongation. At 30 μM concentration, daidzein treatment resulted in:

(1) Increased number of filopodia,
(2) Increased translocation of GAP43 to the tip of elongating neurites,
(3) Increased translocation of ERβ, but not ERα, to the plasma membrane, and
(4) Increased phosphorylation levels of both GAP43 and PKCα-isoform (PKCα).

VI.2. Discussion of Pharmacological Effects

As described above, the pharmacological investigations of isoflavone effects on hippocampal neurons, derived from a number of studies that have been using these molecules in a wide spectrum of concentrations ranging from 5 nM (for daidzein, Wang *et al.* 2008) to 250 μM (for genistein, Casey *et al.* 2002). Although, all the data are interesting to consider, those that did use isoflavone in pharmacological dosing likely reachable by dietary means, will be pointed out.

Concerning the bioelectrical features of isoflavone effects, it was shown that, in both CA1 pyramidal neurons (O'Dell *et al.* 1991) and DG granule cells (Casey *et al.* 2002), genistein could negatively impact the expected induction of LTP that normally follows the appropriate electrical stimulation of the respective afferent pathways. However, these two studies did use very high genistein concentrations (110 μM, for the study by O'Dell and 250 μM for the study by Casey). Although, in the study by Casey and colleagues, the concentration of genistein at the immediate vicinity of DG granule cells was, most probably, much inferior to 250 μM for 4 main reasons:

(1) Genistein-containing solution was injected into the lateral ventricle where the cerebrospinal fluid (CSF) may have diluted the drug,

(2) A very small volume was injected (5 µl), and thereof, the CSF dilution effect of the injected drug solution should not be neglected,
(3) The CSF is a circulating biological fluid, and thereof, some quantity of the drug might have been carried away from the lateral ventricle before having reached the hippocampal formation, and, then, the DG granule cells, and
(4) The waste inherent to such intervention (intracerebro-ventricular injection, icv) in whole animal.

Quite interestingly, using genistein at a much lower concentration (20 µM) in the hippocampal slice culture medium, results in a reduction of induced LTP in the CA1 pyramidal neurons of young rats and an almost complete abolition of induced LTP in CA1 pyramidal neurons of aged rats (Shankar *et al.* 1998).

Nonetheless, the concentrations reported above, most likely, would not be reached in the circulating plasma, under the conditions of nutritional consumption of even a highly dosed isoflavone-containing diet or supplementation. In addition, brain slicing, in an unrealistic fashion, exaggerates the exposure of cells to chemicals, and most probably amplifies the biological cell response. Because blocking LTP induction by blocking the associated NMDA receptor signalization impairs both spatial reference memory and verbal memory (Rowland *et al.* 2005), the argument in favour of our assertion is that animals exposed to highly dosed isoflavone-based diets are still able to learn and to retain spatial information in hippocampus-challenging behavioural tasks (Lund[1,2] and Lephart 2001, Lund *et al.* 2001, Lee *et al.* 2004, Luine *et al.* 2006, Monteiro 2008). Moreover, E_2 has been shown to increase both the CA1 neuronal excitability through a non-NMDA-dependent mechanism (Wong and Moss, 1992) and the magnitude of LTP in the CA3-CA1 synapses through a genomic mechanism involving an upregulation of the NMDA receptor subunit, NR2B (Foy *et al.* 1999, Smith and McMahon 2005, 2006, Smith *et al.* 2009) or, at least, through the translocation of NR2B subunit to sites of active synapses (Snyder *et al.* 2010) and/or through post-transcriptional modulation of NMDAR1 subunit (Gazzaley *et al.* 1996). The use of high doses of genistein in the two studies mentioned above, was, obviously, intended to induce a pharmacological inhibition of hippocampal PTKs in order to gain insights into the mechanisms of LTP induction and, thus, should be viewed as mechanistic investigations. Indeed, the respective authors did report robust genistein-mediated inhibitions of pp60c-rsc(+), protein tyrosine kinase, activity (O'Dell *et al.* 1991) and ERK phosphorylation (Casey *et al.* 2002). In keeping with this, E_2-treatment of hippocampal slices also induced LTD and an ERα-mediated, increase of spine density on CA1 neurons and a decrease of thorn density on CA3 neurons, most likely, through the activation of ERK/MAPK pathway (Ogiue-Ikeda *et al.* 2008). Taken together, these findings may suggest that the inhibition of hippocampal PTKs is unlikely in humans consuming isoflavone-containing foods, since the reported plasma concentrations do not exceed (10 µM) (Mathey *et al.* 2006). Additionally, the pharmacological effects of high doses of genistein on the bioelectrical activity of hippocampal neurons were also seen as:

(1) An inhibition of IGF-1-induced, AMPAR-mediated increased excitability (220µM, Ramsey *et al.* 2005),
(2) An alleviation of mGluR-mediated inhibition of I_{AHP}, (100 µM, Abdul-Ghani *et al.* 1996),

(3) A blockade of 4βPMA-mediated potentiation of NMDA-activated currents (50 μM, Lu *et al.* 1999),
(4) A reduction of depolarization-coupled HVA Ca^{++} currents (100 μM, Potier *et al.* 1999), and
(5) A potentiation of the ACh-evoked currents (100 μM for Charpantier *et al.* 2005, 10 μM for Cho *et al.* 2005).

Quite strikingly, completely opposite effects have been reported for much lower concentrations of E_2, in different intracellular signalizations taking place in brain neurons. Indeed, E_2 was shown:

(1) to elicit induction of IGF-1 mRNA expression in immortalized hippocampal neurons (Shingo and Kito, 2003),
(2) to activate mGlu receptors through an ER-dependent mechanisms (Boulware and Mermelstein 2009) and to decrease the amplitude of I_{AHP} currents, most likely, through an inhibition of L-type Ca^{++} channel (Kumar et Foster 2002),
(3) To enhance an NMDA-dependent induction of hippocampal LTP (Foy *et al.* 1999, Smith and McMahon 2005, 2006, Foy *et al.* 2008, Smith *et al.* 2009)
(4) To bind directly and, consequently, to potentiate L-type voltage-gated Ca^{++} channel (VGCC) currents in embryonic ED18 primary cultured hippocampal neurons (Sarkar *et al.* 2008),
(5) To inhibit nicotine-evoked $\alpha_4\beta_2$nAChR currents (Damaj, 2001) although an allosteric effect-mediated potentiation of ACh-evoked human $\alpha_4\beta_2$nAChR currents was reported for E_2 in both *Xenopus* oocytes and HEK cells (Curtis *et al.* 2002).

Altogether, these results confirm that the findings obtained with high doses of isoflavones, and more particularly of genistein, in hippocampus-derived neuronal systems, are more relevant to PTK-inhibiting signalizations assimilated to anti-estrogenic effects rather than E_2-mimicking signalizations.

We also reported data depicting physio-molecular alterations that were triggered by isoflavone molecules. These data did, for example, document the mechanistic features of channel trafficking in hippocampus-derived cell systems. Here again, a wide range of isoflavone concentrations have been used. More specifically, genistein was capable of blocking insulin-elicited AMPA receptor internalization with relatively a lower concentration than those used to impact the bioelectrical activity (1 μM, Beattie *et al.* 2000). This dose that is, probably, reachable under normal nutritional conditions, might affect some AMPA-related neuronal excitability and/or LTP expression in subjects exposed to dietary/supplemental isoflavones. However, such investigations still have to be conducted under nutritional isoflavone content-controlled experimental conditions.

Finally, concerning their well established characteristics as robust neuroprotective molecules in numerous cell systems, we described the wide extent of isoflavone protective effects upon hippocampal neurons that have been challenged by different cell stressors. These life-saving features of isoflavones upon hippocampal neurons might promote the integrity of hippocampal neuronal networks and, consequently, that of the upon-grafted learning and memory functions.

Furthermore, one interesting aspect of isoflavone effects on hippocampal neurons consisted in their ability to operate as neurotrophic signalling molecules. Indeed, daidzein was capable of triggering axonal outgrowth in dissociated hippocampal neurons at 30 μM concentration (Wang *et al.* 2008). Although such a dose is likely still out of reach to a dietary/supplemental isoflavone exposure, this effect of daidzein resembles those exerted by E_2 on a variety of brain neurons (De Lacalle 2006). Therefore, daidzein should be tested on such outcomes at lower doses for longer exposure-time.

VII. Pharmacological Effects of Isoflavones in non-Hippocampal cells

In the present section, we provide some data describing the effects of isoflavones in non-hippocampal neuronal systems. Because gaining some more insights into the effects of isoflavones in non-hippocampal, neural tissues may be helpful in highlighting further the mechanistic regulations taking place in the hippocampal formation. We also provide data depicting the effects of isoflavones that have been studied in non-neuronal cells. This might be helpful in broadening our view angle on the mechanistic features of isoflavone signalization.

VII.1. Isoflavone Signalization in Neuronal Cells

Paillart *et al.* (1997) have reported that genistein binds directly and inhibits voltage-sensitive sodium (Na^+) channels in rat foetal brain neurons and cerebellar granule cells (IC_{50} = 60 μM). Also, daidzein (IC_{50} = 195 μM) was able to mimic the inhibitory effect of genistein on Na^+ channel activity. Genistein (50 μM), but not daidzein (50 μM), reversibly reduces the slope conductance of the GABA-A-mediated currents in cultured dorsal medulla neurons from Wistar rat foetuses (Wan *et al.* 1997). Using retinal rod photoreceptors and olfactory receptor neurons from tiger salamanders, and transfected *Xenopus* oocytes, Molokanova *et al.* (1999, 2000) were able to show that genistein, at doses as low as 10 μM and up to 1000 μM, dramatically attenuates the apparent affinity of cyclic nucleotide-gated (CNG) channels for cyclic-guanosine-mono-phosphate (cGMP) and inhibits the maximal current through CNG channels. Interestingly in the absence of ATP, the inhibition is already observed, although not complete, from 1 μM exhibiting a nice dose-response effect. Somponpun and Sladek (2002) have reported that genistein at 0.1 μM is an efficient inhibitor of NMDA-stimulated AVP and Oxyt release from explants of the hypothalamo-neurohypophyseal system, probably, through an ER-mediated mechanism.

In a model of βamyloid-induced cell death, genistein (50 μM) was reported to have no neuroprotective effect upon cultured cortical neurons from Sprague-Dawley PND1 rat pups (Wang *et al.* 2001). In human SH-SY5Y neuroblastoma cells, which are known to express both ERα and ERβ, genistein (at 1 and 10 μM) displayed pro-survival effects against βamyloid-induced neuronal death and DNA fragmentation (Bang *et al.* 2004). These effects of genistein, that were abrogated by the ER antagonist, ICI 182,780, did not seem to be mediated through a blockade of PTK activity but rather through an ER-mediated genomic action (Bang *et al.* 2004).

A human embryonic carcinoma-derived cells, namely hNTera-2/CD1 (also called NT-2) cells, exhibit human neuronal progenitor cell-like characteristics after continuous exposure to retinoic acid (under these conditions, these cells undergo neuronal differentiation). In this model, genistein was shown to induce neuronal differentiation and neurite outgrowth at low dose (5 μM) but not at high doses (10 and 15 μM). It also induced ERK1/2 phosphorylation in an irreversible way and a dose dependent manner, and did upregulate protein expression for cyclin-dependent kinase (Cdk5), protein-21 (p21) and N-cadherin, indicating a positive role in initiating neuronal differentiation rather than neuronal proliferation (Hung *et al.* 2005). Bartlett (1997) did demonstrate that under a low concentration of NGF in the culture medium (100 fg/mL) and even used at a low dose (2.5 μM), genistein was able to promote efficiently both neuronal survival and neurite outgrowth in cultured embryonic chick sensory neurons.

Schreihofer and Redmond (2009) addressed the neuroprotective ability of isoflavones, genistein, daidzein and equol, in cultured primary cortical neurons taken from Long-Evans rat ED18 foetuses. In this case, the neuroprotective ability of these isoflavones was investigated under ischemia-like conditions, namely, glutamate-insult, thapsigargin-induced apoptosis, hypoxia and oxygen-glucose deprivation (OGD). The main findings were as follow: (1) genistein, daidzein and equol, at low concentrations (1 μM), were all capable of protecting embryonic cortical neurons from ischemic insult-induced apoptosis through an ER-mediated mechanism, (2) genistein, most likely, operated through a MAPK/PI3K-dependent mechanism, (3) daidzein did act through a PI3K-dependent mechanism and (4) equol, too, may have, possibly, operated through a PI3K-dependent mechanism. On the contrary, in another study, genistein was found to be rather pro-apoptotic on cultured cortical neurons from ED18 foetuses (Linford *et al.* 2001).

VII.2. Isoflavone Signalization in non-Neuronal cells

As a widely expressed neurotrophin, BDNF plays a crucial role in the neuronal signalization throughout the central nervous system and, more specifically, within the hippocampal formation neuronal networks (Cowansage *et al.* 2009). BDNF activates TrKB which is able to trigger intracellular/membrane signalization by initiating a cascade of protein phosphorylations. Rogalski *et al.* (2000) reported that genistein (100 μM) blocks BDNF-activated, TrKB-mediated inhibition of the G protein-activated inwardly rectifying K+ channels (Kir.3) in a transfected Xenopus oocyte model.

Under the paradigm of rapid signalization, the GPR30 is capable of translating, within few seconds-to-minutes time laps, the extracellular E_2 signal into an intracellular one in a number of cells. Indeed, Thomas and Dong (2006) used a human HEK293 cell line devoted of ERα, ERβ and GPR30 to investigate the putative rapid signalization by genistein (0.2 μM) in GPR30-transfected HEK293-cells. They showed that genistein (0.2 μM) binds with high affinity (1/10th of that of E_2) to plasma membranes prepared from GPR30 expression vector-transfected cells. The binding of genistein to the plasma membranes elicited a robust production of cAMP, most likely, throughout the activation of a membrane-associated adenylyl cyclase system. Using ERα-positive MCF7, ER-negative/GPR30-positive SKBR3, GPR30-negative MDA-MB231 and GPR30-negative BT-20 breast cancer cells, Maggiolini and colleagues (2004) did demonstrate that genistein (1 μM) is highly efficient in eliciting c–fos, pS2 and cathepsin D gene expression through both ERα-dependent and ERα-independent mechanisms. The ERα-independent mechanism involved GPR30-mediated transduction of genistein signal through Gβγ-associated activation of ERK1/2 pathway. Vivacqua *et al.* (2006) have investigated genistein signalization in human WRO cell line, human anaplastic

FRO and ARO thyroid tumour cell lines and in HeLa cells. These authors reported that the genistein-enhanced proliferation rate obtained at 0.1 μM was associated with an upregulation of the expression levels of cyclin A, cyclin D1, Bcl-X_L and c-fos mRNAs, while the expression levels of progesterone receptor or pS2 remained unaffected by genistein (1 μM for one hour). A c-fos promoter lacking the ERE-responsive element was still able to mediate transactivation by genistein (1 μM), most probably, through a MEK/ERK/PI3K/MAPK signalling pathway. Interestingly, genistein-elicited responses were completely abrogated when the cells were pretreated with a GPR30 antisense oligodeoxynucleotide (AS-ODN).

Quite interestingly, another aspect of isoflavone signalization is examplified by the fact that these molecules are capable of modulating the endogenous estrogen metabolism. Indeed, in a cell culture model of hepatic HepG2 cells, and oppositely to E_2, genistein was shown to increase, dose-dependently (0.1 to 10 μM), both the enzymatic activity level and the gene expression level of the aromatase, P450-*CYP19* (Ye *et al.* 2009). Albeit, it was not statistically significant at the dose of 0.1 μM, the tendency of the aromatase activity to increase was already obviously seen at this concentration (Ye *et al.* 2009). It became manifest at the concentrations of 1 μM and 10 μM (Ye *et al.* 2009). This effect that could not be triggered by E_2, involved the activation of the intracellular PKCα/P38/ERK1-2 signalling pathways that lead to the ultimate activation of CREB-dependent gene regulation (Ye *et al.* 2009). Similar effect was obtained by genistein (1 nM) in human endometrial stromal cells (Edmunds *et al.* 2005). In contrast, biochanin A was shown to downregulate aromatase P450-*CYP19* activity in *CYP19*-transfected MCF-7 cells and the mRNA expression levels in an ER-negative, breast cancer cell system, SK-BR-3 (Wang *et al.* 2008). Similar inhibitory effects were obtained by genistein (10 μM) and daidzein (10 μM) or by the combination of genistein (100 nM), daidzein (100 nM) and biochanin A (100 nM) on human granulosa-luteal (hGL) cells (Rice *et al.* 2006). In the same context, isoflavone-triggered inhibition of aromatase and 3β-hydroxysteroid dehydrogenase (HSD) activities was obtained by genistein (10 μM), daidzein (10 μM), biochanin A (10 μM) in hGL cells (Lacey *et al.* 2005). The fact that most of these concentrations, that did efficiently modulate the activity of steroid metabolizing enzymes, are likely reachable under nutritional conditions, suggest that dietary/supplemental isoflavones may be capable of triggering some E_2-non-mimicked, own specific effects in different organs of the body and, probably, in some brain neurons, as well. Although, a previous study that did examine the dietary effects of an isoflavone-rich diet (Phyto-600) on the enzymatic activity of aromatase, could not report any significant change in the preoptic area/mediobasal hypothalamus of Phyto-600 fed male and female rat brains (Lephart *et al.* 2001). Therefore, isoflavone-mediated regulations of neurosteroidogenic enzyme activities remain to be addressed in brain cells and, more particularly, in the hippocampal formation neurons.

VIII. Physio-Molecular Impact of Isoflavones on Brain Physiology

The studies presented below deal mainly with dietary exposure of animal models. To help interpreting the respective results, a tentative to correlate animal and human exposure is given here. First according to Adlercreutz *et al.* (1993) a Japanese man consume between 1 and 2 kg of food a day containing between 45 and 60 mg isoflavones. Second a rodent needs between 5 and 10 times more isoflavones to reach the same plasma bioavailable isoflavone levels than man (Xu *et al.* 1994, Izumi *et al.* 2000, Picherit *et al.* 2000). This means that a 10 mg/Kg of body weight isoflavone exposure in rodent is roughly equivalent to 1 or 2 mg/Kg of body weight in humans. Besides impacting the functioning of both hippocampal and non-hippocampal neuronal systems, and non-neuronal cell systems, as well, isoflavones have also been shown to deeply impact the brain physiology. This impact may probably affect indirectly the hippocampal formation function.

Lephart *et al.* (2000) evaluated the impact of a diet highly enriched with soy isoflavones (Phyto-600, *i.e.* 600 µg/g of diet, diet i.e. in man between 60 to 120 mg/Kg of food roughly twice the exposure reported in Japanese people) on brain physiology of Sprague-Dawley male rat. They showed that Phyto-600 fed rats displayed increased levels of isoflavones in circulating plasma. The bioavailable doses were 2465 ng/mL which is about 10 µM and can be reached in certain humans with 50 mg isoflavones twice a day in the morning and the evening (Mathey *et al.* 2006). Lephart *et al.* (2000) did further show that Phyto-600 diet results in (1) increased levels of isoflavones in mediobasal hypothalamus/preoptic area (69,9 ng/g), (2) unaffected levels of enzyme activity of aromatase and 5-α-reductase, two androgen-metabolizing enzymes, in mediobasal hypothalamus/preoptic area, (3) decreased mRNA expression levels of the neuroprotective Ca^{++}-binding proteins, calbindin and calretinin in the amygdala and (4) decreased mRNA expression levels of calbindin, but not calretinin, in mediobasal hypothalamus/preoptic area. In the adult male rat frontal cortex, isoflavone-enriched diet was associated with both decreased expression levels of calbindin mRNA (a pro-survival factor) and increased expression levels of cyclooxygenase-2 (an inducible inflammatory factor). The same research group (Bu and Lephart 2007) did show that Phyto-600 fed male rats displayed 22 % decrease in the number of ERβ-immunopositive neurons in the anteroventral periventricular nucleus (AVPV), most likely due to an increased rate of apoptosis occurrence as evidenced by the increased coexpression of TUNNEL immuno-staining and ERβ-IR within AVPV neuronal population. Pan[1] *et al.* (1999) studied the effects of a soy isoflavone-enriched diet (equivalent for a woman, to a daily intake of 150 mg of isoflavones) in OVX Sprague-Dawley retired breeder rat brains. They showed that, after 8 weeks on diet, dietary isoflavones were able to upregulate the expression levels of BDNF mRNA in the frontal cortex of OVX rats. This effect resembled that of a control diet with added E_2 (equivalent to 2 mg/day) (Pan *et al.* 1999). On the contrary, in adult male hooded Lister rats, a soy isoflavone-containing diet (150 µg isoflavones/g of diet) was associated with a downregulation of BDNF mRNAs in the CA3 and CA4 fields of the hippocampal formation and also in the cerebral cortex, although the expression levels of GAD67 mRNAs and GFAP mRNAs were not affected by the diet in the studied regions (File *et al.* 2003). More specifically, a genistein-enriched (1250 ppm) diet was associated with increased levels of hypothalamic contents in arginin-vasopressin (AVP), but not β-endorphin (Scallet *et al.* 2003). In Long-Evans male rats exposed to Phyto-600 diet, as compared to those fed an isoflavone-free diet, the expression levels of BAD protein (a proapoptotic factor) were (1) increased in the amygdala, (2) decreased in the mediobasal hypothalamus and frontal cortex and (3) unaffected by the diet in the cerebellum and the hippocampus, whereas the expression levels of βIII-tubilin (an early marker of neuronal differentiation/survival) were (1) increased in the amygdala, hippocampus, mediobasal hypothalamus and frontal cortex but (2) unaffected in the cerebellum.

Under the paradigm of aversive conditioning, day-old chicks were tested in a discriminated passive avoidance task. It was found that intra-striatal injection of genistein produces a strong amnesic effect on passive avoidance learning (Whitechurch *et al.* 1997) probably, through an anti-protein tyrosine kinase mechanism.

Lee and colleagues (2004) have compared the effects of 2 isoflavone-dosed diets (0.3 g of isoflavones/Kg of diet and 1.2 g of isoflavones/Kg of diet; one is a plausible dosing for human exposure and the other is about twice the maximum recorded in Asia) to a third isoflavone-free control diet, on age-related impairment of hippocampal functioning and brain cholinergic signalization in aged Sprague-Dawley male rats. These authors have reported that isoflavone-enriched diets decreased acetylcholine esterase (AChE) activity and increased ChAT activity in hippocampal/basal forebrain/cerebral cortical areas, and increased ChAT-IR in both medial septum and hippocampus. All these effects were not seen altogether in the same diet group. In OVX rats, both E_2-treated (3.9 mg/Kg b.w. very high dose) and high soy isoflavone-rich diet (526.9 mg/Kg, largely over the human consumption) fed group did display both lower and higher levels in, respectively, hippocampal AChE activity and basal forebrain ChAT-IR as compared to the two other groups that have been fed with a low soy isoflavone-rich diet (263.4 mg/Kg about 2 or 3 mg/Kg b.w. for a human being) or a soy isoflavone-free control diet (Lee *et al.* 2009). Soy isoflavone-triggered lowering of both hippocampal and frontal cortex AChE activities was also seen in hypercholesterolemic mice (Liu *et al.* 2007). These findings show that at high-dosed soy isoflavones are capable of impacting positively, the cellular metabolism of hippocampus-projecting septomedial-basal forebrain cholinergic neuronal system. Therefore, dietary soy isoflavones might have some inherent potential for the alleviation of age-related degeneration of ACh-synthesizing neurons, and maybe also in human subjects. However, because only E_2 did increase the ChAT enzyme activity in the hippocampus of OVX rats (Lee *et al.* 2009), it seems evident that isoflavones, or at least the soy isoflavone extract used herein, would not trigger all the beneficial effects of native E_2 in the exact same brain areas, not to mention when considering different brain structures. This is a red flag to be waved in the face of those who want to replace estrogen hormone therapy with isoflavones, not to mention the pro-apoptotic/pro-inflammatory effects triggered by these substances in some specific brain regions (see above). Nevertheless, this assertion must be tempered by the fact that we do not know much about the bioavailability/signalizing effects of isoflavones in the human brain and, more particularly, at the level of the hippocampal-projecting cholinergic system.

IX. Effects of Dietary/Supplemental Isoflavones on Cognitive Processes

In the present section, we report a number of data that used isoflavones as dietary and/or supplemental products for their putative ability to impact cognitive brain function. By doing so, we are southing to get the largest view possible on the signalization potential of these molecules under normal conditions of daily nutritional use. Firstly, we are going to summarize these data according to the model examined. Secondly, we are going to discuss the main resulting findings in the light of what is known about the physiology of the hippocampal formation.

IX.1. Description

In the present section, we report a number of data that used isoflavones as dietary and/or supplemental products to deepen our knowledge of their impact on cognitive brain function. Indeed, dietary/supplemental isoflavones have been investigated in (1) different strains of rat (Long-Evans, Sprague-Dawley and Wistar rats), (2) a non-human primate and (3) young, middle-aged or older human subjects. Both genders have been documented in humans in nutritional isoflavone studies. In these studies, the assessment of the cognitive function used well established psychological tasks and mood rating questionnaires. In non-human subjects, however, specific apparatus-based behavioural tasks were called for to gain some objective insights into the learning and memory processes (see Tables 2 for a complete summary).

IX.1.1. Rodents

It is no surprise that, concerning the investigation of isoflavone signalization in whole animals, most studied have been conducted in rodent species (Tables 2.1-3). For example, the impact of a lifelong exposure to a isoflavone-rich diet (Phyto-600, containing 600 μg of isoflavones/g of diet, roughly equivalent in men to 60-to-120mg of isoflavones/day) on memory function was evaluated in Long-Evans male and female rats in the 8 arm-radial-maze. It was reported that, in reference memory-related tasks, the isoflavone-rich diet improves the performance of both female rats and *in-utero* pharmacologically-feminized male rats, whereas it impairs the performance of intact male rats (Lund[1] and Lephart 2001, Lund *et al.* 2001). By contrast, the same diet produced anxiolytic behaviour in both genders when rats were tested in elevated-plus maze (Lund[2] and Lephart 2001.b). Using OVX Sprague-Dawley female rats (2 months of age), Luine and colleagues (2006) did compare the relative effects of a high soy isoflavone-containing diet (810 μg/g of diet, roughly equivalent in men to 80-to-160mg of isoflavones/day) and a low isoflavone-containing diet on the memory function by mean of object placement and object recognition tasks. These authors were able to demonstrate that the animals fed the high-isoflavone-containing diet spent more time exploring the same object at the new location (improved spatial memory), but not the new object at the same location (unaffected object recognition memory). However, as measured in an elevated-plus maze, the instinctively driven exploratory behaviour did not differ between the 2 diet groups.

Using OVX adult female Wistar rats (87-117 days of age), Monteiro *et al.* (2008) compared the relative effects of a soy isoflavone-rich diet (containing 250 μg genistein/g of soy proteins and 230 μg daidzein/g of soy proteins *i.e.* roughly equivalent in men to 50-to-100 mg of isoflavones/day) and a soy isoflavone-free diet on the spatial memory function as assessed in the Morris water maze. The results showed that when animals were put on the isoflavone-rich diet for 30 consecutive days after OVX, all the impaired cognitive parameters seen in the isoflavone-free diet-fed OVX-group were re-adjusted to the levels of those seen in ovary sham-operated group. Indeed, the OVX resulted in (1) an increased latency to find the platform, (2) an increased latency to cross the platform, (3) an increased time spent in the opposite quadrant to target and (4) a decreased time spent in the target quadrant. The isoflavone-enriched diet was able to improve all these 4 parameters and, thus, to rescue the impaired spatial memory function seen in OVX rats. On the contrary, a 60 days pre-treatment (diet initiated 60 days before the rats were OVX) with dietary soy isoflavones could reverse only one cognitive parameter, namely the increased latency to the hidden platform. Lee *et al.* (2004) have compared the effects of 3 diets: (1) a low isoflavone-dosed diet (0.3g of soy

Table 2.1. Effects of dietary/supplemental isoflavones on cognitive processes in rodents.

Diet features	Subject / Target / Mean	Variable parameter	Outcome	Reference
Phyto-600 phytoestrogens-rich diet (600 μ of phyto-estrogens/g of diet) As compared to Phytoestrogen-free diet Lifelong exposure	Long-Evans male and female rats Behavioral testing Diet switching paradigm (40 days after diet switching)	Tests of performance in a 8 arm radial maze 8 arm radial maze task (reference memory) 4/8 arm radial maze task (reference/working memory)	Phyto-600 diet: improves performance in female rats impairs performance in male rats improves performance in feminized male rats (males treated in utero with flutamide, an androgen receptor blocker) In lifelong Phyto-600 fed males, the switching to a Phyto-free diet alleviates the impaired reference/working memory	Lund[2] & Lephart 2001 Lund *et al.* 2001
Phyto-600 and Phyto-free diets Lifelong exposure	Long-Evans male and female rats 50-75 days of age Behavioral testing (anxiety behavior)	Elevated-plus maze Recorded body weight Monitored food and water intake	Phyto-600 diet: increased the number of entries to the open arm area and the time spent on the corresponding platforms (decreased anxiety) increased food and water intake decreased body weight	Lund[1] & Lephart 2001
Isoflavone-rich diet containing 0.3 g of soy isoflavones/Kg diet (S1), Isoflavone-rich diet containing 1.2 g of soy isoflavones/Kg diet (S2), Isoflavone-free control diet 15-16 weeks on diet	Aged Sprague-Dawley male rats age = 10 months at the begining of the exposure to diets Memory function (behavioral testing) Brain cholinergic metabolism in: hippocampus (HPC), cerebral cortex (CCx), basal forebrain (BF) and medial septum (MS)	Cognitive performance: Delayed matching-to-place in a Morris water maze (DMP) task Enzymatic activity measurements: Acetylcholine esterase (AChE) Choline-acetyl transferase (ChAT) Quantitative ChAT-immunoreactivity (ChAT-IR)	Isoflavone diets did not affect food/water consumption or brain weight Isoflavone diets dicreased: AChE activity in HPC and CCx (S1 & S2) and in BF (S1) Isoflavone diets increased: ChAT activity in BF (S1 & S2) and cortex (Cx), but not in HPC Isoflavone diets increased: ChAT-IR in CA1 and MS Isoflavone diets decreased: swimming time in DMP task	Lee *et al.* 2004

Table 2.2. Effects of dietary/supplemental isoflavones on cognitive processes in rodents.

Diet features	Subject / Target / Mean	Variable parameter	Outcome	Reference
Soy phytoestrogen-rich diet containing 263.4 mg/Kg of diet (4.76 mg of isoflavones/day/ rat = S1 diet Soy phytoestrogen-rich diet containing 526.9 mg/Kg of diet (9.48 mg of isflavo-nes/day/rat = S2 diet Isoflavone-free control diet 15-16 weeks on diet	Sprague-Dawley female rats at age = 45 weeks rats were OVX + begining of the exposure to diets Memory function behavioral testing) Cholinergic metabolism in: hippocampus (HPC) cerebral cortex (CCx) diagonal band of Broca (DBB)	Cognitive performance: Reference memory parameters (Morris water maze task): distance to the platform (swimming path length) time to reach the platform (escape latency) Enzymatic activity measurements: Acetylcholine esterase (AChE) Choline-acetyl transferase (ChAT) Quantitative ChAT-immunoreactivity (ChAT-IR)	Isoflavone diets did not affect food/water consumption Isoflavone diets increased: ChAT activity in CCx (S1 & S2) but not in HPC ChAT-IR in DBB (S1 & S2) Isoflavone diets decreased: AChE activity in HPC (S1 & S2) but not in CCx Isoflavone diet (S2) decreased: swimming path length escape latency On the 2nd day of the acquisition	Lee *et al.* 2006
Phytoestrogen-rich diet containing: 810 µg of isoflavones Low phytoestrogen-containing diet 5-7 weeks on diet	Sprague-Dawley female rats 2 month-old OVX Behavioral testing (Anxiety behavior)	The memory testing was performed in an Open-field apparatus using 2 different tasks: Object Placement Task Object Recognition Task The anxiety behavior was monitored in an Eleveted-Plus Maze (EPM) apparatus where the times spent in closed and open arms were recorded	The Phyto-rich diet did affect the performance of rats in the object place-ment test only: the animals did spent more time exploring the object at the new location The 2 groups spent equal amount of time exploring the open and closed arm in the EPM	Luine *et al.* 2006

Table 2.3. Effects of dietary/supplemental isoflavones on cognitive processes in rodents.

Diet features	Subject / Target / Mean	Variable parameter	Outcome	Reference
Soy isoflavone rich diet containing: 250 µg genistein/g of soy proteins 230 µg daidzein/g of as compared to Soy isoflavone-free diet	Adult female Wistar rats 87-117 days of age OVX Pre-treated rats: diet was initiated before OVX or Sham surgery Post-treated rats: diet was initiated after OVX or sham surgery Behavioral testing (spatial memory)	Memory performance in the Moris water maze The following learning parameters were examined: Acquisition phase : (training days 1-5, with 4 trials/day, 15-20 min inter-trial delay): Latency to find the platform Test phase (1 trial, 24h after the last acquisition trial): Latency to cross the platform location Number of target crossings Time spent in the target quadrant Time spent in the opposite quadrant to target	60 days pre-treatment with soy isoflavones prevented the OVX-Induced increase in the latency to the hidden platform 30 days post-treatment with soy isoflavones reversed all the impaired parameters seen in OVX animals: increased latency to find the platform increased latency to cross the platform increased time spent in the opposite quadrant decreased time spent in the target quadrant	Monteiro *et al.* 2008

Table 3. Effects of dietary/supplemental isoflavones on cognitive processes in a non-human primate.

Diet features	Subject / Target / Mean	Variable parameter	Outcome	Reference
Infant Soy Formula Soy Formula + Mn Cow's milk- based Formula Duration: 0-4 months	Nonhumain primates (Rhesus Monkey infants) 8 subjects/group sampling: 0-18 months Assessment of cognitive function and general health parameters Behavioral testing	Tests used: Object discrimination learning and reversal test Delayed nonmatch to sample test Position learning and reversal test Reward delay test	No obvious specific effect of soy isoflavones on cognitive brain function could be reported in the studied non-human primate CSF contents in catecholamines (DA and 5HT) and their metabolites were also unaffected by the given diets	Golub *et al.* 2005

isoflavones/Kg of diet, *i.e.* close to Japanese diet in humans), (2) a high isoflavone-dosed diet (1.2 g of soy isoflavones/Kg of diet, *i.e.* twice or fourth times in the classical Japanese diet) and (3) an isoflavone-free control diet, on age-related impairment of hippocampal-dependent spatial memory function in Sprague-Dawley male rats. These authors addressed the aging hippocampal function in a water Morris delayed matching-to-place (DMP) task. They reported a significant improvement in the performance of middle aged male rats (14 months of age) put on the low isoflavone-dosed diet (0.3 g/Kg) as compared to rats put on the 2 other diets. On five stages of acquisition trials, the animals on the low isoflavone-dosed diet swam faster (on stages 2, 3 and 4) than the animals on the other diets. Interestingly, there was no significant difference in the performance between the 2 groups put on high isoflavone-dosed diet (1.2 g/Kg) or isoflavone-free diet. However, this same research group (Lee et al., 2009) have reported that, in OVX rats, the performance in the Morris water maze was rather improved, along side with the E_2-treated OVX rats (3.9 mg/Kg b.w.), in the OVX rats that had been on a high soy isoflavone-rich diet (526.9 mg/Kg; four or eight times the Japanese diet) as compared to those kept on a low soy isoflavone-rich diet (263.4 mg/Kg; twice or four times the Japanese diet) or a soy isoflavone-free control diet.

IX.1.2. Non-human primates

To our knowledge, only one study did address the cognitive effects of soy in a non-human primate model (Table 3). In this study, Golub and colleagues (2005) investigated the existence of putative effects of a soy-rich infant formula given for a 4 consecutive month duration on general health parameters, including cognitive development, in Rhesus monkey infants. The isoflavone concentration in the soy-based infant formula was not documented. A second experimental group was given a soy-rich infant formula with added manganese (Mn). The third experimental group (control subjects) was given a cow's milk based formula. To assess the cognitive brain function in the developing monkeys, four behavioural tests were regularly administered to subjects, namely (1) object discrimination learning and reversal test, (2) delayed non-match to sample test, (3) position learning and reversal test and (4) reward delay test. Up to 18 months of age, none of the 4 measures resulting from these 4 tests was significantly affected by the soy-rich infant formula in the developing monkeys. Moreover, no alteration of the cerebro-spinal fluid (CSF) contents in catecholamine (dopamine, DA, or serotonin, 5HT) metabolites could be found in soy fed monkeys.

IX.1.3. Human subjects

In the human studies of isoflavone impact on cognitive abilities, neuropsychological tests have been proposed to participants, most of the time, at baseline and at the completion of the dietary/supplemental intervention period. These tests were designed to evaluate the performance of participants in attention and memory tasks. They were also designed to evaluate both the executive functions and the mood states of volunteers. To help interpreting these data, a sample of behavioural evaluations of isoflavone impact on cognitive abilities is summarized in Tables 4.1-5. When available, the information concerning the brain territories activated under those cognitive tests and the source of the tests is also given. In behavioural studies on humans, the duration of exposure to isoflavones ranged from 12 weeks to 12 months. All the studies that did test an active treatment, were designed as double-blind, placebo-controlled, randomized studies (Tables 5.1-8).

Epidemiological studies. A cross-sectional correlation study between dietary intakes of lignans or isoflavones, using a food-frequency questionnaire, and brain functioning was carried-out in postmenopausal Dutch woman (age = 60-75 years). Episodic memory was evaluated by the Rey's Auditory Verbal Learning Tests (Rey immediate recall, Rey delayed recall and Rey recognition tests). Visual memory was assessed by the Doors Test. The appreciation of short-term/working memory materialized as a measure of the performance in the digit span forward/reverse tests. Verbal fluency was addressed by the letter fluency and category generation tests. Attention abilities and mental flexibility were assessed by the trail making (A1/A2/B) tests. The Boston Naming Test was used to evaluate the verbal competence and the semantic retrieval memory. The evaluation of cognitive/perceptual speed materialized as a measure of the digit symbol substitution test. Finally, the Dutch Adult Reading Test was used in order to assess the verbal intelligence quotient. The outcomes in this study did show significant correlation between the daily intake of lignans and improved performance in both attention/mental flexibility tasks (trailmaking A1 and A2) and cognitive/perceptual speed task (digit symbol substitution task). However, the neuropsychological assessment results could not be correlated to any feature of the dietary isoflavone intake of participants (Kreijkamp-Kaspers *et al.* 2007). Nonetheless, in this population the isoflavone intake was shown to be rather low (0.18, 0.34, 2.99 and 14.64 mg/day according to the different quartiles considered). Hogervorst and colleagues (2008) have conducted a cross-sectional correlation study where they did compare the daily consumption of 2 soy-derived products: tofu (a "soy milk" preparation, precipitated with $MgCl_2$) and tempeh (a fermented soy preparation) to the memory performance. The results showed that high tofu intake was associated with low scores in the memory tests, whereas high tempeh intake was independently related to higher scoring in memory tests. Another correlation study carried out by Huang *et al.* (2006) in middle-aged (age = 42-52 years) Asian women looked for any parallel between the estimated daily intake of isoflavones (as assessed by a food frequency questionnaire) and cognitive performance. In this study, the estimated median intake of genistein ranged from 3.5 µg/day-to-6.8 µg/day. No significant link between the isoflavone consumption profiles and the cognitive performances could be found within the experimental groups. Two epidemiological studies now report a higher risk of dementia in men with high daily tofu consumption, more than twice a week (White *et al.* 2000; Hogervorst *et al.* 2008). Tofu was found by Coward *et al.* (1998) to contain 32 mg of isoflavones per 100 g of weigh (one serving). These studies are in accordance with that of Lund[2] and Lephart (2001) and Lund *et al.* (2001) mentioned previously which showed altered memory performances in male rats under phyto-600 diet.

Interventional studies. Casini *et al.* (2006) have conducted a double-blind, placebo-controlled, cross-over study (12 months duration: 6 months on the active treatment or placebo followed by 1 month wash-out and then by 6 months on the alternative treatment) where they provided soy isoflavone-containing tablets (60 mg/day with: 40-45% genistein, 40-45% daidzein and 10-20% glycitein) to postmenopausal woman (age = 44-54 years) and assessed the cognitive function of participants at baseline and after each treatment. The results showed that among the 23 tests administered, 6 tests were found to be positively affected by the isoflavone supplementation. The recalling of pairs was the only test improved that concerned the memory function. The 5 other tests significantly improved by the supplemental isoflavones, concerned the mood states of the participants. File and colleagues (2001)

Table 4.1. Behavioural Evaluation of Isoflavone Impact on Brain Cognitive Function in Human subjects:

Tests of Attention	Description/Authors	Assessed Cognitive Function/Authors Source/Authors Activated Brain Region/Authors
Digit Symbol Substitution test (DSS)	A sheet of paper with the digits 1-9 in random order is given to participants. Using a key containing a corresponding symbol for each digit, participants have to place the correct symbol in the empty box below each digit. 90s to complete the task. The number of correct substitutions is scored. File *et al.* 2001. Kreijkamp-Kaspers *et al.* 2004, 2007.	Assessment of attention and perceptual/psychomotor speed. From the Wechsler Adult Intelligence Scale-Revised (WAIS-R). By Wechsler 1981.
Digit Concellation test (DC)	Participants are asked to cross out all the occurrences of a specific digit from a sheet of random numbers. The number of correct cancellations is scored. 90s to complete the test. File *et al.* 2001.	Assessment of focused attention. By Curran *et al.*, 1991.
Serial digit addition test	Adding together successive pairs of digits read from a list of 61numbers presented at different speeds from one digit every 2.4s (tape A) to one digit every 1.2s (tape D). The total number of correct responses is scored (maximum 60). File *et al.* 2001.	Assessment of sustained attention. By Spreen and Strauss 1991. From the Paced Auditory Serial Addition Test. (PASAT).
Trailmaking tests	Pseudorandomly placed circles with numbers (Trailmaking A1), with letters (Trailmaking A2) and with both letters and numbers (Trailmaking B) have to be connected with a line as fast as possible in a fixed order. Kreijkamp-Kaspers *et al.* 2004, 2007.	Assessment of attention and mental flexibility. From Reitan and Wolfson 1985.

Table 4.2. Behavioural Evaluation of Isoflavone Impact on Brain Cognitive Function in Human subjects.

Tests of Memory	Description/Authors	Assessed Cognitive Function/Authors Source/Authors Activated Brain Region/Authors
Story recalling test	A short story with 25 units of information is read to participants at the rate of one unit/s. Participants have to recall the story as closely as possible when asked immediately after listening. The number of correctly recalled units is scored. File *et al.* 2001. Maki *et al.* 2009.	Assessment of short-term verbal memory. From the Wechsler Memory Scale-Revised (WMS-R). By Wechsler 1987.
Word list recalling/ recognition tests	A list of 15 words is presented to participants. Tey are asked to recall it immediately (immediate recall) for 5 times consecutively (maximum score 75) or 25/30 min (delayed recall, maximum score 15). Tey are also asked to recognize the words in that list out of a 30 words list. Kreijkamp-Kaspers *et al.* 2004, 2007.	Assessment of short-term/long-term verbal memory and of visual/semantic recognition memory. From the Rey Auditory Verbal Learning Test. By Lezak 1995.
Word list recalling tests	A target list of 16 words from 4 semantic categories, is read aloud to participants on three successive verbal learning trials. Tey are asked to recall as many words as possibe. Another list (interference list) of 16 words from overlapping and different semantic categories is read. Participants are asked to recall the words on the second list. Next, participants are asked to recall the words on the original target list, first, without (short-delay free recall), and then with, categorical cues. After 20 min, they are asked to recall the words on the original target list, first, without (long-delay free recall), and then with, categorical cues. Maki *et al.* 2009.	Assessment of short-term/longterm verbal memory, semantic/logical verbal memory and mnemonic interferences. From the California Verbal Learning Test.

Table 4.3. Behavioural Evaluation of Isoflavone Impact on Brain Cognitive Function in Human subjects:

Tests of Memory	Description/Authors	Assessed Cognitive Function/Authors Source/Authors Activated Brain Region/Authors
Delayed Matching To Sample test (DMTS).	A sample complex pattern is displayed on a computer screen. Four similar patterns are also displayed at different delays of 0s (simultaneous presentation), 4s or 12s. The median times to make a correct response are scored for 5 patterns at each delay. File *et al.* 2001.	Assessment of short-term visual memory. From the Cambridge Neuropsychological Test Automated Battery (CANTAB). Activates the temporal cortex. By Owen *et al.* 1995.
The doors test	A sample of 2 series of 12 photographs of doors is shown to participants. Tey are asked to recognize that doors from 4 pictures of doors. Maximum score, 24. Kreijkamp-Kaspers *et al.* 2004, 2007.	Assessment of visual memory. From Davis *et al.* 1999.
Delayed picture recalling test	A set of Pictures (22) of common objects is presented to participants at the pace of 5s per picture. 20 min later, participants are asked to recall as many pictures as possible. The number of items correctly recalled is scored. File *et al.* 2001.	Assessment of longterm non-verbal memory.
Digit span forward (DSF) test Digit span reversed (DSR) test	Participants are asked to repeat a string of digits in the original order. Maximum score, 8. Participants are asked to repeat the string of digits in the reverse order. Maximum score, 7. Kreijkamp-Kaspers *et al.* 2004, 2007.	Assessment of short-term memory and short-term working memory. From the Wechsler Adult Intelligence Scale (WAIS). By Wechsler 1945.

Table 4.4. Behavioural Evaluation of Isoflavone Impact on Brain Cognitive Function in Human subjects.

Tests of Memory	Descriptions/Authors	Assessed Cognitive Functions/Authors Sources/Authors Activated Brain Regions/Authors
Category generation task	Participants are given 20s to name as many names of items as possible within a specific category (drinks, animals, etc.) of items. The number of generated names is scored. File *et al.* 2001.	Assessment of semantic memory. Activates the left medial temporal lobe. By Pihlajamaki *et al.* 2000.
Drawing recognition test	Participants are shown 60 line drawings that have to be properly named. Maximum score, 180 (3 points for a correct answer). Kreijkamp-Kaspers *et al.* 2004, 2007.	Assessment of verbal and semantic memory. From the Boston Namming Task (BNT). By Kaplan and Goodglass 1982.
Visual reproduction copy	Participants are allowed to view a geometric line drawing for 10s. Immediately after, they are asked to attemp to reproduce the drawing on a sheet of paper. 10 trials are administered to each participant with increasing complexity of drawing. Maki *et al.* 2009.	Assessment of short-term figural memory. From the Benton Visual Retention Test (BVRT). By Benton 1968.
Card rotation test	Participants are allowed to view a target line drawing of a geometric figure and eight alternatives of 2- or 3-dimensional rotation of the target. They are asked to find links between the 2-dimensional (picture plane) and 3-dimensional (mirror image) target drawing-rotated image, and the alternative images. The score is the number of correct minus the number of incorrect responses. Maximum score, 160. Maki *et al.* 2009.	Assessment of visual short-term memory and visuo-spatial ability. From the Modified Card Rotation Test. By Wilson *et al.* 1975.

Table 4.5. Behavioural Evaluation of Isoflavone Impact on Brain Cognitive Function in Human subjects.

Frontal Lobe Function	Description/Authors	Assessed Cognitive Function/Authors Source/Authors Activated Brain Region/Authors
Word generation task	Participants are given 60s to name as many words as possible starting with a given letter within the alphabet list. The number of generated words is scored. File *et al.* 2001. Kreijkamp-Kaspers *et al.* 2004, 2007.	Assessment of verbal (letter) fluency. Activates the frontal cortex. By Phelps *et al.* 1997, Schlosser *et al.* 1998.
Rule shifting and reversal test (IDED)	A series of paired patterns is presented to participants on a computer screen. Participants are asked to learn the rule that determined which pattern is correct. After accurate learning, the rule is either reversed or shifted to another rule within the same dimension (Intradimensional Shift, ID) or to a different dimension (Extradimensional shift, ED). The number of trials to complete the series is recorded. File *et al.* 2001.	Assessment of mental flexibility. From the Cambridge Neuropsychological Test Automated Battery (CANTAB). Activates the frontal cortex. By Owen *et al.* 1991.
Planning ability test	Two sets of three coloured balls are shown on a computer screen. The balls could be housed in three stockings. The task is to move the balls in the lower part of the screen to reach the exact pattern matching to the one in the upper part of the screen. Two to five moves are necessary to complete the matching. File *et al.* 2001.	Assessment of planning ability. From the Cambridge Neuropsychological Test Automated Battery (CANTAB). (The Stockings of Cambridge test). Activates the frontal cortex. By Owen *et al.* 1990.

Table 5.1. Effects of dietary/supplemental isoflavones on cognitive processes in human subjects.

Diet features	Subject / Target / Mean	Variable parameter	Outcome	Reference
A cross-sectional correlation study of: Dietary intake of lignans and isoflavones (as assessed by a food-frequency questionnaire) And brain functionning	Postmenopausal Dutch women 60-75 years of age N = 301 Cognitive function Psychological assessment	Tests of memory: - Episodic memory: Rey's Auditory Verbal Learning Tests Rey Immediate Recall Rey Delayed Recall Rey Recognition - Visual memory: Doors test - Short-term/working memory: Digit Span Forward Digit Span Reverse Tests of Verbal Fluency: Verbal fluency "N" Verbal fluency "A" Verbal test with animal-category Verbal test with occupation-category - Tests of attention and mental flexibility: Trail-Making A1 Trail-Making A2 Trail-Making B - Verbal competence and semantic retrieval: Boston Naming Test - Cognitive and perceptuel speed: Digit Symol Substitution (DSS) - Verbal intelligence quotient: Dutch Adult Reading Test	Daily consumption of phytoestrogens, expressed as median intake values, was found to be: For isflavones: 1st quartile: 0.18 mg/day 2nd quartile: 0.34 mg/day 3rd quartile: 2.99 mg/day 4th quartile: 14.64 mg/day For lignans: 1st quartile: 0.65 mg/day 2nd quartile: 1.39 mg/day 3rd quartile: 1.80 mg/day 4th quartile: 2.29 mg/day The neuropsychological test results could not be correlated to any noticeable feature parameter of their dietary intake of isoflavones However, the performance of participants in both attention/mental flexibility tasks (trailmaking A1 and A2) and cognitive/perceptuel speed task (DSS) was significantly correlated to the daily intake of lignans. The relative scores were noticeably better in subjects consuming the highest levels of lignans	Kreijkamp-K *et al.* 2007

Table 5.2. Effects of dietary/supplemental isoflavones on cognitive processes in human subjects.

Diet features	Subject / Target / Mean	Variable parameter	Outcome	Reference
A cross-sectional correlation study of: Soy (tofu or tempe) consumption (assessed using a food-frequency questionnaire) and memory function	Elderly Indonesian men and women 52-98 years of age N = 791 Psychological assessment of memory	Test of memory Hopkins Verbal Learning Test (HVLT) Immediate recall Delayed recall	Few participants consumed no tofu or no tempeh at all (5%) A majority consumed both products (65%) High tofu intake was associated with low memory performance High tempeh intake was independently related to higher scoring in memory tests	Hogervorst *et al.* 2008
A correlation study between: Estimated daily intake of isoflavones (from a Food Frequency Que-stionnaire) and cognitive performance	Middle-aged Asian women 42-52 years of age 195 Japanese 185 Chinese Cognitive function Psychological assessment	Tests of memory: - Episodic memory: East Boston Memory Tests (EBMT) with immediate and delayed recall - Working memory: Digit Span Backward (DBS) test - Psychomotor ability: Symbol Digit Modalities Test (SDMT)	Estimated median intakes of genistein: 3.5-6.8 (µg/day) No correlative connection between the daily soy-isoflavone consumption and the cognitive performance could be found in the individuals within the studied cohort	Huang *et al.* 2006
Soy isoflavone tablets (60 mg/day): 40-45% genistein 40-45% daidzein 10-20% glycitein Isoflavone-free placebo Six months on isoflavones or placebo One month washout Six months on the opposite treatment	Postmenopausal women 44-54 years of age N = 78 39 women in each group Cognitive function Psychological assessment at baseline and after each 6 month period	Cognitive evaluation: Digit Symbol Test (DST) Digit Span Test Visual Scaning Test Mood states Beck Depression Inventory (BDI) Spielberg State-Trait Anxiety Inventory (SSTI) Profile of Mood States (PMS) Visual Analogue Scales (VAS)	Among the 23 tests exploring the cognitive function, 6 tests were found to to be positively affected by the isoflavone supplementation: the recalling of pairs test (in the DST) was the only test improved that concerned the memory function the 5 other tests improved, concerned the mood states of the participants	Casini *et al.* 2006

Table 5.3. Effects of dietary/supplemental isoflavones on cognitive processes in human subjects.

Diet features	Subject / Target / Mean	Variable parameter	Outcome	Reference
Soy diet containing 100 mg of soy isoflavones As compared to 0.5 mg of soy isoflavone-containing diet 10 weeks on the diets	Young men (N = 12) and women (N = 15) 20-30 years of age Cognitive function Psychological assessment at baseline and after treatment	- Tests of attention: Digit Symbol Substitution (DSS) test Digit Concellation (DC) test Paced Auditory Serial Addition Test - Tests of memory: Episodic memory: Immediate story recalling test Delayed matching to sample test Delayed (> 20 min) recalling of common object pictures Semantic memory: Category generation task - Tests of frontal lobe function: Verbal fluency test Rule shifting and reversal test Planning ability test - Mood ratings: Visual Analogue Scale (from The Bond and Lader Scales)	Dietary isoflavones did not affect attention They did improve, however, the performance of participants in those tests that were related to: Short-term memory (immediate recall and delayed matching to sample) Longterm memory (pictures recall) Mental flexibility (males + females) Planning ability (females only) Letter fluency performance was improved in women whereas it was impaired in men Subjects on isoflavone supplementation rated themeselves as more restrained and more relaxed than the control subjects	File *et al.* 2001
Soy isoflavone supplement: 60 mg of isoflavones/day As compared to isoflavone-free placebo 12 week-duration	Postmenopausal women 50-65 years of age N = 36 Cognitive function Psychological assessment at baseline and after treatment	Episodic/semantic memory Verbal fluency and frontal lobe function Sustained attention (see File *et al.* 2001 for details) For mood ratings: Greene Climacteric Scale Stanford/Epworth Sleepiness Scales	Soy isoflavone supplement improved the performance in: Delayed picture-recalling test Paced Auditory Serial Addition Test Rule learning and reversal test All other parameters tested remained unaffected by soy isflavones	Duffy *et al.* 2003

Table 5.4. Effects of dietary/supplemental isoflavones on cognitive processes in human subjects.

Diet features	Subject / Target / Mean	Variable parameter	Outcome	Reference
Isoflavone-rich (99 mg) soy protein-based diet: 52 mg genistein 41 mg daidzein 6 mg glycitein As compared to Milk protein-based diet 12 months on dietary supplement	Postmenopausal women 60-75 years of age Placebo group: N = 87 Soy group: N = 88 Cognitive function Psychological assessment at baseline and after treatment	- Tests of episodic memory: The Rey Auditory Verbal Learning Test: Immediate/Delayed Recall tests Recognition test - Test of visual memory: The Doors test Tests of short-term/working memory: Digit span forward/reversed tests - Tests of semantic memory (verbal fluency): Category generation task The Boston Naming Task - Tests of frontal lobe function: Digit Symbol Substitution Test Trailmaking tests	12 months of dietary intake of soy-isflavones did not affect significantly any one of the following parameters: Hippocampus-dependent memories Prefrontal cortex-dependent cognition Bone mineral density or Plasma lipids	Kreijkamp-K *et al.* 2004
Soy milk/placebo supplement: 72 mg of isoflavones/day: 30.9 ± 1.5 mg daidzein 37.4 ± 1.3 mg genistein 3.6 ± 0.5 mg glycitein Cow's milk/isoflavone supplement: 70 mg isoflavones/day: 30 mg daidzein 33 mg genistein 7 mg glycitein Cow's milk and a placebo supplement 16 weeks on diet	Postmenopausal women 48-65 years of age N = 79 Cognitive function Psychological assessment at baseline and after 16 weeks on diet	- Selective Attention: the Stroop Task - Tests of memory: Working Memory: Digit Ordering Task Forward Digit Span Spatial working memory: Corsi Block-Tapping Color Matching Task Short-term memory: The Benton Visual Retention Test Longterm memory: Visual Pattern Recognition - Mood rating: The Beck Depression Inventory (BDI)	Urine contents in isoflavones were measured in 45 woman (N = 15/group): Soy milk group = 36.5 ± 6.3 μmol/L Supplement group = 52.2 ± 6.3 μmol/L Control group = undetectable No significant differences in the urine contents in isoflavones were found between groups The performance in the digit ordering task (a measure of verbal working memory) was impaired in the soy milk group All other tasks were not affected by the isoflavone intake	Fournier *et al.* 2007

Table 5.5. Effects of dietary/supplemental isoflavones on cognitive processes in human subjects.

Diet features	Subject / Target / Mean	Variable parameter	Outcome	Reference
Soy isoflavone supplement: 80 mg of isoflavones/day A placebo Six months on supplementation	Postmenopausal Chinese women 55-76 years of age N = 168 Cognitive function Psychological assessment at baseline and after 6 months on supplemental isoflavones	- Test of attention: Digit symbol (Wechsler Adult Intelligence Scale-Revised) - Tests of motor function: Finger tapping-right Finger tapping-left - Tests of memory: Hong Kong List Learning Test (HKLLT): Trials 1-5 Short delay recall Long delay recall - Executive function: Color Trail 1 Color Trail 2 Verbal Fluency - Visual Reproduction: Wechsler Memory Scale-Revised I (WMSR-I) Wechsler Memory Scale-Revised II (WMSR-II) - Visual perception and constructional ability: Visual reproduction copy Composite cognitive score - Language: Boston Naming Test - Global cognitive level: Modified Mini-Mental State Examination (MMSE)	The neuropsychological test results showed that supplemental isoflavones did not have any major measurable effect and that the performance scores at baseline and post-treatment did not differ significantly No noticeable adverse effects could neither be diagnosed among the participants in this study	Ho *et al.* 2007

Table 5.6. Effects of dietary/supplemental isoflavones on cognitive processes in human subjects.

Diet features	Subject / Target / Mean	Variable parameter	Outcome	Reference
Soy isoflavone supplement in tablet form: 100 mg of isoflavones as a daily intake (about 85% daidzein and genistein Placebo Six months on supplementation	Elderly healthy men and women 62-89 years of age N = 30 No previous use of hormone therapy for women Cognitive function Psychological assessment at baseline and after 6 months on treatment Plasma isoflavones were also analyzed at baseline (0 mo.) and after 1month (1mo.) and 6 months (6 mo.) on supplementation	- Verbal and visuospatial memory: Buschke Selective Reminding Test Paragraph recalling test Rey complex figure test Visual spatial learning test - Language and fluency: Boston Naming Test Letter fluency test Category fluency test - Visual motor function: Rey complex figure test-copy points Grooved Pegboard - Executive function: Stroop Color Word Test Trailmaking test B Mazes	Total plasma isoflavones nmol/L: * Placebo group: 0 mo. = 16.0 ± 5.5 1 mo. = 8.4 ± 2.7 6 mo. = 8.6 ± 3.8 * Isoflavone group: 0 mo. = 21.1 ± 15.7 1 mo. = 334.6 ± 188.6 6 mo. = 345.8 ± 183.5 The neuropsychological test results showed a significant improvement of subjects under isoflavone treatment in 8 tests examined and concerning: visuospatial memory mental construction ability verbal fluency speeded dexterity Also, placebo-treated subjects outperformed isoflavone-treated subjects in 2 tests concerning executive function	Gleason *et al.* 2007

Table 5.7. Effects of dietary/supplemental isoflavones on cognitive processes in human subjects:

Diet features	Subject / Target / Mean	Variable parameter	Outcome	Reference
Red clover extract standarized to 120 mg of isoflavones/day: genistein = 1.6 mg daidzein = 0.9 mg biochanin = 57.5 mg formononetin = 56.6 mg Black cohosh extract standarized to 7.27 mg triterpene glycosides CEE/MPA daily intake : 0.625 mg/2.5 mg Placebo 12 months on treatment CEE: conjugated equine estrogen MPA: medroxy-progesterone	Postmenopausal women with moderate to severe vasomotor symptoms (>35 hot flashs/week) Mean age: 53 years N = 66 Red clover: N = 14 Black cohosh: N = 18 CEE/MPA: N = 17 Placebo: N = 17 Cognitive function Psychological assessment at baseline and after 12 months on treatment	- Verbal memory: California Verbal Learning Test: Total learning Short-delay free recall Long-delay free recall - Logical memory: Immediate recall Delayed recall Benton Visual Rentention Test (BVRT) - Visuo-spatial Memory: Modified Card Rotation Test - Attention and working memory: Digit span forward Digit span backward Auditory attention tetst - Visuoperceptual Speed (Finding As test) - Verbal Intelligence: Primary Mental Abilities Vocabulary Test Memory functioning questionnaire: Frequency/seriousness of forgetting Mnemonics usage Retrospective functioning - Positive And Negative Affect Scale (PANAS): positive/negative affect	The results revealed that supplemental isoflavone-containing red clover extracts, as well as black cohosh extracts, have no significant impact on cognitive brain function Also, neither plant extracts did improve objectively vasomotor symptoms Oppositely, CEE/MPA seemed to impair the performance in verbal fluency test, although, it did releaved hot flashe-related complaints	Maki *et al.* 2009

Table 5.8. Effects of dietary/supplemental isoflavones on cognitive processes in human subjects:

Diet features	Subject / Target / Mean	Variable parameter	Outcome	Reference
Supplemental iso-flavones isoflavones/day: genistein = 12 mg daidzein = 68 mg glycitein = 36 mg Placebo 12 weeks on supplementation	Healthy men 49 ± 10 years of age N = 34 Non-usual soy consumers Cognitive function Psychological assessment at baseline and after 12 weeks on supplementation	- Paired Associate Learning Task: First Recall Total Recall Delayed Recall - Novel Spatial Working Memory Task: Pairs viewed Memory errors Time to complete task - Rey Auditory Verbal Learning Test: First recall Total recall Immediate recall after distraction Delayed recall Initial letter fluency task Backwards digit span - Trailmaking test: Trail A Trail B Total (B – A) Letter-number sequencing task - Mental rotation task: Number correct Time to complete task	Results showed that supplemental isoflavones were significantly associated with an overall improvement of spatial working memory and cognitive processing: → 18% fewer pairs viewed → 23% fewer memory errors → 17% faster processing Statistical analysis revealed that there was no other cognitive task improved or impaired by the daily supplemention with isoflavones	Thorp *et al.* 2009

examined the effect of a 10-week-treatment with a soy diet containing 100 mg of soy isoflavones on cognitive brain function in young men and women aged 20-30 years. The authors assessed the brain function of participants at baseline and after treatment, by using psychological tests of (1) attention (digit symbol substitution test, digit cancellation test and paced auditory serial addition test), (2) episodic memory (immediate story recalling test, delayed recalling test and delayed recalling of pictures), (3) semantic memory (category generation tests), (4) frontal lobe function (verbal fluency test, rule shifting and reversal test and planning ability test) and (5) mood rating (visual analogue scale, from the Bond and Lader scales). In this study, dietary isoflavones did not affect the attention abilities of participants. They did improve, however, the performance of participants in those tests that were related to short-term memory (immediate recall test and delayed matching to sample test) and to longterm memory (delayed recalling (> 20 min) of common object pictures). In addition, both mental flexibility (in both male and female subjects) and planning ability (in female subjects, only) were also among the cognitive functions significantly improved by isoflavones. By contrast, the performance scored in the letter fluency test, was improved in women, whereas it was impaired in men. Finally, the subjects under the active treatment did rate themselves as more restrained and more relaxed than the subjects that were under the control diet. A double-blind, placebo-controlled study conducted by Duffy and colleagues (2003) did investigate, in postmenopausal women (age = 50-60 years), the effects of soy isoflavones supplements (60 mg/day, for 12 weeks) on cognition. Episodic memory, semantic memory, frontal lobe function, sustained attention and mood were all rated using appropriate psychological tests. The outcomes of this study revealed that soy isoflavone supplements improved the performance in the following tests: (1) the test of delayed recall of pictures, (2) the paced auditory serial addition test and (3) the test of rule shifting and reversal. All the other parameters arising from the battery of tests used in the study remained unchanged between the group under the active treatment and the group under placebo. Kreijkamp-Kaspers and colleagues (2004) did address the impact of 12 month-intake of an isoflavone-rich soy protein-based diet (containing: genistein = 52 mg, daidzein = 41 mg, glycitein = 6 mg) on the cognitive abilities of a postmenopausal women group (60-75 years old). The assessment of the cognitive function that was conducted as described above in Kreijkamp-Kaspers *et al.* (2007), did not show any significant effect on cognitive performance. Fournier *et al.* (2007) did report that introducing soy milk (with approximately, 31 mg of daidzein, 37 mg of genistein and 4 mg of glycitein) into the daily diet of postmenopausal women (between 48 and 65 years old), resulted in impaired verbal working memory (digit ordering task), whereas adding the same amount of isoflavones (30 mg of daidzein, 33 mg of genistein and 7 mg of glycitein) to cow's milk daily intake did not impair nor improve the memory function. Moreover, Ho *et al.* (2007) who assessed the impact of daily intake of supplemental soy isoflavones (containing 80 mg of isoflavones) on brain function in postmenopausal Chinese women (age = 55-76 years) did not found any significant effects. A double blind, placebo-controlled supplemental (100 mg of isoflavones, about 85% daidzein and genistein) study, conducted by Gleason *et al.* (2007) in elderly healthy men (n = 15) and women (n = 15) (62-89 years old) randomized to two groups (placebo, n = 15, active treatment, n = 15) did report that, within the battery of tests administered to participants at baseline and after 6 months on supplementation, 8 tests were significantly improved by isoflavones. These tests concerned visuo-spatial memory, mental construction ability, verbal fluency, and speeded dexterity. Also, this study reported that the subjects under placebo out-performed those under active treatment in 2 tests concerning the executive function. In addition, after a 12 month-long supplementation of postmenopausal women (mean age = 53 years and the last periods having occurred between 35 to 50 months earlier) with a daily intake of isoflavone-containing red clover extracts, Maki *et al.* (2009) could not attribute any noticeable alteration of the memory function to the supplemental isoflavones. In a double blind, placebo-controlled, cross-over study, Thorp and colleagues (2009) examined the effects of isoflavone supplements (12 weeks duration with daily intake: genistein = 12 mg, daidzein = 68 mg, glycitein = 36 mg) on the cognitive functions of middle-aged (age = 49 ± 10 years) healthy men, non habitual consumers of soy-derived products. The results showed that supplemental isoflavones were significantly associated with an overall improvement of the performance in a spatial memory-related task. Indeed, in the novel spatial working memory task used by the authors, there were 18% fewer pairs viewed, 23% fewer memory errors and 17% faster processing within the group of subjects under the active treatment. Statistical analysis revealed that there was no other cognitive task improved or impaired by the daily intake of supplemental isoflavones. A 6-month, double-blind, randomized, placebo-controlled study, that was documented by Kritz-Silverstein *et al.* (2003) in healthy postmenopausal women, investigated the effects of soy isoflavones (110 mg /day) on cognitive function (SOPHIA study). The authors reported a significant improvement in category fluency, verbal memory, trails B tasks in the group of women who had been on the active treatment.

IX.2. Discussion

The compiled data reported above establish that dietary Isoflavones can exert some significant modulatory actions on cognitive function in both humans and rodent models, but not in a non-human primate model (Rhesus monkey) in the condition tested. Some of these effects are measurably reachable processes through adequately administered behavioural tasks. Sometimes, they do correlate with specific physiological aspects of brain functioning. It is noteworthy to observe that, when it comes to their outcomes, the studies depicted above are highly conflicting with each others, especially human studies, and that the results were depending on the task examined, the age of the studied group, the gender of the considered subjects, the time of administration from last period, the duration of isoflavone treatment, the isoflavone compound administered and the form of the dietary intake (supplements *vs.* soy based food).

In rodent studies, dietary isoflavones did show some capacity to impact the cognitive function in intact males (Lund and Lephart 2001a, Lund *et al.* 2001, Lee *et al.* 2004), intact females (Lund and Lephart 2001a, Lund *et al.* 2001) and OVX females (Lee *et al.* 2009, Luine *et al.* 2006, Monteiro 2008). Dietary isoflavones did also impact cognitive behaviours in Sprague-Dawley (Luine *et al.* 2006), Long-Evans (Lund[2] and Lephart 2001, Lund *et al.* 2001) and Wistar (Monteiro, 2008) rats. In radial arm maze tasks addressing the visual spatial memory, a life-long isoflavone-enriched diet did improve performance of both intact female rats and *in utero*-feminized, flutamide-treated male rats, whereas it did impair performance in control, oil-treated male rats. This implies that dietary isoflavones are:

(1) Capable of reaching and impacting the brain areas that are devoted to the accomplishment of higher cognitive functions such as the hippocampus, prefrontal cortex and associative cerebral cortices,

(2) Efficiently, capable of modulating the neuronal network(s) sub-serving the visual spatial memory function in the upper-mentioned brain areas,
(3) Sex-dependently, able to alter the behavioural expression of learning and memory processes, and
(4) Capable of counter-acting a developmental sex hormone-related cognitive defect.

Additionally, a much shorter time exposure (15-16 weeks) to a low-dose of dietary isoflavones was also able to improve the defective aging-related spatial memory performance in the Morris water maze in middle-aged (14 months old) male Sprague-Dawley rats (Lee *et al.* 2004). Hence, the administration of dietary isoflavones can provide a beneficial relieving treatment against age-associated impairment of reference memory. However, a group of rats who had been on a high-dose of dietary isoflavones did not differ from a control group who had been on a free-isoflavone diet (Lee *et al.* 2004) evidencing that the highly-dosed isoflavone diets or the highly-dosed isoflavone supplementations may not display necessarily beneficial effects on memory function. This is, most likely, due to the fact that a complex cognitive task must activate different cerebral territories that are not at the same levels of estrogen signalization. For example, such territories may differ by their:

(1) Isoflavone-mediated modulation of the expression levels of ER genes,
(2) Number of ER sites available for the molecular interaction with circulating isoflavone molecules,
(3) Relative abundance of ER subtypes that may putatively lead to some differential ER subtype-specific signalizations,
(4) Specific ultrastructural distribution of ERs that could lead to a hypothetical cell-compartmentalized isoflavone signalizations, and/or
(5) Task-linked regulation of blood flow within the given brain area(s).

Elsewhere, under the paradigm of hormone replacement therapy using isoflavones, the memory performance in OVX rats was improved in the spatial reference memory by an isoflavone-rich diet in the sense that it was associated with significant decreases in both path length and escape latency monitored on the 2nd day of acquisition (Lee *et al.* 2009). Furthermore, an isoflavone-enriched diet did increase the amount of time spent by OVX rats at exploring an old object displaced to a new position within the exploration maze (Luine *et al.* 2006). This finding may reflect an isoflavone-mediated improvement of the object placement memory in surgically menopausal young animals. In the same study by Luine *et al.* (2006), the object placement memory performance of isoflavone-fed OVX rats was ameliorated in the 7th week of treatment but not in the 5th week of treatment advocating for the likelihood of duration-specific effects of isoflavones on cognitive processes. Moreover, isoflavones did not impact the memory performance in a new object recognition task administered to OVX rats at 6, 8 and 9 weeks, after the diet introduction (Luine *et al.* 2006) implying that the modulation of the cognitive function by isoflavones depends on the type of memory needed for the resolution of the task. In the same context, an isoflavone-rich soy-based diet provided to OVX rats soon after the surgical intervention was efficiently able to rescue a hippocampus-dependent cognition, namely reference memory function that was monitored in a Morris water maze (Monteiro *et al.* 2008). Indeed, OVX rats exposed to isoflavones did acquire the maze faster than those who were not. Also, on the day of probing, 3 parameters that can be gained from the analysis of the platform-searching behaviour of rats (occupation time of the target quadrant, occupation time of the opposite quadrant and latency to cross the platform) were all improved (up-graded to the levels of parameters recorded in sham-operated rats) by the immediate exposure to isoflavones after OVX. Interestingly, in this same study, sham-operated rats that have been exposed to isoflavones for the same duration as those who were OVX, did not show any impairment in both the acquisition phase or the probing session, evidencing that nutritional adjunction of isoflavones to normal circulating levels of estrogens does not impair the hippocampal formation functioning in the ovaries retaining animals. In addition, it has been shown that soy isoflavones dose-dependently, improve spatial working memory in OVX retired breeder rats with or without added E_2, implying that dietary isoflavones do not antagonize the beneficial effects of circulating estrogens (Pan *et al.* 2000). Taken altogether, the findings described above highlights the beneficial effects of an isoflavone-rich diet in females on hippocampal-activating mnemonic tasks that have been impaired by experimental estrogen withdrawal. None of the dietary interventions resulted into an anti-estrogenic effect of isoflavones. Therefore, dietary and/or supplemental isoflavones may be of interest in the ongoing efforts for the search of a hormonal substitute for the surgically-induced or age-associated estrogen withdrawal and the resulting impairment of hippocampus-dependent cognition experienced by aged/menopausal women. In males the discrepancy observed between some results may reflect the plausible existence of a tricky dose-dependent response effect.

In non-human primates, and to our knowledge, the only study that did address the question of soy impact on the cognitive function is the one published by Golub *et al.* (2005), although the isoflavone exposure was not documented in this trial. This research group did investigate the effects of a soy formula-based diet on brain functioning in the developing Rhesus monkeys until their 18th month of age. They reported no significant impact of soy-based infant formula on cognitive abilities. However, to ascertain this possibility, the isoflavone content should have been measured and soy-fed monkey brains should have been examined at the neuroanatomical level in order to rule out any adverse occurrence such as, for example, a neurological defect.

In human studies, dietary isoflavones did improve the performance in tasks related to short-term memory (immediate recall and delayed matching to sample), longterm memory (pictures recall) and mental flexibility in young (aged, 20-30 years) subjects of both genders (File *et al.* 2001). These results are in agreement with those reported by Gleason *et al.* (2009) in elderly men and women (aged, 62-89 years) in whom visuo-spatial memory, mental construction ability, verbal fluency and speeded dexterity were improved by a 6-months intake of a soy isoflavone supplement. However, dietary isoflavones displayed some sex-specific effects on planning abilities that were improved in female subjects but unaffected in male subjects (File *et al.* 2001). By contrast, some adverse effects of isoflavones can also be seen with respect to the gender of the group examined as it was the case for a letter fluency task that was improved in female subjects but impaired in male subjects (File *et al.* 2001). In a slightly different context, supplemental isoflavones did cause a remarkable improvement of spatial working memory in healthy men, aged 49 ± 10 years (Thorp *et al.* 2009). Similarly to the young human subjects under dietary isoflavones, short-term memory, longterm memory and cognitive processing speed were improved by an isoflavone supplementation in postmenopausal (aged, 44-65 years) women (Casini *et al.* 2006, Duffy *et al.* 2003). A 12 month-long daily supplementation with an isoflavone-rich red clover extract did show a null effect on cognitive function in elderly persons. However, in this specific case the bioavailable isoflavone plasma levels were very low indicating, probably, a modification in the absorption, the metabolism and/or the excretion rates of isoflavones with ageing (Maki *et al.* 2009). Similarly, a 6 month-long isoflavone supplementation did not affect cognitive function in postmenopausal women (Ho *et al.* 2007). However, in this study the menopausal parameters were highly variable: age, time since menopause, HRT treatment or not; and all these parameters would have influenced the cognitive response leading to non-significant differences. Considering the nature of the isoflavone compound administered to participants, Fournier *et al.* (2007) did report an impaired verbal working memory in postmenopausal women (aged 48-65 years) who have been on a diet containing soy milk (72 mg isflavones/day) + a placebo supplement as compared to those who have been on a diet containing a cow's milk + an isoflavone supplement (70 mg isoflavones/day) or a diet containing a cow's milk + placebo supplement. These findings suggest that the isoflavone-containing matrix (herein, soy milk) or the preparation procedure of the isoflavone-derived product, may have some detrimental actions on the brain function, regardless of the active compound tested. In addition, in cross-sectional correlation studies, high tofu intake was associated with low performance in the memory testing (Hogervorst *et al.* 2008) and with deleterious brain functioning in old men (White *et al.* 2000) when tempeh was associated with memory improvement (Hogervorst *et al.* 2008). In addition, high isoflavone-rich diets could not be associated with any improvement of, or with any adverse effect on memory testing (Kreijkamp-Kaspers *et al.* 2007, Huang *et al.* 2006). Similarly, an isoflavone-rich soy protein based diet did not impact cognitive functioning in postmenopausal women aged 60-75 years (Kreijkamp-kaspers *et al.* 2004). Nonetheless, the effects observed in the tofu consumer group appear to corroborate the one obtained in soy milk group. With respect to hippocampal-associated memory, discrepant results could be noticed depending on the study considered. Indeed, the performance in such memory subtype, for example, among supplemental-designed investigations, was improved in Duffy *et al.* (2003), Casini *et al.* (2006) and Gleason *et al.* (2009) studies but was unaffected in Ho *et al.* (2007) study. In keeping with this, and considering soy-based isoflavone-containing diets, hippocampus-activating mnemonic tasks were improved in File *et al.* study (2001) but were unaffected in Kreijkamp-Kaspers *et al.* (2004) study. Still, the large differences in (1) the age of participants (20-30 years in File's study and 60-75 years in Kreijkamp-Kaspers's study, with no cycles for several years), (2) the treatment durations (10 weeks in File's study and 12 months in Kreijkamp-Kaspers's study) and (3) the number of participants (n = 27, in File's study and n = 175, in Kreijkamp-Kaspers's study) may offer plausible explanations to the apparent discrepancy in the outcomes. Besides, the participants in File's study were represented by men (n = 12) and by women (n = 15) and the small subject number did not allow a separate statistical investigation, whereas in Kreijkamp-Kaspers's study, the participants were all postmenopausal women. Altogether, these observations suggest that dietary isoflavones might exert some (1) exposure time-specific, (2) exposure nature-specific, (3) age-specific and/or (4) sex-specific effects on the hippocampal-dependent memory function in humans.

X. Concluding Remarks

Cellular signalization by estrogens is a multifaceted and a widespread physiological mechanism among all and every single organ within the body including the brain. Therefore, this signalization arose as both a highly challenging research topic and a highly critical issue for human health. Consequently, due to their structural resemblance to estrogen molecules and to their natural occurrence in food along side with high quality proteins (for example, in soy), isoflavones have erupted into the spotlighted arena of human curiosity, having succeeded in capturing a large part of scientific, media, industrial and public attention throughout the world. Thus, in the present review, we provided *in vitro*, animal and clinical studies that did describe a plethora of effects arising from the use of isoflavones as signalling molecules. Each one of these studies could be classified in one of the following 2 main ensembles: pharmacological studies or dietary/supplemental studies. Below we are going to recall the main resulting findings according to the relative nature of the study: pharmacological studies *versus* nutritional studies. In the reported effects, we will be focusing on brain neurons, and more specifically, on hippocampal neurons. We will also try to draw the main conclusions and plausible consequences as of the isoflavone impact on human cognition. Of course, keeping in mind the upper levels of plasma concentrations that can be reasonably reached by dietary/supplemental isoflavones. In parallel to our closing conclusions, we will be evoking some findings falling into the grey, shadowy area of highly controversial reporting and also some research topics that are still highly suffering from the complete lack of investigative research.

X.1. Pharmacological Studies

Herein, isoflavone molecules have been revealed as potent pharmacological substances modulating intracellular transduction pathways in various cell systems, and more specifically, in hippocampal formation neurons. The pharmacological studies carried out using high doses of isoflavones (> 10 μM) that are unlikely occurring concentrations in circulating plasma under the normal nutritional status, have produced signalizing effects reflecting the powerful inhibitory action of isoflavones (genistein mainly) upon the phosphorylation activity of cellular/neuronal PTKs. This inhibitory action was most likely due, at least in part, to the competitor property of this isoflavone (genistein) at the ATP-binding site of PTKs. Nonetheless, high concentrations of isoflavones produced manifest alterations of the bioelectrical neuronal activity (LTP, ion channel-sustained currents and neurotransmitter-mediated signalizations).

Albeit in a significant number of data where the concentrations of isoflavones used would not be relevant for nutritional strategies, isoflavone molecules, nonetheless, remain robust signalling substances that must be handled with extreme caution. More specifically, when it comes to the matter of supplemental isoflavones, overdosing must be adversely forbidden and elderly people must be extremely sensitized against such practice. Also, such recommendations must be clearly advertised in supplemental product leaflets accompanying the over-the-counter products available in drug stores for public sale.

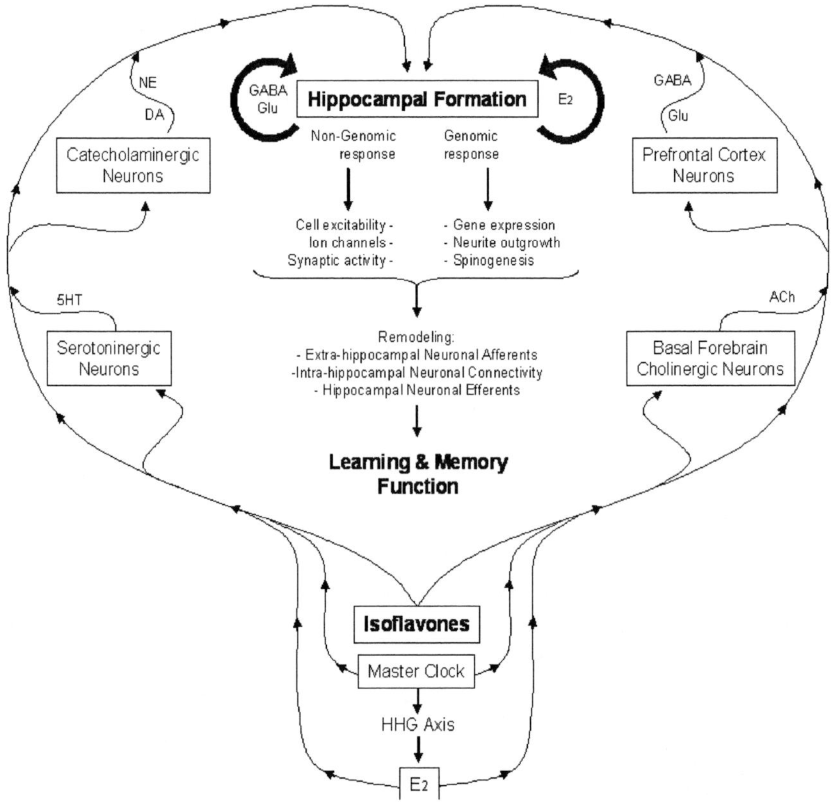

5HT: 5-hydroxy-tryptamine (serotonin), ACh: acetylcholine, DA: dopamine, E2: estradiol, GABA: γ-aminobutyric acid, Glu: glutamate, HHG axis: hypothalamo-hypophyso-gonadal axis, NE: norepinephrine.

Figure 2. Putative signalling mechanisms by dietary/supplemental isoflavones in mnemonic brain structures. Dietary/supplemental isoflavones may impact hippocampal formation neurons directly and indirectly, through monoaminergic systems and prefrontal cortex neurons. Nutritional isoflavones may act through genomic and non-genomic mechanisms to re-modulate intra/extra-hippocampal neuronal network activity and, thereof, to impact the learning and memory function. The master biological clock located in the suprachiasmatic nucleus exerts both neural and endocrine (through the HHG axis-generating E2) controls upon brain neurons sub-serving the hippocampus-dependent learning and memory function.

X.2. Nutritional Studies

As largely developed in the present chapter, dietary/supplemental isoflavones are powerful substances that display highly potent signalizing characteristics in various cell systems including brain neuronal systems. In particular, such isoflavones showed substantial leverage with the neuronal physiology of the hippocampal formation. In humans, the effects reported here would indicate that isoflavones at doses ranging from 60 to 100 mg a day may improve cognitive features in women when administered soon after the menopause occurrence. Although not consistently recurrent, similar outcomes have been identified for the hormonal replacement therapy using variously designed estrogens or different combinatorial configurations with co-administered therapeutical hormone candidates (for example, progesterone) along side with estrogens. At lower doses (< 60 mg/day), however, no effect on human cognitive abilities could be, directly or indirectly, inferred to nutritional isoflavones.

In men, too, isoflavones did exhibit positive effects when administered in supplements or in soy-deriving foods, such as tempeh. Nevertheless, daily intake of high quantities of a soy-arising meal, herein tofu, seems to produce deleterious effects as those seen in two epidemiological studies. It must be mentioned here that, in the *Western* recipes, the 7 deepings, that are traditionally applied in the Asian recipe, are most often omitted. These Deepings associated to heating, most probably, reduce the content of the soya matrix in antinutritional compounds (urease, trypsin inhibitors, phytase, saponins, hemaglutinins, stigmasterol) and in isoflavone of traditional tofu, since isoflavones in their glycosylated forms, are soluble in water. It is unclear from the animal/human studies if high intake of nutritional isoflavones would produce positive or rather deleterious effects, since in all the studies that have shown a positive impact, either administered doses or reported plasma levels were low or medium. Furthermore, the effects of isoflavones on both the cognitive behaviours (in animal and human studies) and the brain physiology indicate that not only the hippocampus would be influenced by dietary isoflavones. Most probably, brain areas like the pre-frontal/frontal cortices, amygdala, striatum, preoptic/mediobasal hypothalamic structures, midbrain and brainstem, that are well known to be both highly enriched in ERs and functionally interconnected, are plausible target sites for some isoflavone-mediated signalizations (Figure 2). These signalizations, therefore, would transpire behaviourally as hippocampus-involving improvement/impairment of the memory function.

Isoflavones raised the hope of being used as alternative brain-protecting drugs to hormone replacement therapy and in aging/menopausal-associated dysfunctions, but an eye still have to be pointed out on their proliferative effects in estrogen-dependent cancers. However, as reported enclosed, isoflavones do trigger some alarming effects on neuronal survival, neuronal excitability, brain physiology, cognitive function and, thus, these molecules still have to be considered with wise caution. Hence, little skepticism must and is well still around as has been reportedly published in the scientific literature and more specifically in expert opinion-reporting reviews.

All along the present chapter, we provided data that have reported some highly interesting effects reflecting both the neuroprotective and the neurotrophic properties of isoflavones. However, the respective studies should be redesigned with lower concentrations of isoflavones and under dietary/supplemental configurations, as well, in order to get full acceptance as beneficial effects arising from nutritional isoflavones. Moreover, and regardless of the administration route/form of isoflavones, these substances did produce a variety of biological responses that have been shown to depend specifically on the sex, dose, age, animal model, hormonal status, exposure time, cell type and behavioural task. These occurrences make it evident that isoflavones, definitely, will not suite all the human beings nor every single health or aging-related condition. Consequently, we could not close this chapter without raising the question of the code of good practice, concerning the consumption of nutritional isoflavone-enriched products. Indeed, and to our knowledge, the only study that did address the question of isoflavone-mediated modulation of the activity of cerebral steroid-metabolizing enzymes under the dietary isoflavone intake status is the one previously published in isoflavone-rich diet fed rats and where the enzymatic activity of aromatase P450-*CYP19* was measured in the preoptic/mediobasal hypothalamic regions. In that study, the lack of effect of isoflavones on the aromatase activity, in the single brain area examined to date, must underlies, rather than occludes, the need of investigating the impact of dietary/supplemental isoflavones on the acitivity of neurosteroidogenic enzymes in multiple brain regions including the hippocampal formation. More particularly, this impact should be evaluated in the contexts of both normal and pathophysiological brain aging, as well as in that of experimental E_2 withdrawal with and without estrogen replacement strategies. As a consequence of all these observations, it makes it evident that, beyond the tempting advertisement of isoflavone supplemental makers, the common sense of a safe use of isoflavones remains the seeking of physician advices.

Dietary/supplemental isoflavones, obviously, displayed some beneficial effects on mood/anxiety aspects in both animal models and human subjects. The findings that isoflavones are efficient modulators of behaviourally/verbally expressed emotional states do agree with reports on the impact of estrogens on the mood function. Could it be possible that dietary/supplemental isoflavones induce some mild sedation-like effects that made some human subjects feel more relaxed/restrained and less tense (as they did express) and others, including animal subjects under isoflavone diets, better performers in behavioural cognitive tasks? A number of data we did report in this chapter, such as the downregulation of both cell excitability and $[Ca^{++}]_i$, would substantiate this suggestion. Once again, the corresponding studies remain to be redesigned with the nutritional constraints of isoflavone bio-availablity in mind.

The beneficial effects of dietary/supplemental isoflavones on mood-related cognition may be the iceberg tip of the impact of these molecules on the cerebral signalization involving neurotransmitter neuronal systems, namely GABAergic, glutamatergic, catecholaminergic, serotoninergic and basal forebrain cholinergic neurons. More specifically, dietary/supplemental isoflavones did show interesting and beneficial leverage with both central nervous system cholinergic signalization and neuroprotection of basal forebrain neurons against age-triggered defects. Could it be possible that this same leverage with the modulation of cerebral cholinergic signalization is the one that did allow isoflavones to impact positively the expression of both the affective states and the hippocampus-linked mnemonic behaviours? If such positive potentials are really substantiated by future studies, then, sparing forebrain cholinergic neurons from age-triggering neurodegeneration may participate in rescuing the aging hippocampal formation and, thereof, in relieving age-related impairments of the memory function that are widespread among elderly men and women, and more particularly among those victims of neurodegenerative diseases.

Abbreviations:

$[^{35}S]$-GTPγ-S	: sulphate 35-radio-labeled guanosine-triphosphate gamma-stimulating	GDP	: guanosine-di-phosphate
		GFAP	: glial fibrillary acidic protein
$[^{125}I]$estrogen	: iode 125-radio-labelled estrogen	GAD67	: glutamic acid decarboxylase-67
$[Ca^{++}]_i$	: intracellular calcium concentration	GAP43	: growth cone-associated protein-43
α_7nAchR	: α_7 nicotinic acetylcholine receptor	GSK-3β	: glycogen synthase kinase-3β
5HT	: 5-hydroxy-tryptamine (serotonin)	GPCR	: G protein-coupled receptor
AChR	: acetylcholine receptor	GPCRs	: G protein-coupled receptors
ACh	: acetylcholine	GPR30	: G protein-coupled estrogen receptor
AChE	: acetylcholine esterase	GPR30-IR	: G protein-coupled estrogen receptor-immunoreactivity
AD	: Alzheimer's disease		
ADP	: adenosine-di-phosphate	hERα	: human estrogen receptor α
ATP	: adenosine-tri-phosphate	hERβ	: human estrogen receptor β

AMPA	: α-amino-3-hydroxyl-5-methyl-4-isoxazolepropionate	IGF-1	: insulin-like growth factor-1
AMPAR	: α-amino-3-hydroxyl-5-methyl-4-isoxazolepropionate receptor	i.p.	: intraperitoneal
AS-ODN	: antisense oligodeoxynucleotide	K^+	: potassium
AVP	: arginin-vasopressin	KA	: kainic acid
AVPV	: anteroventral periventricular nucleus	KCl	: potassium chloride
BDNF	: brain-derived neurotrophic factor	LDH	: lactate dehydrogenase
b.w.	: body weight	LTP	: long-term potentiation
Ca^{++}	: calcium	MAP-2	: microtubule-associated protein 2
CA1/2/3/4	: cornu ammonis fields 1/2/3/4	MAPK	: mitogen-activated protein kinase
cAMP	: cyclic-adenosine mono-phosphate	mGluR	: metabotropic glutamate receptor
cGMP	: cyclic-guanosine-mono-phosphate	mRNA	: messenger ribonucleic acid
ChAT	: choline acetyltransferase	MTT	: 3-[4,5-di-Methyl-Thiazol-2-yl]-2,5-diphenyl Tetrazolium bromide
ChAT-IR	: choline acetyltransferase-immunoreactivity	Na^+	: sodium
CNG	: cyclic nucleotide-gated	nAchR	: nicotinic acetylcholine receptor
CSF	: cerebro-spinal fluid	NGF	: nerve growth factor
DA	: dopamine	NMDA	: N-methyl-D-aspartate
DG	: dentate gyrus	OVX	: ovariectomy
DMP	: delayed matching-to-place	PCR	: polymerase chain reaction
DNA	: deoxyribonucleic acid	PELP1	: praline/glutamic acid/leucine-rich protein-1
DRG	: dorsal root ganglion	Phyto-600	: soy phytoestrogen-rich diet containing 600 µg/kg of diet
E_2	: 17-β-estradiol	PI3K	: phosphate-inositol-3 kinase
EB	: estradiol benzoate	PND	: post-natal day
ED	: embryonic day	pp60c-src$^{(+)}$	: cellular counterpart of Rous sarcoma virus protein tyrosine kinase
ER	: estrogen receptor	PSD95	: postsynaptic density protein-95
ERα	: estrogen receptor α	PTK	: protein tyrosine kinase
ERα-IR	: estrogen receptor α-immunoreactivity	RNA	: ribonucleic acid
ERβ	: estrogen receptor β	RT	: reverse transcriptase
ERβ-IR	: estrogen receptor β-immunoreactivity	SERCA	: sarco-endoplasmic calcium ATPase
ERK1/2	: extracellular signal-regulated kinase1/2	SOCCs	: store-operated calcium channels
GABA	: γ-amino-butyric-acid	TrKA/B	: tropomyosin-receptor-kinase A/B
GABA-A	: γ-amino-butyric acid receptor-A	VDCC	: voltage-dependent calcium channel

References

Abdul-Ghani, MA; Valiante, TA; Carlen, PL; Pennefather, PS. Tyrosine kinase inhibitors enhance a Ca(2+)-activated K+ current (IAHP) and reduce IAHP suppression by a metabotropic glutamate receptor agonist in rat dentate granule neurones. *J Physiol.*, 1996, 496(Pt 1), 139-44.

Adams, MM; Shah, RA; Janssen, WG; Morrison, JH. Different modes of hippocampal plasticity in response to estrogen in young and aged female rats. *Proc Natl Acad Sci*, U S A. 2001, 98(14), 8071-6.

Adlercreutz, H; Markkanen, H; Watanabe, S. Plasma concentrations of phyto-oestrogens in Japanese men. *Lancet.*, 1993, 342(8881), 1209-10.

Agís-Balboa, RC; Pinna, G; Zhubi, A; Maloku, E; Veldic, M; Costa, E; Guidotti, A. Characterization of brain neurons that express enzymes mediating neurosteroid biosynthesis. *Proc Natl Acad Sci*, U S A. 2006, 103(39), 14602-7.

Albertazzi, P; Purdie, DW. Reprint of The nature and utility of the phytoestrogens: a review of the evidence. *Maturitas.*, 2008, 61(1-2), 214-26.

Anderson, RL; Wolf, WJ. Compositional changes in trypsin inhibitors, phytic acid, saponins and isoflavones related to soybean processing. *J. Nutr.*, 1995, 125, 581S-588S.

Aydin, M; Yilmaz, B; Alcin, E; Nedzvetsky, VS; Sahin, Z; Tuzcu, M. Effects of letrozole on hippocampal and cortical catecholaminergic neurotransmitter levels, neural cell adhesion molecule expression and spatial learning and memory in female rats. *Neuroscience.*, 2008, 151(1), 186-94.

Azcoitia, I; Moreno, A; Carrero, P; Palacios, S; Garcia-Segura, LM. Neuroprotective effects of soy phytoestrogens in the rat brain. *Gynecol Endocrinol.*, 2006, 22(2), 63-9.

Bang, OY; Hong, HS; Kim, DH; Kim, H; Boo, JH; Huh, K; Mook-Jung, I. Neuroprotective effect of genistein against beta amyloid-induced neurotoxicity. *Neurobiol Dis.*, 2004, 16(1), 21-8.

Barrett-Connor, E; Laughlin, GA. Endogenous and Exogenous Estrogen, Cognitive Function and Dementia in Postmenopausal Women: *Evidence from Epidemiologic Studies and Clinical Trials Semin Reprod Med.*, 2009, 27(3), 275–82.

Bartlett, SE. Protein tyrosine kinase inhibitors synergize with nerve growth factor in embryonic chick sensory neuronal cell survival. *Neurosci Lett.*, 1997, 227(2), 87-90.

Beattie, EC; Carroll, RC; Yu, X; Morishita, W; Yasuda, H; von Zastrow, M; Malenka, RC. Regulation of AMPA receptor endocytosis by a signaling mechanism shared with LTD. *Nat Neurosci.*, 2000, 3(12), 1291-300.

Belcher, SM; Zsarnovszky, A. Estrogenic actions in the brain: estrogen, phytoestrogens, and rapid intracellular signaling mechanisms. *J Pharmacol Exp Ther.*, 2001, 299(2), 408-14.

Bellino, FL; Wise, PM. Nonhuman primate models of menopause workshop. *Biol Reprod.*, 2003, 68(1), 10-8.

Bennetau-Pelissero, C; Arnal-Schnebelen, B; Lamothe, V; Sauvant, P; Sagne, JL; Verbruggen, MA; Mathey, J; Lavialle, O. ELISAs as new methods to measure isoflavones in food and human fluids. *Food Chem.*, 2003, 82, 645-658.

Bennetau-Pelissero, C; Latonnelle, K; Séqueira, A; Lamothe, V. Phytœstrogens, possible endocrine disrupters from food. *Analusis.*, 2000, 28(9), 763-775.

Bennetts, HW; Underwood, EJ; Shier, FL. A specific breeding problem of sheep on subterranean clover pastures in Western Australia. *Aust Vet J.*, 1946, 22, 2-12.

Benton, AL. Differential behavioural effects in frontal lobe disease. *Neuropsychologia.*, 1968, 6:53-60.

Billard JM. Ageing, hippocampal synaptic activity and magnesium. *Magnes Res.* 2006, 19(3):199-215.

Blomquist CH; Lima PH; Hotchkiss JR. Inhibition of 3alpha-hydroxysteroid dehydrogenase (3alpha-HSD) activity of human lung microsomes by genistein, daidzein, coumestrol and C(18)-, C(19)- and C(21)-hydroxysteroids and ketosteroids. *Steroids.* 2005, 70(8).507-14.

Boon WC; Diepstraten J; van der Burg J; Jones ME; Simpson ER; van den Buuse M. Hippocampal NMDA receptor subunit expression and watermaze learning in estrogen deficient female mice. *Brain Res Mol Brain Res.* 2005, 140(1-2):127-32.

Boulware MI; Mermelstein PG. Membrane estrogen receptors activate metabotropic glutamate receptors to influence nervous system physiology. *Steroids.* 2009, 74(7):608-13.

Braden AWH; Hart NK; Lamberton JA. The œstrogenic activity and metabolism of certain isoflavones in sheep. *Aust J Agric Res.* 1967, 18:335-348.

Brailoiu E; Dun SL; Brailoiu GC; Mizuo K; Sklar LA; Oprea TI; Prossnitz ER; Dun NJ. Distribution and characterization of estrogen receptor G protein-coupled receptor 30 in the rat central nervous system. *J Endocrinol.* 2007, 193(2):311-21.

Brake WG; Alves SE; Dunlop JC; Lee SJ; Bulloch K; Allen PB; Greengard P; McEwen BS. Novel target sites for estrogen action in the dorsal hippocampus: an examination of synaptic proteins. *Endocrinology.* 2001, 142(3):1284-9.

Brann DW; Dhandapani K; Wakade C; Mahesh VB; Khan MM. Neurotrophic and neuroprotective actions of estrogen: basic mechanisms and clinical implications. *Steroids.* 2007, 72(5):381-405.

Brann DW; Zhang QG; Wang RM; Mahesh VB; Vadlamudi RK. PELP1--a novel estrogen receptor-interacting protein. *Mol Cell Endocrinol.* 2008, 290(1-2):2-7.

Bu L; Lephart ED. AVPV neurons containing estrogen receptor-beta in adult male rats are influenced by soy isoflavones. *BMC Neurosci.* 2007, 8:13.

Carlson S; Peng N; Prasain JK; Wyss JM. Effects of botanical dietary supplements on cardiovascular, cognitive, and metabolic function in males and females. *Gend Med.* 2008, 5 (SupplA):76-90.

Casey M; Maguire C; Kelly A; Gooney MA; Lynch MA. Analysis of the presynaptic signalling mechanisms underlying the inhibition of LTP in rat dentate gyrus by the tyrosine kinase inhibitor, genistein. *Hippocampus.* 2002, 12(3):377-85.

Casini ML; Marelli G; Papaleo E; Ferrari A; D'Ambrosio F; Unfer V. Psychological assessment of the effects of treatment with phytoestrogens on postmenopausal women: a randomized, double-blind, crossover, placebo-controlled study. *Fertil Steril.* 2006, 85(4):972-8.

CDC: The Centers for Disease Control and Prevention. Public health and aging: trends in aging--United States and worldwide. *JAMA.* 2003, 289(11):1371-3.

Charpantier E; Wiesner A; Huh KH; Ogier R; Hoda JC; Allaman G; Raggenbass M; Feuerbach D; Bertrand D; Fuhrer C. Alpha7 neuronal nicotinic acetylcholine receptors are negatively regulated by tyrosine phosphorylation and Src-family kinases. *J Neurosci.* 2005, 25(43):9836-49.

Cho CH; Song W; Leitzell K; Teo E; Meleth AD; Quick MW; Lester RA. Rapid upregulation of alpha7 nicotinic acetylcholine receptors by tyrosine dephosphorylation. *J Neurosci.* 2005, 25(14):3712-23.

Choi JM; Romeo RD; Brake WG; Bethea CL; Rosenwaks Z; McEwen BS. Estradiol increases pre- and postsynaptic proteins in the CA1 region of the hippocampus in female rhesus macaques (Macaca mulatta). *Endocrinology.* 2003, 144(11):4734-8.

Christensen K; Doblhammer G; Rau R; Vaupel JW. Ageing populations: the challenges ahead. *Lancet.* 2009, 374(9696):1196-208.

Chung WC; Pak TR; Suzuki S; Pouliot WA; Andersen ME; Handa RJ. Detection and localization of an estrogen receptor beta splice variant protein (ERbeta2) in the adult female rat forebrain and midbrain regions. *J Comp Neurol.* 2007, 505(3):249-67.

Clarkson TB; Anthony MS; Hughes CL Jr. Estrogenic soybean isoflavones and chronic disease Risks and benefits. *Trends Endocrinol Metab.* 1995, 6(1):11-6.

Cornwell T; Cohick W; Raskin I. Dietary phytoestrogens and health. *Phytochemistry.* 2004, 65(8):995-1016.

Cowansage KK; Ledoux JE; Monfils MH. Brain-Derived Neurotrophic Factor: A Dynamic Gatekeeper of Neural Plasticity. *Curr Mol Pharmacol.* 2009 Dec 23. [Epub ahead of print].

Coward L; Smith M; Kirk M; Barnes S. Chemical modification of isoflavones in soyfoods during cooking and processing. *Am J Clin Nutr.* 1998, 68(6 Suppl):1486S-1491S.

Crews DE. Senescence; aging, and disease. J Physiol Anthropol. 2007, 26(3):365-72.

Curran HV; Schifano F; Lader M. Models of memory dysfunction? Acomparison of the effects of scopolamine and lorazepam on memory, psychomotor performance and mood. *Psychopharmacology.* 1991, 103:83-90.

Curtis L; Buisson B; Bertrand S; Bertrand D. Potentiation of human alpha4beta2 neuronal nicotinic acetylcholine receptor by estradiol. *Mol Pharmacol.* 2002, 61(1):127-35.

Damaj MI. Influence of gender and sex hormones on nicotine acute pharmacological effects in mice. *J Pharmacol Exp Ther.* 2001, 296(1):132-40.

Daniel JM; Hulst JL; Berbling JL. Estradiol replacement enhances working memory in middle-aged rats when initiated immediately after ovariectomy but not after a long-term period of ovarian hormone deprivation. *Endocrinology.* 2006, 147(1):607-14.

Davis C; Bradshaw CM; Szabadi E. The doors and People Memory Test: validation of norms and some new correction formulae. *Br J Clin Psychol.* 1999, 38(pt3):305-314.

De Lacalle S. Estrogen effects on neuronal morphology. Endocrine. 2006, 29(2):185-90. dehydrogenase and aromatase in human granulosa-luteal cells. *J Steroid Biochem Mol Biol.* 2005, 96(3-4):279-86.

Downs JL; Urbanski HF. Neuroendocrine changes in the aging reproductive axis of female rhesus macaques (Macaca mulatta). *Biol Reprod.* 2006, 75(4):539-46.

Downs JL; Wise PM. The role of the brain in female reproductive aging. *Mol Cell Endocrinol.* 2009, 299(1):32-8.

Driscoll I; Hamilton DA; Petropoulos H; Yeo RA; Brooks WM; Baumgartner RN; Sutherland RJ. The aging hippocampus: cognitive, biochemical and structural findings. *Cereb Cortex.* 2003, 13(12):1344-51.

Duffy R; Wiseman H; File SE. Improved cognitive function in postmenopausal women after 12 weeks of consumption of a soya extract containing isoflavones. *Pharmacol Biochem Behav.* 2003, 75(3):721-9.

Dumas J; Hancur-Bucci C; Naylor M; Sites C; Newhouse P. Estradiol interacts with the cholinergic system to affect verbal memory in postmenopausal women: evidence for the critical period hypothesis. *Horm Behav.* 2008, 53(1):159-69.

Edmunds KM; Holloway AC; Crankshaw DJ; Agarwal SK; Foster WG. The effects of dietary phytoestrogens on aromatase activity in human endometrial stromal cells. *Reprod Nutr Dev.* 2005, 45(6):709-20.

Favit A; Fiore L; Nicoletti F; Canonico PL. Estrogen modulates stimulation of inositol phospholipid hydrolysis by norepinephrine in rat brain slices. *Brain Res.* 1991, 555(1):65-9

Fernandez SM; Lewis MC; Pechenino AS; Harburger LL; Orr PT; Gresack JE; Schafe GE, Frick KM. Estradiol-induced enhancement of object memory consolidation involves hippocampal extracellular signal-regulated kinase activation and membrane-bound estrogen receptors. *J Neurosci.* 2008 Aug 27, 28(35):8660-7.

Ferrari CK. Functional foods, herbs and nutraceuticals: towards biochemical mechanisms of healthy aging. *Biogerontology.* 2004, 5(5):275-89.

Ferrucci L; Giallauria F; Guralnik JM. Epidemiology of aging. *Radiol Clin North Am.* 2008, 46(4):643-52, v.

Filardo E; Quinn J; Pang Y; Graeber C; Shaw S; Dong J; Thomas P. Activation of the novel estrogen receptor G protein-coupled receptor 30 (GPR30) at the plasma membrane. *Endocrinology.* 2007, 148(7):3236-45.

File SE; Hartley DE; Alom N; Rattray M. Soya phytoestrogens change cortical and hippocampal expression of BDNF mRNA in male rats. *Neurosci Lett.* 2003, 338(2):135-8.

File SE; Jarrett N; Fluck E; Duffy R; Casey K; Wiseman H. Eating soya improves human memory. *Psychopharmacology.* 2001, 157(4):430-6.

Findlay JK; Buckmaster JM; Chamley WA; Cumming IA; Hearnshaw H et Goding JR. Release of luteinising hormone by œstradiol 17β and a gonadotrophin-releasing hormone in ewes affected with clover disease. *Neuroendocrinology.* 1973, 11:57-66.

Fitzpatrick LA. Soy isoflavones: hope or hype? *Maturitas.* 2003, 44 (suppl 1):S-21-S29.

Foster TC. Involvement of hippocampal synaptic plasticity in age-related memory decline. *Brain Res Brain Res Rev.* 1999 Nov, 30(3):236-49. Review. PubMed PMID: 10567726.

Fournier LR; Ryan Borchers TA; Robison LM; Wiediger M; Park JS; Chew BP; McGuire MK; Sclar DA; Skaer TL; Beerman KA. The effects of soy milk and isoflavone supplements on cognitive performance in healthy, postmenopausal women. *J Nutr Health Aging.* 2007, 11(2):155-64.

Foy MR; Baudry M; Diaz Brinton R; Thompson RF. Estrogen and hippocampal plasticity in rodent models. *J Alzheimers Dis.* 2008, 15(4):589-603.

Foy MR; Xu J; Xie X; Brinton RD; Thompson RF; Berger TW. 17beta-estradiol enhances NMDA receptor-mediated EPSPs and long-term potentiation. *J Neurophysiol.* 1999, 81(2):925-9.

Frick KM; Fernandez SM; Bulinski SC. Estrogen replacement improves spatial reference memory and increases hippocampal synaptophysin in aged female mice. *Neuroscience.* 2002, 115(2):547-58.

Frick KM. Estrogens and age-related memory decline in rodents: what have we learned and where do we go from here? *Horm Behav.* 2009, 55(1):2-23.

Frye CA; Rhodes ME; Dudek B. *Estradiol to aged female or male mice improves learning in inhibitory avoidance and water maze tasks. Brain Res.* 2005, 1036(1-2):101-8.

Frye CA. Estrus-associated decrements in a water maze task are limited to acquisition. *Physiol Behav.* 1995, 57(1):5-14.

Garcia-Segura LM; Azcoitia I; DonCarlos LL. Neuroprotection by estradiol. *Prog Neurobiol.* 2001, 63(1):29-60.

Garcia-Segura LM. Aromatase in the brain: not just for reproduction anymore. *J Neuroendocrinol.* 2008, 20(6):705-12.

Gazzaley AH; Weiland NG; McEwen BS; Morrison JH. Differential regulation of NMDAR1 mRNA and protein by estradiol in the rat hippocampus. *J Neurosci.* 1996, 16(21):6830-8.

Geller SE; Studee L. Soy and red clover for mid-life and aging. *Climacteric.* 2006, 9(4):245-63.

Genazzani AR; Pluchino N; Luisi S; Luisi M. Estrogen; cognition and female ageing. *Hum Reprod Update.* 2007, 13(2):175-87.

Giovannini MG; Pazzagli M; Malmberg-Aiello P; Della Corte L; Rakovska AD; Cerbai F; Casamenti F; Pepeu G. Inhibition of acetylcholine-induced activation of extracellular regulated protein kinase prevents the encoding of an inhibitory avoidance response in the rat. *Neuroscience.* 2005, 136(1):15-32.

Giovannini MG. The role of the extracellular signal-regulated kinase pathway in memory encoding. Rev *Neurosci.* 2006, 17(6):619-34.

Gleason CE; Carlsson CM; Barnet JH; Meade SA; Setchell KD; Atwood CS; Johnson 1 SC; Ries ML; Asthana S. A preliminary study of the safety, feasibility and cognitive efficacy of soy isoflavone supplements in older men and women. *Age Ageing.* 2009, 38(1):86-93.

Golub MS; Hogrefe CE; Germann SL; Tran TT; Beard JL; Crinella FM; Lonnerdal B. Neurobehavioral evaluation of rhesus monkey infants fed cow's milk formula, soy formula, or soy formula with added manganese. *Neurotoxicol Teratol.* 2005, 27(4):615-27.

Gruber CJ; Gruber DM; Gruber IML; Wieser F; Huber JC. Anatomy of the estrogen response element. *Trends Endocrinol Metab.* 2004, 15(2):73-78.

Halbreich U; Lumley LA; Palter S; Manning C; Gengo F; Joe SH. Possible acceleration of age effects on cognition following menopause. *J Psychiatr Res.* 1995, 29(3):153-63.

Hall JE. Neuroendocrine changes with reproductive aging in women. *Semin Reprod Med.* 2007, 25(5):344-51.

Hazell GG; Yao ST; Roper JA; Prossnitz ER; O'Carroll AM; Lolait SJ. Localisation of GPR30, a novel G protein-coupled oestrogen receptor, suggests multiple functions in rodent brain and peripheral tissues. *J Endocrinol.* 2009, 202(2):223-36.

Hellström-Lindahl E; Mousavi M; Zhang X; Ravid R; Nordberg A. Regional distribution of nicotinic receptor subunit mRNAs in human brain: comparison between Alzheimer and normal brain. *Brain Res Mol Brain Res.* 1999, 66(1-2):94-103.

Hill AJ. First occurrence of hippocampal spatial firing in a new environment. *Exp Neurol.* 1978, 62(2):282-97.

Ho SC; Chan AS; Ho YP; So EK; Sham A; Zee B; Woo JL. Effects of soy isoflavone supplementation on cognitive function in Chinese postmenopausal women: a double-blind, randomized, controlled trial. *Menopause.* 2007, 14(3 Pt 1):489-99.

Hogervorst E; Sadjimim T; Yesufu A; Kreager P; Rahardjo TB. High tofu intake is associated with worse memory in elderly Indonesian men and women. *Dement Geriatr Cogn Disord.* 2008, 26(1):50-7.

Hojo Y; Murakami G; Mukai H; Higo S; Hatanaka Y; Ogiue-Ikeda M; Ishii H; Kimoto T; Kawato S. Estrogen synthesis in the brain--role in synaptic plasticity and memory. *Mol Cell Endocrinol.* 2008, 290(1-2):31-43.

Hooper L; Ryder JJ; Kurzer MS; Lampe JW; Messina MJ; Phipps WR; Cassidy A. Effects of soy protein and isoflavones on circulating hormone concentrations in pre- and post-menopausal women: a systematic review and meta-analysis. *Hum Reprod Update.* 2009, 15(4):423-40.

Huang MH; Luetters C; Buckwalter GJ; Seeman TE; Gold EB; Sternfeld B; Greendale GA. Dietary genistein intake and cognitive performance in a multiethnic cohort of midlife women. *Menopause.* 2006, 13(4):621-30.

Hung SP; Hsu JR; Lo CP; Huang HJ; Wang JP; Chen ST. Genistein-induced neuronal differentiation is associated with activation of extracellular signal-regulated kinases and upregulation of p21 and N-cadherin. *J Cell Biochem.* 2005, 96(5):1061-70.

Ishii H; Tsurugizawa T; Ogiue-Ikeda M; Asashima M; Mukai H; Murakami G; Hojo Y; Kimoto T; Kawato S. Local production of sex hormones and their modulation of hippocampal synaptic plasticity. *Neuroscientist.* 2007, 13(4):323-34.

Ishunina TA; Fischer DF; Swaab DF. Estrogen receptor alpha and its splice variants in the hippocampus in aging and Alzheimer's disease. *Neurobiol Aging.* 2007, 28(11):1670-81.

Izumi T; Piskula MK; Osawa S; Obata A; Tobe K; Saito M; Kataoka S; Kubota Y; Kikuchi M. Soy isoflavone aglycones are absorbed faster and in higher amounts than their glucosides in humans. *J Nutr.* 2000, 130(7):1695-9.

Jefremov V; Rakitin A; Mahlapuu R; Zilmer K; Bogdanovic N; Zilmer M; Karelson E. 17beta-Oestradiol stimulation of G-proteins *J Neuroendocrinol.* 2008, 20(5):587-96.

Jamali K; Naylor BR; Kelly MJ; Rønnekleiv OK. Effect of 17β-estradiol on mRNA expression of large-conductance, voltage-dependent, and calcium-activated potassium channel alpha and beta subunits in guinea pig. Endocrine. 2003, 20(3):227-37.

Kajta M; Domin H; Grynkiewicz G; Lason W. Genistein inhibits glutamate-induced apoptotic processes in primary neuronal cell cultures: an involvement of aryl hydrocarbon receptor and estrogen receptor/glycogen synthase kinase-3beta intracellular signaling pathway. *Neuroscience*. 2007, 145(2):592-604.

Kalita K; Szymczak S; Kaczmarek L. Non-nuclear estrogen receptor beta and alpha in the hippocampus of male and female rats. *Hippocampus.* 2005, 15(3):404-12.

Kaplan E; Goodglass H; Weintraub S. *The Boston Naming Test.* Philadelphia, Pa: Lea & Febiger. 1982.

Katzenellenbogen JA; Katzenellebogen BS; Tatee T; Robertson DW; Landvatter SW.. The chemistry of estrogens and antiestrogens: relationships between structure, receptor binding ans biological activity. In: *Estrogens in the Environment.* JA McLachlan Ed, Elsevier Inc. Pub, 1980, 33 - 51.

Kelly MJ; Levin ER. Rapid actions of plasma membrane estrogen receptors. *Trends End Metab*. 2001, 12:152-156.

Kelly MJ; Rønnekleiv OK. Membrane-initiated estrogen signalling in hypothalamic neurons. *Molec Cel Endocrinol.* 2008, 290:14-23.

Kelly MJ; Wagner EJ. Estrogen modulation of G-protein-coupled receptors. *Trends End Metab.* 1999, 10:369-374.

Kim JS; Kim HY; Kim JH; Shin HK; Lee SH; Lee YS; Son H. Enhancement of rat hippocampal long-term potentiation by 17 beta-estradiol involves mitogen-activated protein kinase-dependent and -independent components. *Neurosci Lett.* 2002, 332(1):65-9.

Kitamura N; Araya R; Kudoh M; Kishida H; Kimura T; Murayama M; Takashima A; Sakamaki Y; Hashikawa T; Ito S; Ohtsuki S; Terasaki T; Wess J; Yamada M. Beneficial effects of estrogen in a mouse model of cerebrovascular insufficiency. *PLoS One*. 2009, 4(4):e5159.

Knaus HG; Schwarzer C; Koch RO; Eberhart A; Kaczorowski GJ; Glossmann H; Wunder F; Pongs O; Garcia ML; Sperk G. Distribution of high-conductance Ca(2+)-activated K+ channels in rat brain: targeting to axons and nerve terminals. *J Neurosci.* 1996, 16(3):955-63.

Koss DJ; Riedel G; Platt B. Intracellular Ca2+ stores modulate SOCCs and NMDA receptors via tyrosine kinases in rat hippocampal neurons. *Cell Calcium*. 2009, 46(1):39-48.

Kreijkamp-Kaspers S; Kok L; Grobbee DE; de Haan EH; Aleman A; Lampe JW; vanderSchouw YT. Effect of soy protein containing isoflavones on cognitive function, bone mineral density and plasma lipids in postmenopausal women: a randomized controlled trial. *JAMA*. 2004, 292(1):65-74.

Kreijkamp-Kaspers S; Kok L; Grobbee DE; de Haan EH; Aleman A; vanderSchouw YT. Dietary phytoestrogen intake and cognitive function in older women. *J Gerontol A Biol Sci Med Sci*. 2007, 62(5):556-62.

Kretz O; Fester L; Wehrenberg U; Zhou L; Brauckmann S; Zhao S; Prange-Kiel J; Naumann T; Jarry H; Frotscher M; Rune GM. Hippocampal synapses depend on hippocampal estrogen synthesis. *J Neurosci*. 2004, 24(26):5913-21.

Kritz-Silverstein D; Von Mühlen D; Barrett-Connor E; Bressel MA. Isoflavones and cognitive function in older women: the SOy and Postmenopausal Health In Aging (SOPHIA) Study. *Menopause*. 2003, 10(3):196-202.

Kuiper GC; Carlsson B; Grandien K; Enmark E; Haggblad J; Nilsson S; Gusfasson JA. Comparison of the ligand binding specificity and transcript tissue distribution of estrogen receptors alpha and beta. *Endocrinology*. 1997, 138(3):863-70.

Kuiper GG; Lemmen JG; Carlsson B; Corton JC; Safe SH; van der Saag PT; van der Burg B; Gustafsson JA. Interaction of estrogenic chemicals and phytoestrogens with estrogen receptor beta. *Endocrinology*. 1998, 139(10):4252-63.

Kumar A; Foster TC. 17beta-estradiol benzoate decreases the AHP amplitude in CA1 pyramidal neurons. *J Neurophysiol*. 2002, 88(2):621-6.

Lacey M; Bohday J; Fonseka SM; Ullah AI; Whitehead SA. Dose-response effects of phytoestrogens on the activity and expression of 3beta-hydroxysteroid dehydrogenase and aromatase in human granulosa-luteal cells. *J Steroid Biochem Mol Biol*. 2005, 96(3-4):279-86.

Lacreuse A. Effects of ovarian hormones on cognitive function in nonhuman primates. *Neuroscience*. 2006, 138(3):859-67.

Lâm TT; Leranth C. Role of the medial septum diagonal band of Broca cholinergic neurons in oestrogen-induced spine synapse formation on hippocampal CA1 pyramidal cells of female rats. *Eur J Neurosci*. 2003, 17(10):1997-2005.

Lee YB; Lee HJ; Sohn HS. Soy isoflavones and cognitive function. *J Nutr Biochem*. 2005, 16(11):641-9.

Lee YB; Lee HJ; Won MH; Hwang IK; Kang TC; Lee JY; Nam SY; Kim KS; Kim E; Cheon SH; Sohn HS. Soy isoflavones improve spatial delayed matching-to-place performance and reduce cholinergic neuron loss in elderly male rats. *J Nutr*. 2004, 134(7):1827-31.

Lee YB; Lee KH; Sohn HS; Lee SJ; Cho KH; Kang IJ; Kim DW; Shin YK; Pai T; Hwang IK; Won MH. Effects of soy phytoestrogens on reference memory and neuronal cholinergic enzymes in ovariectomized rats. *J Med Food*. 2009, 12(1):64-70.

Lee[1] SJ; Campomanes CR; Sikat PT; Greenfield AT; Allen PB; McEwen BS. Estrogen induces phosphorylation of cyclic AMP response element binding (pCREB) in primary hippocampal cells in a time-dependent manner. *Neuroscience*. 2004, 124(3):549-60.

Lephart ED; Adlercreutz H; Lund TD. Dietary soy phytoestrogen effects on brain structure and aromatase in Long-Evans rats. *Neuroreport*. 2001, 12(16):3451-5.

Lephart ED; Setchell KD; Handa RJ; Lund TD. Behavioral effects of endocrine-disrupting substances: phytoestrogens. *ILAR J.* 2004, 45(4):443-54.

Lephart ED; Setchell KD; Lund TD. Phytoestrogens: hormonal action and brain plasticity. *Brain Res Bull.* 2005, 65(3):193-8.

Lephart ED; Thompson JM; Setchell KD; Adlercreutz H; Weber KS. Phytoestrogens decrease brain calcium-binding proteins but do not alter hypothalamic androgen metabolizing enzymes in adult male rats. *Brain Res.* 2000, 859(1):123-31.

Levin ER. Plasma membrane estrogen receptors. *Trends Endocrinol Metab.* 2009, 20(10):477-82.

Levitzki A; Mishani E. Tyrphostins and other tyrosine kinase inhibitors. *Annu Rev Biochem.* 2006, 75:93-109.

Lezak MD. *Neuropsychological assessment.* New York: Oxford University Press. 1995.

Linford NJ; Yang Y; Cook DG; Dorsa DM. Neuronal apoptosis resulting from high doses of the isoflavone genistein: role for calcium and p42/44 mitogen-activated protein kinase. *J Pharmacol Exp Ther.* 2001, 299(1):67-75.

Liu F; Day M; Muñiz LC; Bitran D; Arias R; Revilla-Sanchez R; Grauer S; Zhang G; Kelley C; Pulito V; Sung A; Mervis RF; Navarra R; Hirst WD; Reinhart PH; Marquis KL; Moss SJ; Pangalos MN; Brandon NJ. Activation of estrogen receptor-beta regulates hippocampal synaptic plasticity and improves memory. *Nat Neurosci.* 2008, 11(3):334-43.

Liu YQ; Xin TR; Lü XY; Ji Q; Jin Y; Yang HD. Memory performance of hypercholesterolemic mice in response to treatment with soy isflavones. *Neurosci Res.* 2007, 57(4):544-49.

Lu WY; Xiong ZG; Lei S; Orser BA; Dudek E; Browning MD; MacDonald JF. G-protein-coupled receptors act via protein kinase C and Src to regulate NMDA receptors. *Nat Neurosci.* 1999, 2(4):331-8.

Luine V; Attalla S; Mohan G; Costa A; Frankfurt M. Dietary phytoestrogens enhance spatial memory and spine density in the hippocampus and prefrontal cortex of ovariectomized rats. *Brain Res Rev.* 2006, 1126(1):183-7.

Lund TD; West TW; Tian LY; Bu LH; Simmons DL; Setchell KD; Adlercreutz H; Lephart ED. Visual spatial memory is enhanced in female rats (but inhibited in males) by dietary soy phytoestrogens. *BMC Neurosci.* 2001, 2:20.

Lund[1] TD; Lephart ED. Dietary soy phytoestrogens produce anxiolytic effects in the elevated plus-maze. *Brain Res.* 2001, 913(2):180-4.

Lund[2] TD; Lephart ED. Manipulation of prenatal hormones and dietary phytoestrogens during adulthood alter the sexually dimorphic expression of visual spatial memory. *BMC Neurosci.* 2001, 2:21.

Lutz W; Qiang R. Determinants of human population growth. *Philos Trans R Soc Lond B Biol Sci.* 2002, 357(1425):1197-210.

Maggiolini M; Picard D. The unfolding stories of GPR30, a new membrane-bound estrogen receptor. *J Endocrinol.* 2010, 204(2):105-14.

Maggiolini M; Vivacqua A; Fasanella G; Recchia AG; Sisci D; Pezzi V; Montanaro D; Musti AM; Picard D; Andò S. The G protein-coupled receptor GPR30 mediates c-fos upregulation by 17beta-estradiol and phytoestrogens in breast cancer cells. *J Biol Chem.* 2004, 279(26):27008-16.

Maki PM; Dumas J. Mechanisms of action of estrogen in the brain: insights from human neuroimaging and psychopharmacologic studies. *Semin Reprod Med.* 2009, 27(3):250-9.

Maki PM; Rubin LH; Fornelli D; Drogos L; Banuvar S; Shulman LP; Geller SE. Effects of botanicals and combined hormone therapy on cognition in postmenopausal women. *Menopause.* 2009, 16(6):1167-77.

Marrian GF; Haslewood GAD. Equol a new inactive phenol isolated from the ketohydroxy-œstrin fraction of mares' urine. *Biochem. J.* 1932, 26:1227-1232.

Mathey J; Lamothe V; Coxam V; Potier M; Sauvant P; Bennetau-Pelissero C. Concentrations of isoflavones in plasma and urine of post-menopausal women chronically ingesting high quantities of soy isoflavones. *J Pharm Biomed Anal.* 2006, 41:957-965.

Mathey J; Mardon J; Fokialakis N; Puel C; Kati-Coulibaly S; Mitakou S; Bennetau-Pelissero C; Lamothe V; Davicco MJ; Lebecque P; Horcajada MN; Coxam V. Modulation of soy isoflavones bioavailability and subsequent effects on bone health in ovariectomized rats: the case for equol. *Osteoporos Int.* 2007, 18(5):671-9.

Matsuda K; Sakamoto H; Mori H; Hosokawa K; Kawamura A; Itose M; Nishi M; Prossnitz ER; Kawata M. Expression and intracellular distribution of the G protein-coupled receptor 30 in rat hippocampal formation. *Neurosci Lett.* 2008, 441(1):94-9.

McEwen B. Estrogen actions throughout the brain. Recent Prog Horm Res. 2002, 57:357-84.

McEwen BS; Alves SE; Bulloch K; Weiland NG. Ovarian steroids and the brain: implications for cognition and aging. *Neurology.* 1997, 48(5 Suppl 7):S8-15.

McEwen BS; Milner TA. Hippocampal formation: shedding light on the influence of sex and stress on the brain. *Brain Res Rev.* 2007, 55(2):343-55.

McNaughton BL; Barnes CA; O'Keefe J. The contributions of position, direction, and velocity to single unit activity in the hippocampus of freely-moving rats. *Exp Brain Res.* 1983, 52(1):41-9.

Mehra RD; Sharma K; Nyakas C; Vij U. Estrogen receptor alpha and beta immunoreactive neurons in normal adult and aged female rat hippocampus: a qualitative and quantitative study. *Brain Res.* 2005, 1056(1):22-35.

Mendelson CR; Jiang B; Shelton JM; Richardson JA; Hinshelwood MM.Transcriptional regulation of aromatase in placenta and ovary. *J Steroid Biochem Mol Biol.* 2005, 95(1-5):25-33.

Messinis IE. Ovarian feedback, mechanism of action and possible clinical implications. *Hum Reprod Update.* 2006, 12(5):557-71.

Mhyre AJ; Dorsa DM. Estrogen activates rapid signaling in the brain: role of estrogen receptor alpha and estrogen receptor beta in neurons and glia. *Neuroscience.* 2006, 138(3):851-8.

Mielke JG; Mealing GA. Cellular distribution of the nicotinic acetylcholine receptor alpha7 subunit in rat hippocampus. *Neurosci Res.* 2009, 65(3):296-306.

Milner TA; McEwen BS; Hayashi S; Li CJ; Reagan LP; Alves SE. Ultrastructural evidence that hippocampal alpha estrogen receptors are located at extranuclear sites. *J Comp Neurol.* 2001, 429(3):355-71.

Milner TA; Ayoola K; Drake CT; Herrick SP; Tabori NE; McEwen BS; Warrier S; Alves SE. Ultrastructural localization of estrogen receptor beta immunoreactivity in the rat hippocampal formation. *J Comp Neurol.* 2005, 491(2):81-95.

Milner TA; Lubbers LS; Alves SE; McEwen BS. Nuclear and extranuclear estrogen binding sites in the rat forebrain and autonomic medullary areas. *Endocrinology.* 2008, 149(7):3306-12.

Mitra SW; Hoskin E; Yudkovitz J; Pear L; Wilkinson HA; Hayashi S; Pfaff DW; Ogawa S; Rohrer SP; Schaeffer JM; McEwen BS; Alves SE. Immunolocalization of estrogen receptor beta in the mouse brain: comparison with estrogen receptor alpha. *Endocrinology.* 2003, 144(5):2055-67. Erratum in: *Endocrinology.* 2003, 144(7):2844.

Molokanova E; Savchenko A; Kramer RH. Interactions of cyclic nucleotide-gated channel subunits and protein tyrosine kinase probed with genistein. *J Gen Physiol.* 2000, 115(6):685-96.

Molokanova E; Savchenko A; Kramer RH. Noncatalytic inhibition of cyclic nucleotide-gated channels by tyrosine kinase induced by genistein. *J Gen Physiol.* 1999, 113(1):45-56.

Montague D; Weickert CS; Tomaskovic-Crook E; Rothmond DA; Kleinman JE; Rubinow DR. Oestrogen receptor alpha localisation in the prefrontal cortex of three mammalian species. *J Neuroendocrinol.* 2008, 20(7):893-903.

Monteiro SC; de Mattos CB; Ben J; Netto CA; Wyse AT. Ovariectomy impairs spatial memory: prevention and reversal by a soy isoflavone diet. *Metab Brain Dis.* 2008, 23(3):243-53.

Moutsatsou P. *The spectrum of phytoestrogens in nature: our knowledge is expanding.* Hormones (Athens). 2007, 6(3):173-93.

Nomura M; Akama KT; Alves SE; Korach KS; Gustafsson JA; Pfaff DW; Ogawa S. Differential distribution of estrogen receptor (ER)-alpha and ER-beta in the midbrain raphe nuclei and periaqueductal gray in male mouse: Predominant role of ER-beta in midbrain serotonergic systems. *Neuroscience.* 2005, 130(2):445-56.

O'Dell TJ; Kandel ER; Grant SG. Long-term potentiation in the hippocampus is blocked by tyrosine kinase inhibitors. *Nature.* 1991, 353(6344):558-60.

Ogiue-Ikeda M; Tanabe N; Mukai H; Hojo Y; Murakami G; Tsurugizawa T; Takata N; Kimoto T; Kawato S. Rapid modulation of synaptic plasticity by estrogens as well as endocrine disrupters in hippocampal neurons. *Brain Res Rev.* 2008 Mar, 57(2):363-75.

Ohlsson A; Ullerås E; Oskarsson A. A biphasic effect of the fungicide prochloraz on aldosterone, but not cortisol, secretion in human adrenal H295R cells--underlying mechanisms. *Toxicol Lett.* 2009, 191(2-3):174-80.

Olton DS; Branch M; Best PJ. Spatial correlates of hippocampal unit activity. *Exp Neurol.* 1978, 58(3):387-409.

Owen AM; Downes JJ; Sahakian BJ; Polkey CE; Robbins TW. Planning and spatial working memory following frontal lobe lesions in man. *Neuropsychologia.* 1990, 28:1021-1034.

Owen AM; Roberts AC; Polkey CE; Sahakian BJ; Robbins TW. Extra-dimensional versus intra-dimensional set shifting performance following frontal lobe excisions, temporal lobe excisions or amygdalo-hyppocampectomy in man. *Neuropsychologia.* 1991, 29:993-1006.

Owen AM; Sahakian BJ; Semple J; Polkey CE; Robbins TW. Visuospatial short-term recognition memory and learning after temporal lobe excisions, frontal lobe excisions or amygdalo-hyppocampectomy in man. *Neuropsychologia.* 1995, 33:1-24.

Pacheco Otalora LF; Hernandez EF; Arshadmansab MF; Francisco S; Willis M; Ermolinsky B; Zarei M; Knaus HG; Garrido-Sanabria ER. Downregulation of BK channel expression in the pilocarpine model of temporal lobe epilepsy. *Brain Res.* 2008, 1200:116-31.

Paillart C; Carlier E; Guedin D; Dargent B; Couraud F. Direct block of voltage-sensitive sodium channels by genistein, a tyrosine kinase inhibitor. *J Pharmacol Exp Ther*. 1997, 280(2):521-6.

Pan Y; Anthony M; Clarkson TB. Effect of estradiol and soy phytoestrogens on choline acetyltransferase and nerve growth factor mRNAs in the frontal cortex and hippocampus of female rats. *Pro Soc Exp Biol Med.* 1999, 221(2):118-25.

Pan Y; Anthony M; Watson S; Clarkson TB. Soy phytoestrogens improve radial arm maze performance in ovariectomized retired breeder rats and do not attenuate benefits of 17beta-estradiol treatment. *Menopause.* 2000, 7(4):230-5.

Pan[1] Y; Anthony M; Clarkson TB. Evidence for upregulation of brain-derived neurotrophic factor mRNA by soy phytoestrogens in the frontal cortex of retired breeder female rats. *Neurosci Lett.* 1999 , 12 :261(1-2):17-20.

Pelissero C; Lenczowski M; Chinzi D; Sumpter JP; Fostier A. Effects of flavonoids on aromatase activity, an *in vitro* study. *J Steroid Biochem Molec Biol.* 1996, 57(3/4):215-223.

Pelletier G; Dupont E; Simard J; Luu-The V; Bélanger A; Labrie F. Ontogeny and subcellular localization of 3 beta-hydroxysteroid dehydrogenase (3 beta-HSD) in the human and rat adrenal, ovary and testis. *J Steroid Biochem Mol Biol.* 1992, 43(5):451-67.

Phelps EA; Hyder F; Blamire AM; Shulman RG. fMRI of the prefrontal cortex during overt verbal fluency. *Neuroreport.* 1997, 8:561-565.

Picherit C; Coxam V; Bennetau-Pelissero C; Kati-Coulibaly S; Davicco MJ; Lebecque P; Barlet JP. Daidzein is more efficient than genistein in preventing ovariectomy-induced bone loss in rats. *J Nutr.* 2000, 130(7):1675-81.

Pihlajamaki SM; Tanila H; Hanninen T; Kononen M; Laakso M; Partanen K; Soininen H; Aronen HJ. Verbal fluency activates the left medial temporal lobe: a functional magnetic resonance imaging study. *Ann Neurol.* 2000, 47:470-476.

Potier B; Rovira C. Protein tyrosine kinase inhibitors reduce high-voltage activating calcium currents in CA1 pyramidal neurones from rat hippocampal slices. *Brain* 1999, 816(2):587-97.

Prange-Kiel J; Rune GM; Leranth C. Median raphe mediates estrogenic effects to the hippocampus in female rats. *Eur J Neurosci.* 2004, 19(2):309-17.

Prange-Kiel J; Rune GM. Direct and indirect effects of estrogen on rat hippocampus. *Neuroscience.* 2006, 138(3):765-72.

Prange-Kiel[1] J; Fester L; Zhou L; Lauke H; Carrétero J; Rune GM. Inhibition of hippocampal estrogen synthesis causes region-specific downregulation of synaptic protein expression in hippocampal neurons. *Hippocampus.* 2006, 16(5):464-71.

Prossnitz ER; Arterburn JB; Sklar LA. GPR30: A G protein-coupled receptor for estrogen. *Mol Cell Endocrinol.* 2007, 265-266:138-42.

Qiu J; Bosch MA; Jamali K; Xue C; Kelly MJ; Rønnekleiv OK. Estrogen upregulates T-type calcium channels in the hypothalamus and pituitary. *J Neurosci.* 2006, 26(43):11072-82.

Ramsey MM; Adams MM; Ariwodola OJ; Sonntag WE; Weiner JL. Functional characterization of des-IGF-1 action at excitatory synapses in the CA1 region of rat hippocampus. *J Neurophysiol.* 2005, 94(1):247-54.

Reitan RK; Wolfson D. *The Halestead-Reitan neuropsychological test battery.* Tucson, AZ: Neuropsychological press. 1985.

Rice S; Mason HD; Whitehead SA. Phytoestrogens and their low dose combinations inhibit mRNA expression and activity of aromatase in human granulosa-luteal cells. *J Steroid Biochem Mol Biol.* 2006, 101(4-5):216-25.

Rogalski SL; Appleyard SM; Pattillo A; Terman GW; Chavkin C. TrkB activation by brain-derived neurotrophic factor inhibits the G protein-gated inward rectifier Kir3 by tyrosine phosphorylation of the channel. *J Biol Chem.* 2000, 275(33):25082-8.

Rowland I; Faughnan M; Hoey L; Wähälä K; Williamson G; Cassidy A. Bioavailability of phyto-oestrogens. *Br J Nutr.* 2003, 89 Suppl 1:S45-58.

Rowland LM; Astur RS; Jung RE; Bustillo JR; Lauriello J; Yeo RA. Selective cognitive impairments associated with NMDA receptor blockade in humans. *Neuropsychopharmacology.* 2005, 30(3):633-9.

Rune GM; Frotscher M. Neurosteroid synthesis in the hippocampus: role in synaptic plasticity. *Neuroscience*s. 2005, 136(3):833-42.

Rune GM; Wehrenberg U; Prange-Kiel J; Zhou L; Adelmann G; Frotscher M. Estrogen upregulates estrogen receptor alpha and synaptophysin in slice cultures of rat hippocampus. *Neuroscience.* 2002, 113(1):167-75.

Sakamoto H; Matsuda K; Hosokawa K; Nishi M; Morris JF; Prossnitz ER; Kawata M. Expression of G protein-coupled receptor-30, a G protein-coupled membrane estrogen receptor, in oxytocin neurons of the rat paraventricular and supraoptic nuclei. *Endocrinology.* 2007, 148(12):5842-50.

Sarkar SN; Huang RQ; Logan SM; Yi KD; Dillon GH; Simpkins JW. Estrogens directly potentiate neuronal L-type Ca2+ channels. *Proc Natl Acad* Sci USA. 2008, 105(39):15148-53.

Scallet AC; Wofford M; Meredith JC; Allaben WT; Ferguson SA. Dietary exposure to genistein increases vasopressin but does not alter beta-endorphin in the rat hypothalamus. *Toxicol Sci.* 2003, 72(2):296-300.

Scharfman HE; MacLusky NJ. Estrogen and brain-derived neurotrophic factor (BDNF) in hippocampus: complexity of steroid hormone-growth factor interactions in the adult CNS. *Front Neuroendocrinol.* 2006, 27(4):415-35.

Schlosser R; Hutchinson M; Joseffer S; Rusinek H; Saarimaki A; Stevenson J; Dewey SL; Brodie JD. Functional magnetic resonance imaging of human brain activity in a verbal fluency task. *J Neurol Neurosurg Psychiatry.* 1998, 64:492-498.

Schreihofer DA; Redmond L. Soy phytoestrogens are neuroprotective against stroke *Neuroscience.* 2009, 158(2):602-9.

Schreihofer DA. Transcriptional regulation by phytoestrogens in neuronal cell lines. *Mol Cell Endocrinol.* 2005, 231(1-2):13-22.

Setchell KD; Brown NM; Lydeking-Olsen E. The clinical importance of the metabolite equol-a clue to the effectiveness of soy and its isoflavones. *J Nutr.* 2002, 132(12):3577-84.

Setchell KD. Soy isoflavones—benefits and risks from nature's selective estrogen receptor modulators (SERMs). *J Am Coll Nutr.* 2001, 20(5 Suppl):354S-362S

Setchell KDR; Cassidy A. Dietary isoflavones: Biological effects and relevance to human health. *J Nutr.* 1999, 129:758S-767S.

Shankar S; Teyler TJ; Robbins N. Aging differentially alters forms of long-term potentiation in rat hippocampal area CA1. *J Neurophysiol.* 1998 Jan, 79(1):334-41.

Sherwin BB. Estrogen and cognitive aging in women. *Neuroscience.* 2006, 138(3):1021-6.

Sherwin BB. The clinical relevance of the relationship between estrogen and cognition in women. *J Steroid Biochem Mol Biol.* 2007 Aug-Sep, 106(1-5):151-6.

Shingo AS; Kito S. Estrogen induces insulin-like growth factor-1 mRNA expression in the immortalized hippocampal cell: determination by quantitative real-time polymerase chain reaction. *Neurochem Res.* 2003, 28(9):1379-83.

Shrestha LB. Population aging in developing countries. *Health Aff (Millwood).* 2000, 19(3):204-12.

Shughrue PJ; Merchenthaler I. Evidence for novel estrogen binding sites in the rat hippocampus. *Neuroscience.* 2000, 99(4):605-12.

Smith CC; McMahon LL. Estradiol-induced increase in the magnitude of long-term potentiation is prevented by blocking NR2B-containing receptors. *J Neurosci.* 2006, 26(33):8517-22.

Smith CC; McMahon LL. Estrogen-induced increase in the magnitude of long-term potentiation occurs only when the ratio of NMDA transmission to AMPA transmission is increased. *J Neurosci.* 2005, 25(34):7780-91.

Smith CC; Vedder LC; McMahon LL. Estradiol and the relationship between dendritic spines, NR2B containing NMDA receptors, and the magnitude of long-term potentiation at hippocampal CA3-CA1 synapses. Psychoneuro*Endocrinology.* 2009, 34 Suppl 1:S130-42.

Smith MJ; Jennes L. Neural signals that regulate GnRH neurones directly during the oestrous cycle. *Reproduction.* 2001, 122(1):1-10.

Snyder MA; Cooke BM; Woolley CS. Estradiol potentiation of NR2B-dependent EPSCs is not due to changes in NR2B protein expression or phosphorylation. *Hippocampus.* 2010. [Epub ahead of print].

Somponpun S; Sladek CD. Role of estrogen receptor-beta in regulation of vasopressin and oxytocin release *in vitro*. *Endocrinology.* 2002, 143(8):2899-904.

Son JH; Winzer-Serhan UH. Expression of neuronal nicotinic acetylcholine receptor subunit mRNAs in rat hippocampal GABAergic interneurons. *J Comp Neurol.* 2008, 511(2):286-99.

Spencer JL; Waters EM; Romeo RD; Wood GE; Milner TA; McEwen BS. Uncovering the mechanisms of estrogen effects on hippocampal function. *Front Neuroendocrinol.* 2008, 29(2):219-37.

Spreen O; Strauss E. *A compendium of neuropsychological tests: administrarion norms and commentary.* Oxford University Press, Oxford. 1991.

Takanami K; Sakamoto H; Matsuda K; Hosokawa K; Nishi M; Prossnitz ER; Kawata M. Expression of G protein-coupled receptor 30 in the spinal somatosensory system. *Brain Res.* 2010, 1310:17-28.

Talboom JS; Williams BJ; Baxley ER; West SG; Bimonte-Nelson HA. Higher levels of estradiol replacement correlate with better spatial memory in surgically menopausal young and middle-aged rats. *Neurobiol Learn Mem.* 2008, 90(1):155-63.

Thijssen JH; Daroszewski J; Milewicz A; Blankenstein MA. Local aromatase activity in human breast tissues. *J Steroid Biochem Mol Biol.* 1993, 44(4-6):577-82.

Thomas P; Dong J. Binding and activation of the seven-transmembrane estrogen receptor GPR30 by environmental estrogens: a potential novel mechanism of endocrine disruption. *J Steroid Biochem Mol Biol.* 2006, 102(1-5):175-9.

Thompson MA; Lasley BL; Rideout BA; Kasman LH. Characterisation of the estrogenic properties of a nonsteroidal estrogen equol, extracted from the urine of pregnant macaques. *Biol Reprod* 1984, 31:705-713.

Thorp AA; Sinn N; Buckley JD; Coates AM; Howe PR. Soya isoflavone *Br J Nutr.* 2009, 1:1-7.

Tohgi H; Utsugisawa K; Yoshimura M; Nagane Y; Mihara M. Age-related changes in nicotinic acetylcholine receptor subunits alpha4 and beta2 messenger RNA expression in postmortem human frontal cortex and hippocampus. *Neurosci Lett.* 1998, 245(3):139-42.

Usui T. Pharmaceutical prospects of phytoestrogens. *Endocr J.* 2006, 53(1):7-20.

Valverde MA; Rojas P; Amigo J; Cosmelli D; Orio P; Bahamonde MI; Mann GE; Vergara C; Latorre R. Acute activation of Maxi-K channels (hSlo) by estradiol binding to the beta subunit. *Science.* 1999, 285(5435):1929-31.

VanderHorst VG; Gustafsson JA; Ulfhake B. Estrogen receptor-alpha and -beta immunoreactive neurons in the brainstem and spinal cord of male and female mice: relationships to monoaminergic, cholinergic, and spinal projection systems. *J Comp Neurol.* 2005, 488(2):152-79.

VanderHorst VG; Terasawa E; Ralston HJ 3rd. Estrogen receptor-alpha immunoreactive neurons in the brainstem and spinal cord of the female rhesus monkey: species-specific characteristics. *Neuroscience.* 2009, 158(2):798-810.

Vergne S; Titier K; Bernard V; Asselineau J; Durand M; Lamothe V; Potier M; Perez P; Demotes-Mainard J; Chantre P; Moore N; Bennetau-Pelissero C; Sauvant P. Bioavailability and urinary excretion of isoflavones in humans: effects of soy-based supplements formulation and equol production. *J Pharm Biomed Anal.* 2007, 43(4):1488-94.

Vergne S; Bennetau-Pelissero C; Lamothe V; Chantre P; Potier M; Asselineau J; Perez P; Durand M; Moore N; Sauvant P. Higher bioavailability of isoflavones after a single ingestion of a soya-based supplement than a soya-based food in young healthy males. *Br J Nutr.* 2008, 99(2):333-44.

Vivacqua A; Bonofiglio D; Albanito L; Madeo A; Rago V; Carpino A; Musti AM; Picard D; Andò S; Maggiolini M. 17beta-estradiol, genistein, and 4-hydroxytamoxifen induce the proliferation of thyroid cancer cells through the g protein-coupled receptor GPR30. *Mol Pharmacol.* 2006, 70(4):1414-23.

VonSchassen C; Fester L; Prange-Kiel J; Lohse C; Huber C; Böttner M; Rune GM. Oestrogen synthesis in the hippocampus: role in axon outgrowth. *J Neuroendocrinol.* 2006, 18(11):847-56.

Wan Q; Man HY; Braunton J; Wang W; Salter MW; Becker L; Wang YT. Modulation of GABAA receptor function by tyrosine *J Neurosci.* 1997, 17(13):5062-9.

Wang[1] Y; Man Gho W; Chan FL; Chen S; Leung LK. The red clover (Trifolium pratense) isoflavone biochanin A inhibits aromatase activity and expression. *Br J Nutr.* 2008, 99(2):303-10.

Wang[2] P; Jeng CJ; Chien CL; Wang SM. Signaling mechanisms of daidzein-induced axonal outgrowth in hippocampal neurons *Biochem Biophys Res Commun.* 2008, 366(2):393-400.

Wang CN; Chi CW; Lin YL; Chen CF; Shiao YJ. The neuroprotective effects of phytoestrogens on amyloid beta protein-induced toxicity are mediated by abrogating the activation of caspase cascade in rat cortical neurons. *J Biol Chem.* 2001, 276(7):5287-95.

Waters EM; Mitterling K; Spencer JL; Mazid S; McEwen BS; Milner TA. Estrogen receptor alpha and beta specific agonists regulate expression of synaptic proteins in rat hippocampus. *Brain Res*. 2009, 1290:1-11.

Watson CS; Alyea RA; Jeng YJ; Kochukov MY. Nongenomic actions of low concentration estrogens and xenoestrogens on multiple tissues. *Mol Cell Endocrinol.* 2007, 274(1-2):1-7.

Wechsler D. Wechsler *Adult Intelligence Scale-Revised.* Chicago, IL: Psychological Corporation, Harcourt Brace Jovanovich, 1981.

Wechsler D. Wechsler Memory Scale. A standarized memory scale for clinical use. *Psychology.* 1945, 19:87-95.

Wechsler D. Wechsler Memory Scale-Revised. New York, NY: Psychological Corporation, 1987.

Weiland NG; Orikasa C; Hayashi S; McEwen BS. Distribution and hormone regulation of estrogen receptor immunoreactive cells in the hippocampus of male and female rats. *J Comp Neurol.* 1997, 388(4):603-12.

White LR; Petrovitch H; Ross GW; Masaki K; Hardman J; Nelson J; Davis D; Markesbery W. Brain aging and midlife tofu consumption. *J Am Coll Nutr.* 2000, 19(2):242-55.

Whitechurch RA; Ng KT; Sedman GL. Tyrosine kinase inhibitors impair long-term memory formation in day-old chicks. *Brain Res* Cogn *Brain Res*. 1997, 6(2):115-20.

Williams BJ; Bimonte-Nelson HA; Granholm-Bentley AC. ERK-mediated NGF signalling in the rat septo-hippocampal pathway diminishes with age. *Psychopharmacology.* 2006, 188(4):605-18.

Wilson J; DeFries J; McLearn G; Vandenberg S; Johson R; Rashad M. Cognitive abilities: use of family data as a control to assess sex and age differences in two ethnic groups. *Int J Aging Hum Dev.* 1975, 6:261-276.

Wise PM; Dubal DB; Wilson ME; Rau SW; Böttner M; Rosewell KL. Estradiol is a protective factor in the adult and aging brain: understanding of mechanisms derived from *in vivo* and *in vitro* studies. *Brain Res Brain Res Rev*. 2001, 37(1-3):313-9.

Woclawek-Potocka I; Piskula MK; Bah M; Siemieniuch MJ; Korzekwa A; Brzezicka E; Skarzynski DJ. Concentrations of isoflavones and their metabolites in the blood of pregnant and non-pregnant heifers fed soy bean. *J Reprod Dev.* 2008, 54(5):358-63.

Wong M; Moss RL. Long-term and short-term electrophysiological effects of estrogen on the synaptic properties of hippocampal CA1 neurons. *J Neurosci*. 1992, 12(8):3217-25.

Wuttke W; Jarry H; Becker T; Schultens A; Christoffel V; Gorkow C; Seidlová-Wuttke D. Phytoestrogens: endocrine disrupters or replacement for hormone replacement therapy? *Maturitas.* 2003, 44(Suppl 1):S9-S20.

Xiao CW. Health effects of soy protein and isoflavones in humans. *J Nutr.* 2008, 138(6):1244S-9S.

Xu X; Wang HJ; Murphy PA; Cook L; Hendrich S. Daidzein is a more bioavailable soymilk isoflavone than is genistein in adult women. *J Nutr*. 1994, 124(6):825-32.

Xu XW; Shi C; He ZQ; Ma CM; Chen WH; Shen YP; Guo Q; Shen CJ; Xu J. Effects of phytoestrogen on mitochondrial structure and function of hippocampal CA1 region of ovariectomized rats. *Cell Mol Neurobiol.* 2008, 28(6):875-86.

Yague JG; Wang AC; Janssen WG; Hof PR; Garcia-Segura LM; Azcoitia I; Morrison JH. Aromatase distribution in the monkey temporal neocortex and hippocampus. *Brain Res*. 2008, 1209:115-27.

Yanagihara N; Toyohira Y; Shinohara Y. Insights into the pharmacological potential of estrogens and phytoestrogens on catecholamine signaling. *Ann N Y Acad Sci.* 2008, 1129:96-104.

Ye L; Chan MY; Leung LK. The soy isoflavone genistein induces estrogen synthesis in an extragonadal pathway. *Mol Cell Endocrinol.* 2009, 302(1):73-80.

Zeng H; Chen Q; Zhao B. Genistein ameliorates beta-amyloid peptide (25-35)-induced hippocampal neuronal apoptosis. *Free Radic Biol Med.* 2004, 36(2):180-8.

Zhao L; Chen Q; Diaz Brinton R. Neuroprotective and neurotrophic efficacy of phytoestrogens in cultured hippocampal neurons. *Exp Biol Med (Maywood).* 2002, 227(7):509-19.

Zhao L; O'Neill K; Diaz Brinton R. Selective estrogen receptor modulators (SERMs) for the brain: current status and remaining challenges for developing NeuroSERMs. *Brain Res Brain Res Rev*. 2005, 49(3):472-93.

Zhao^1 L; Brinton RD. Estrogen receptor alpha and beta differentially regulate intracellular Ca(2+) dynamics leading to ERK phosphorylation and estrogen neuroprotection in hippocampal neurons. *Brain Res*. 2007, 1172:48-59.

Zhao^2 L; Brinton RD. WHI and WHIMS follow-up and human studies of soy isoflavones on cognition. *Expert Rev Neurother.* 2007, 7(11):1549-64.

In: Isoflavones Biosynthesis, Occurence and Health Effects ISBN: 978-1-61728-113-6
Editor: Melanie J. Thompson, pp. 83-110 © 2010 Nova Science Publishers, Inc.

Chapter II

Prevention of Cardiovascular Diseases by Soy Protein and Isoflavones

***M. A. Rostagno**[*1]**, N. Manchón**[1]**, M. D`Arrigo**[1]**, A. García- Lafuente**[1]**, A. Villares**[1]**, E. Guillamón**[1]**, A. Ramos**[1]**, J. A. Martínez**[1,2]*

[1]Centro para la Calidad de los Alimentos, Instituto Nacional de Investigación y Tecnología Agraria y Alimentaria (INIA), Campus Universitario "Duques de Soria", Soria, 42004, Soria, Spain

[2]Universidad de Navarra, Dpto. Fisiología y Nutrición, Edificio de Investigación, C/Irunlarrea, 1, 31008, Pamplona, Spain

Keywords: Cardiovascular diseases; soy protein; isoflavones;

Abstract

The use of functional foods in the diet is one of the latest trends in a strategy of intervention to reduce risk factors for the development of cardiovascular diseases. Soybeans and derived foods has attracted much attention of health professionals since its consumption is associated with a reduction of cholesterol levels in blood and other risk factors. In spite of the mechanisms and components responsible for the potential effect of soy foods for the prevention of cardiovascular diseases are not yet completely identified, the most recent studies suggest that several factors are inter-related, such as the functional composition of soybeans, the synergism between components in food, individual metabolism and genetics as well as other risk factors

* Corresponding author: E-mail: rostagno.mauricio@inia.es, Telephone: (+34) 975 233204, Fax: (+34) 975 233205

1. Introduction

Cardiovascular disease (CVD) is a general term used to describe a wide range of unhealthy conditions, from arteriosclerosis to disorders of the peripheral vascular system. Currently, CVDs are considered to be responsible for approximately 30% of deaths worldwide. It is also estimated that more than 17.5 million people die annually from CVDs, and the number of deaths are projected to increase to 20 million by 2015 [1].

In the last decades several studies tried to identify the prevalence of recognized biological and behavioral risk factors associated with development of CVDs in order to improve its prevention. The most important factors that have been identified to show a significant increase in the risk of developing a CVD were divided into non-modifiable and modifiable risk factors. The major non-modifiable risk factors include heredity, gender, and age. Since these are non-modifiable risk factors, prevention has focused mainly on modifiable risk factors that include elevated blood pressure, dyslipidemia (triglycerides, total cholesterol, LDL-cholesterol and HDL cholesterol levels), smoking, obesity, physical inactivity and diabetes mellitus. At the present time, there is increasing evidence that reducing or controlling these factors can significantly reduce morbidity and mortality caused by CVDs. In fact, identification and control of these conditions are the basis of several risk assessment and reduction strategies [2-4].

Despite the fact that there is correlation between these risk factors and CVDs incidence, the search for additional risk factors is an active field of research and a number of novel risk factors or markers have been proposed (Table 1). Although the mechanisms that produce the development of atherosclerosis are not entirely clear, there are evidences indicating that there is a direct relationship between development and oxidation of LDL cholesterol, chronic inflammation, endothelial dysfunction and thrombosis [5,6].

Inflammation is increasingly being recognized to play an important role in the pathogenesis of several CVDs [2]. Inflammatory markers, such as cell adhesion molecules, cytokines, pro-atherogenic enzymes and C-reactive protein, have been found to be associated with future cardiovascular risk. With the identification of additional CVD risk markers it is possible to improve prediction of a CVD event by measuring combinations of different potential makers rather than lipid testing alone [7].

However, among identified risk factors associated with an increased incidence of CVDs, high total cholesterol, low-density lipoprotein (LDL) cholesterol, and triglycerides, and low high-density lipoprotein (HDL) cholesterol concentrations are the ones most studied as significant predictors of CVDs [8,9]. Nowadays, the use of healthy diets with the objective of reducing these risk factors is one of the main strategies for the prevention of CVDs. This strategy is increasing its importance with the evidence that certain foods and/or food components can be particularly efficient in the modulation of these specific targets (total, LDL and HDL cholesterols and triglycerides). In fact, one of the latest innovations in the food industry is the development of new products enhanced with these foods or components specifically aimed to provide them with a cholesterol lowering effect.

Among potential food components or naturally occurring compounds that may be potentially helpful in improving cholesterol and triglycerides levels and profiles are polyunsaturated fatty acids (Omega 3 and 6), phytosterols and antioxidants. These and other potentially bioactive compounds are present in a wide variety of foods, including fruits, vegetables and whole grains [10-12]. Soybeans and derived products (soy protein, soy milk, tofu, etc.) are attracting a lot of attention since there are several epidemiological studies indicating that in Asian countries; where soybeans are consumed regularly in considerable quantities, present a lower incidence of CVDs in relation to western countries [13-17]. Although there are plausible epidemiological indications that soy foods can potentially help reduce the risk of developing CVDs, the specific components occurring in these foods that are responsible for the potential beneficial effects remain unknown. In the last decades, identification of phytochemicals that may be related to the possible lipid lowering effects of soybeans has been, and continues to be, an active research area. In this context, several phytochemicals were characterized in soybeans, including phytosterols, protease inhibitors, inositol hexafosfate, saponins and isoflavones, among others [18-21].

Table 1. Main novel risk factors for the development of CVDs*.

Inflammatory Markers	C-reactive protein
	Interleukins
	Serum amiloid A
	Vascular and cellular adhesion molecules
	Other markers
Platelet-Related Factors	Platelet aggregation
	Platelet activity
	Platelet size and volume
Lipid-Related Factors	Small dense low-density lipoprotein (LDL)
	Lipoprotein (a)
	Apolipoproteins A1 and B
	High-density lipoprotein subtypes
	Oxidized LDL
Other Factors	Homocysteine
	Lipoprotein-associated phospholipase A(2)
	PAI-1 genotype
	ApoE genotype
	Psychosocial factors

* Adapted from ref. 2

Isoflavones are a subclass of flavonoids that are capable of exerting estrogen-like effects. For this reason, they are also termed phytoestrogens—plant-derived compounds with estrogenic activity. Isoflavone structures are characterized by a flavone nucleus, composed by two benzene rings (A and B) linked to a heterocyclic ring C (Figure 1).

There are three main soybean isoflavones: genistein (4,5,7-trihidroxyisoflavone), Daidzein (4,7-dihidroxyisoflavone), and Glycitein (4,7-dihidroxy-6-metoxi-isoflavone). These compounds are naturally present in the soybean and non-fermented soyfoods primarily in their beta glycoside forms: genistin, daidzin, and glycitin. Malonyl glucoside derivatives are also relatively abundant while their acetyl glucoside derivatives are usually found in low concentrations. Biochanin A and formononetin (which are derivatives of genistein and daidzein) are generally less rich in soy than the 12 main forms and which are found mostly in clover and alfalfa sprouts [22-25].

Isoflavone content in soybeans can range from 0.4 mg to 9.5 mg of total isoflavones per gram, which can be influenced by genetics, crop year and growth location, among other factors [26-29]. Isoflavones are also present in significant amounts on derived soy products, such as soy protein, tofu and soy milk. Isoflavone content of the different soy foods available in several countries been intensively investigated. Isoflavone quantitation in soybeans and soy foods consumed in the USA, Japan, Italy, UK, Singapore, Australia, Indonesia, Brazil, and Canada have been published in the last decade. Besides of individual reports, there are food composition databases and compilations from these values specifically focusing on isoflavone distribution [30-52].

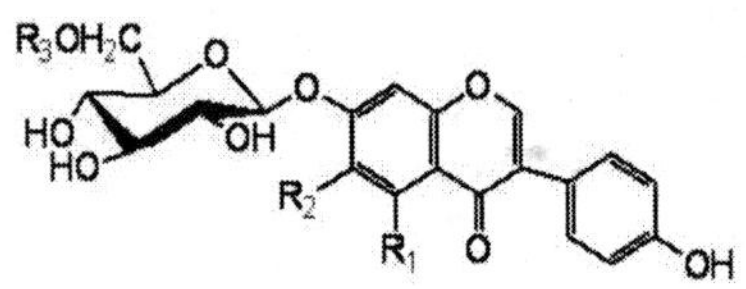

Glucosides

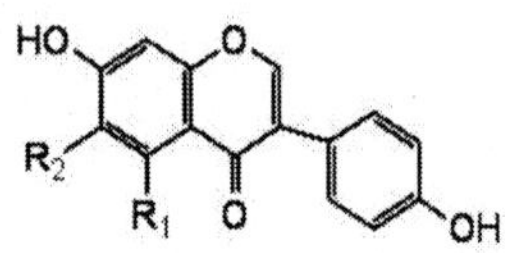

Aglycones

Isoflavone	MW	R1	R2	R3
Malonyl Daidzin	502	H	H	$COCH_2COOH$
Malonyl Glycitin	532	H	OCH_3	$COCH_2COOH$
Malonyl Genistin	518	OH	H	$COCH_2COOH$
Acetyl Daidzin	458	H	H	$COCH_3$
Acetyl Glycitin	488	H	OCH_3	$COCH_3$
Acetyl Genistin	474	OH	H	$COCH_3$
Daidzin	416	H	H	H
Glycitin	446	H	OCH_3	H
Genistin	432	OH	H	H

Isoflavone	MW	R1	R2
Daidzein	254	H	H
Glycitein	284	H	OCH_3
Genistein	270	OH	H

Figure 1. Chemical structure of soybean isoflavones

The hypothesis that soy consumption may reduce CVD risk has focused on isoflavones because soybeans are the only significant dietary source of these compounds [53]. More importantly, these compounds have shown several in vitro and in vivo beneficial properties consistent with the potential soybean effects on health. As a result, enormous efforts are being directed worldwide for the evaluation of isoflavone composition in foods and to identify their relationship with the possible beneficial effects of soybeans and derived products consumption on health. These efforts has increased especially after the United States Food and Drug Administration (FDA) approved a health claim in 1999, allowing products containing a certain amount of soy protein state that they may help prevent heart diseases [54].

2. Meta-Analysis of the Effects of Soy and Isoflavones on Lipids

The interest in the use of soybeans and products to reduce the CVDs risks gave place to several intervention studies that were published in the last decades. In an attempt to evaluate the possible cholesterol lowering potential of soy and its suitability for the prevention of CVDs, different meta-analyses were published. One of the first published meta-analysis evaluated the results obtained in studies conducted until 1995 about the effects of soy protein intake on serum lipid concentration [55]. In total, 29 randomized controlled trials (either a crossover or a parallel design) that used soy protein (either isolated or textured soy protein) were included in the analysis. Ingestion of soy protein (mean intake of 47 g per day) was associated with an average reduction of 23.2 mg/dL (13.5-32.9 mg/dL) in total cholesterol, equivalent to 9.3%, of 21.7 mg/dL in LDL cholesterol (11.2-31.7 mg/dL), equivalent to 12.9% and of 13.3 mg/dL (0.3-25.7 mg/dL) in triglycerides, equivalent to 10.5%. The net changes in total and LDL cholesterol concentration when compared with the control diet were directly related to the initial cholesterol concentration ($P<0.001$) and a small, nonsignificant, increase (2.4%) in serum concentrations of HDL cholesterol was observed.

Another meta-analysis of studies conducted between 1995 and 2002 evaluated specific effects of soy-associated isoflavones on cholesterol concentrations in well-controlled trials substituting soy protein with dairy or animal protein. It was found that a daily intake of 36g of soy protein with 52 mg of associated isoflavones (average values) reduced LDL cholesterol (0.17 mmol/L, equivalent to 4%) and increased HDL cholesterol concentrations (0.03 mmol/L, equivalent to 3%), but did not report a dose-response relation between soy-associated isoflavones and net changes, leading to the conclusion that these compounds are not related with the effects of soy protein in serum lipids [56].

In contrast, the evaluation of eight randomized controlled trials by Zhuo et al [57] indicated that consumption of isolated soy protein (mean of 50g/day) with high amount of isoflavones (mean ingestion of 96 mg/day) reduced the concentration of LDL cholesterol in 0.15 mmol/L when compared with individual that consumed the amount of isolated soy protein with low amount of isoflavones (mean ingestion of 6 mg/day). The reduction on LDL cholesterol of hypercholesterolemic and normochoelsterolemic individuals was 0.18 and 0.14 mmol/L, respectively. With the ingestion of identical amounts of soy protein and higher concentration of isoflavones led to a significant higher reduction of LDL cholesterol when compared to a lower isoflavone concentration, indicating that isoflavones have an effect on the reduction of LDL cholesterol independently of soy protein. However, no linear relation between the effects and soy protein or isoflavones intake was observed.

Another meta-analysis [58], with randomized controlled trials conducted between 1995 and 2002, observed that soy protein with isoflavones are associated with a significant reduction of total cholesterol (0.22 mmol/L equivalent to 3.77%), LDL cholesterol (0.21 mmol/L equivalent to 5.25%) and triglycerides (0.10 mmol/L) and a significant increase of HDL cholesterol (0.04 mmol/L equivalent to 3.03%), although a dose-response relation was not identified. Also, this meta-analysis indicated that reduction of total and LDL cholesterol were higher on men than on women and that the initial total cholesterol concentration has an important effect on the changes on total and HDL cholesterol, especially on hypercholesterolemic individuals. Studies with more than 80 mg of isoflavones per day demonstrated higher effect on lipid profiles. A higher effect of soy protein with isoflavones on reduction of total and LDL cholesterol and triglycerides was observed on a short initial intervention period while an improvement on HDL cholesterol was only observed on studies longer than 12 weeks. Therefore, soy protein with isoflavones may significantly reduce total and LDL cholesterol and triglycerides and increase HDL cholesterol depending of the amount and duration of the intake, gender and initial individual lipid concentration.

However, in this meta-analysis, it was not observed that isoflavones extract supplements had a significant effect on reduction of total cholesterol. The same lack of overall statistical effect of isoflavones tablets, isolated or mixtures with up to 150 mg per day on serum lipids was also reported by Yeung and Yu [59]. Moreover, a specific meta-analysis of the effects of extracted soy isoflavones alone also did not report any significant effect of the supplementation of the diet with 70 mg/day (27-132 mg of aglycone equivalents) for 1-3 months on total and LDL cholesterol in normocholesterolemic menopausal women [60].

On a more recent meta-analysis of a greater number of studies conducted since 1966 specifically about the effects of soy protein on blood lipids observed an association between intake of soy protein and reduction on mean concentration of total cholesterol (5.26 mg/dL), LDL (4.25 mg/dL) and triglycerides (6.26 mg/dL) and a slight increase on cholesterol HDL (0.77 mg/dL). Evaluation of the studies indicated a slightly higher reduction of lipids on subjects with total cholesterol lower than 240 mg/dL and on subjects with LDL cholesterol lower than 160 mg/dL. It was also observed a slightly higher reduction on pre and peri-menopausal women (vs post-menopausal) and a higher increase on HDL cholesterol on hypercholesterolemic individuals, as well as a higher effect of dietary soy protein on longer intervention studies. The meta-regression analysis demonstrated a dose-response relationship between soy protein and isoflavones on the net alteration of serum lipids, with a statistically significant inverse relationship between the amount of soy protein and isoflavones on the modification of total and LDL cholesterol and triglycerides. It was observed a statistically significant positive relation between soy protein and isoflavones and HDL cholesterol modification [61].

Another meta-analysis of studies with soy protein with high and low isoflavones observed that soy isoflavones significantly reduce total (0.10 mmol/L, 3.9 mg/dL or 1,77%) and LDL (0.13 mmol/L, 5 mg/dL or 3.58%) cholesterol, but did not affect concentration of HDL cholesterol. Independently of isoflavone concentration, soy protein significantly improved lipidic profiles. Reduction of LDL cholesterol was higher on hypercholesterolemic individuals than on normocholesterolemic, although there was not a linear correlation between initial cholesterol levels and reduction [62].

On a recent review about the effect of a mean intake of 25g (15-40g) of soy protein on blood lipids in adults with normal and slightly high cholesterol levels indicated that the introduction of soy protein in the diet results on a reduction of 0.23, 0.22 and 0.08 mmol/L of total and LDL cholesterol and triglycerides, respectively. It was not observed a significant effect on apolipoprotein A but there was a reduction of 0.021 mmol/L of apolipoprotein B. An analysis of meta-regression did not indicate a dose response relationship between soy intake (15-40g) and differences in LDL and HDL cholesterol. The authors concluded that the inclusion of modest amounts of soy protein (25g) on the diet of adults results in a highly significant reduction of 6% on total and LDL cholesterol [63].

In view of the conclusions of these meta-analyses, it can be assumed that there is sufficient evidence to suggest a plausible association between the consumption of soy protein and a modest improvement of lipid profiles. It also appears to have evidences that isoflavones, at least in part and depending of several factors, may play a role on the effects of soy protein on improvement of lipid profiles and therefore on the reduction of CVD risk. Taken together, the results of these meta-analyses indicate a reduction in cholesterol concentrations between 3–6% (63-65). Despite these effects may seem small initially, the influence on the risk of developing CVDs can be important. For example, it has been suggested that a reduction of 1% in the concentration of LDL cholesterol may reduce the risk of CVD by 2% and a 1% increase in the concentration of HDL cholesterol may reduce the risk of CVD in 3% (66-67). Therefore, the combined potential effect on LDL and HDL cholesterols may result in a 20% reduction of risk which can be considered important or even decisive for the development of CVDs.

Nevertheless, it is important to highlight that results obtained from the analysis of a wide range of different designed studies are mean values and a more critical evaluation indicates inconsistencies and contradictory evidences. In the same meta-analysis, for instance, we can find studies evaluating the effect of the "same" product that reported no change, a decrease or even an increase in cholesterol concentrations. Moreover, there are even contradictory results obtained in different meta-analysis, such as regarding the effects (or lack of effects) of isoflavones on lipid profiles and to dose-response relationship between soy protein and isoflavones.

3. Effects of Soy and Isoflavones on other Risk Factors

Although the potential lipid lowering effect of soy foods and isoflavones has been the main focus of research in the last decades, there are other important aspects for the reduction of CVD risk. The metabolic syndrome groups together the most important CVD risk factors, such as diabetes and prediabetes, abdominal obesity, high cholesterol and high blood pressure. Therefore, the metabolic syndrome allows assessing the clustering of risk factors associated with an increased risk of cardiovascular disease in addition to the risk associated with the individual components and identifies subjects with increased cardiovascular morbidity and mortality [68,69].

In this context, recent studies suggested that soy foods and isoflavones may improve features of the metabolic syndrome. Azadbakht et al. [70], reported that inclusion of soy nuts in the diet decreased the homeostasis model assessment-insulin score in postmenopausal women with metabolic syndrome, improving glycemic control. Also, it has been reported that Genistein and isolated soy protein supplements may be beneficial for correcting the hyperglycemia and preventing diabetic complications, therefore reducing the risk of developing CVDs [71]. Another study revealed that an isoflavone-containing soy diet improved the diabetic phenotype in male and female rats, exerting a beneficial effect on lipid and glucose metabolism, possibly through activation of the PPAR receptors. In vitro studies indicated that treating cells with Daidzein or Genistein doubled PPAR α-directed gene expression and PPAR γ -directed gene expression was increased by 200–400% [72].

Recently, it has been reported that supplementation of the diet with Genistein for six months induced a significant improvement in glycemic control and endothelial function of postmenopausal women compared to a placebo group [75]. Nevertheless, response to isoflavone supplementation was suggested to be affected by individual's pre-existing metabolic status [73]. On the other hand, long-term consumption of soy protein in a high fat diet significantly decreased serum glucose, free fatty acids, leptin, and the insulin:glucagon ratio compared with animals fed a casein high fat diet. It was also suggested that soy protein with associated isoflavones consumption stimulated insulin secretion to a lower extent despite high fat diet consumption. Furthermore, there was lower pancreatic islet area and insulin, SREBP-1, PPARα, and GLUT-2 mRNA abundance in comparison with rats fed the casein high fat diet. Therefore, soy protein and isoflavones may prove to be useful for reducing the risk of CVDs by ameliorating hyperinsulinemia observed during obesity [74].

Soy foods and associated isoflavones may also act on obesity since there are reports of positive effects on food and water intake, on the adipose tissue and on body weight [75-78]. Although the effects of isoflavones on body weight and food intake it is still a matter of debate and controversy, there is supporting evidence that all three main isoflavones present in soy foods (Genistein, Daidzein and Glycitein) decrease the deposit of adipose tissue. Thus, isoflavone can also potentially reduce CVD risk by influencing adiposity and the development of metabolic syndrome [77,78].

Yet another risk factor that may be affected by soy foods and associated isoflavones is hypertension. Isoflavones may influence blood pressure regulation because they can act as an estrogen receptor agonist and as an inhibitor of tyrosine kinase, which may be involved in the control of vascular tone and generation of hypertension [79-81]. It has been reported that consumption of 40 g of soy protein containing 68 mg of isoflavones reduced blood pressure in perimenopausal women [82]. On another intervention study, consumption of 40g of soy protein containing 118 mg of isoflavones during 3 months by healthy men and postmenopausal women reduced systolic blood pressure, diastolic blood pressure and mean blood pressure [83]. On another study, reduction of blood pressure in men and women with mild-to-moderate hypertension by consumption of soy milk was correlated with urinary excretion of the isoflavone genistein [84]. Sagara et al., also reported similar findings of the supplementation of the diet for 5 weeks with at least 20 g soy protein and isoflavones (at least 80 mg) in high risk middle-aged men [85]

However, the influence of soybeans derived products and components on the risk of CVDs can be determined not only by direct effect on established risk factors, but also different mechanisms related to the development of the disease. The components of soybeans, for instance, may act on the inflammatory processes, thus indirectly preventing the onset of the disease itself. Symptoms of artherosclerosis and inflammatory response are associated by an enzyme, Tyrosine kinase [86,87]. This enzyme may promote muscle cell multiplication, causing the initial obstruction of cardiac arteries. Tyrosine kinase may also activate platelet, causing platelet aggregation [88]. Genistein has been identified as a tyrosine kinase inhibitor and have been reported to reduce platelet activation and to prevent platelet aggregation and to affect platelet derived growth factor (PDGF) [89-91]. Genistein may also prevent platelet activation in response to other stimulants, such as collagen, thromboxane A2, Thrombin and adenosine diphosphate (ADP) [92]. Furthermore, it have been reported that both genistein and Daidzein prevent thromboxane A2 from binding with its receptor, inhibiting platelet aggregation [90]. Genistein may also suppress activation of GPIIb/ III, a receptors present on platelets which bind with fibrinogen, and inhibits monocyte adhesion, via inhibition of tyrosine kinase [16]. Adhesion of blood monocytes to arterial lining may lead to migration into the arterial wall and their posterior conversion into macrophages. Since macrophages may promote cholesterol deposit into the arterial wall and cause arteriosclerosis, inhibiting adhesion of monocytes by some isoflavones may reduce the risk of CVDs.

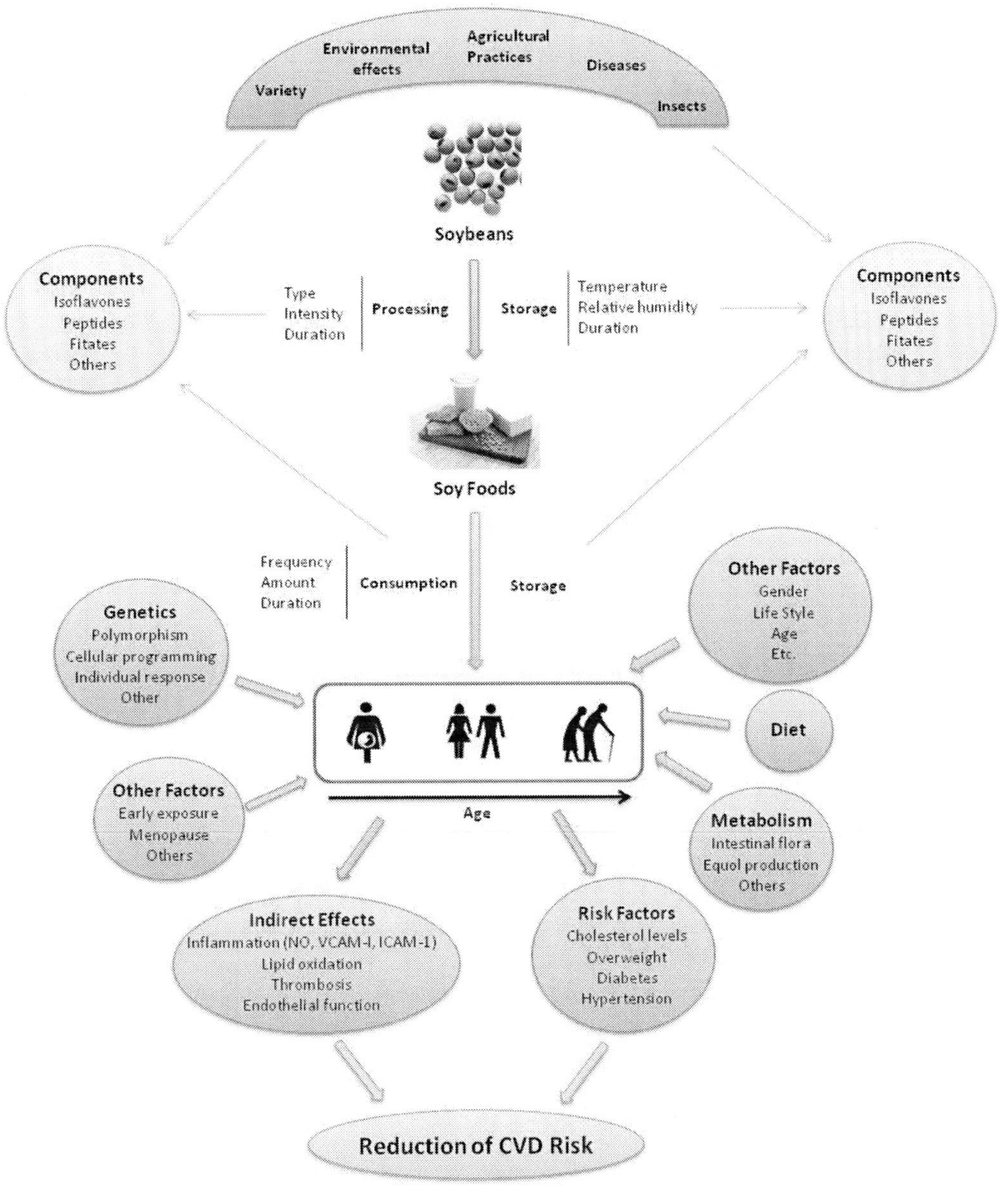

Figure 2. Factors that may influence the reduction of the risk of developing a CVD by consumption of soy products and associated isoflavones

The intake of soy protein with a high content of isoflavones may also promote anti-atherogenic effects by modulation of pro-inflammatory cytokines and cell adhesion proteins, the induction of endothelial nitric oxide formation, the protection against the oxidation of LDL cholesterol, the inhibition of platelet aggregation and the improvement of vascular reactivity [5]. Isoflavones may promote the activation of membrane estrogen receptor in triggering the signaling cascades leading to an increased activity of eNOS (endothelial nitric oxide synthase) and on the induction of genes of antioxidant defense pathway mediated by kinase modulation of transcription factors [93].

One of the earliest events in the arterial wall in the initiation of atherosclerosis is the adherence of mononuclear cells to endothelium, which is triggered by a number of adhesion molecules such as integrins (VCAM-1: vascular adhesion molecule and ICAM-1: intercellular adhesion molecule). Although the available information on the role of soy protein and isoflavones on vascular inflammatory markers and endothelial function is quite contradictory, there is evidence that the intake of isoflavones may affect the endothelium dependent arterial vasodilation and can cause a significant reduction of VCAM-1andICAM-1 [94,95]. However, the positive effects on these integrins were observed only in a subgroup of individuals (menopausal women), with a polymorphism of the ERβ gene [95,96]. This information indicates the need for further research directed to other polymorphisms present on individuals participating in intervention studies and that possibly these polymorphisms are related to the contradictory results observed in different assessments of the consumption of isoflavones on the risk of developing a CVD. For example, on another report, no significant effect on these and other different inflammatory markers (ICAM-1, VCAM-1, E-selectin and Interleukin receptor) was observed [98]. In contrast, feeding menopausal women during 8 week with soy protein or soy nuts may significantly affect concentration of NO, E-selectin and interleukin 18 [93]. NO is a crucial vasodilator, which is produced by the constitutively expressed enzyme endothelial nitric oxide synthase (eNOS). Also prevents intravasal coagulation and platelet aggregation, leukocyte adhesion to and immigration of inflammatory cells into the vessel wall as well as the release of chemokines. The development of atherosclerosis can be diagnosed by a reduced bioavailability of endothelium-derived NO and impaired reactivity of the vessel wall to the vasodilator acetylcholine [99].

In this context, Isoflavones may reduce of CVD risk by affecting vascular tone by regulation of either eNOS expression and/or bioavailability of NO [100,101]. . There is evidence that Genistein may improve flow-mediated endothelium dependent vasodilation in healthy postmenopausal women through a direct effect on the vascular function resulting from an increased ratio of nitric oxide to endothelin [102]. It has also been recently reported that soy isoflavones increase gene expression of endothelial nitric-oxide synthase (eNOS) and antioxidant defense enzymes, resulting in improved endothelial function and lower blood pressure in vivo [93].

Recent advances in basic science have established a major role for inflammation in mediating all disease stages from initiation through progression of atherosclerosis. The most recent trend in the search for the mechanism responsible for the potential reduction of CVD risk by soy derived products and isoflavones is focused on inflammatory markers. There are several different reports of the effects of soy derived products and isoflavones specifically aimed on most important inflammatory markers. Main isoflavones present in soy foods, for instance, have been reported to dose-dependently suppress NO production in supernatants of LPS-activated macrophages [103]. Also, isoflavone aglycones may down-regulate inflammatory mediators, such as TNF-α at endothelial level, thus reducing the risk of inflammation and associated CVDs 104.

On the other hand, there are also evidences that the intake of isoflavones may be inversely associated with serum levels of C-reactive protein, another biomarker for chronic inflammation and a sensitive risk factor for CVD [94,105]. C-reactive protein has been suggested to be associated with the incidence of CVDs and related mortality which could not be explained by conventional risk factors of apparently healthy populations [105,106]. Regarding the effects of soy products and associated isoflavones on C-reactive protein, reports are also controversial since in some studies, isoflavones significantly increased concentration of this inflammation marker only on equol producer individuals [105], while on other studies, no significant effect of isoflavones was observed, independently of equol producer subgroup [96,97]. The difference between these reports is the subjects used in the studies: while on the first case subjects were menopausal women, in the later they were hypercholesterolemic subjects of both gender. This is indication that other aspects of the subjects taking part on intervention studies may also be relevant for a potential positive effect of soy products and isoflavones on CVD risk factors.

4. Influencing Factors on the Reduction of CVD Risk by Soy Products and Isoflavones

Although further investigations are still needed to elucidate the potential use of soy products for the reduction of CVD risk, the involved components, their respective mechanisms of action and suitable applications, it is clear that the outcome of this type of complex research can be influenced by a wide range of factors, which may explain inconsistencies between different studies. Some of the several possible factors that may influence the reduction of the risk of developing a CVD by consumption of soy products and associated isoflavones are graphically illustrated in Figure 2.

The reduction of the risk of development of CVDs by soy protein and isoflavones may be determined by direct improvement of established risk factors as well as by indirect effects on mechanisms associated with onset of the disease itself. On the other hand, the potential reduction of risk factors and other mechanisms may be dependent of other several factors, such as the concentration and profile of isoflavone and other components present in soybeans, derived foods and intervention diets; the type, amount, frequency and duration of intervention diets; genetics; the age that individuals are exposed to soy isoflavones and individual conditions (cholesterol levels, menopause, etc…) and metabolism as well as due to potential synergisms with other components of the diet.

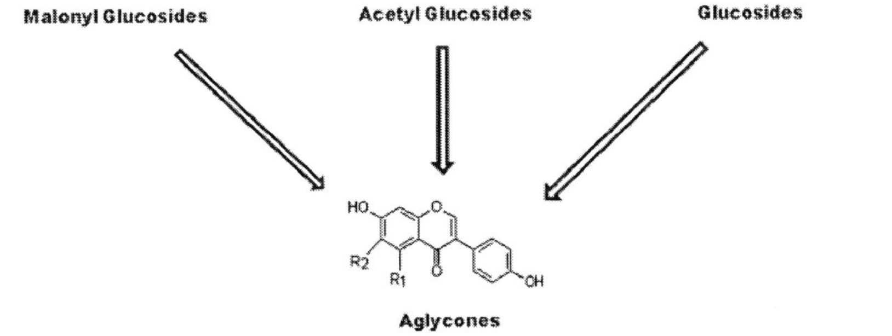

Figure 3. Main degradation path of soybean isoflavones (Reproduced from ref. 53 with permission from Elsevier).

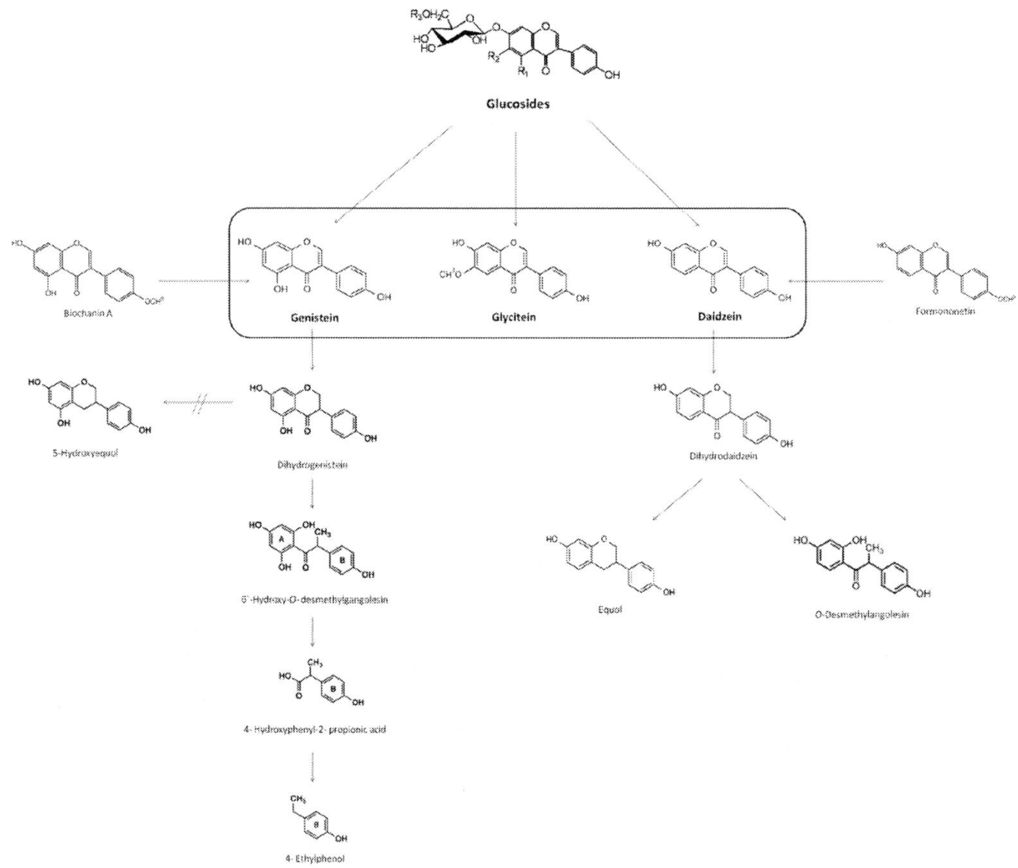

Figure 4. Main metabolism route of isoflavones by human gut microbiota

4.1. Soybean Components

One of the most important factors that may explain the variability and inconsistency reported of the effects of soy products and isoflavone intake on CVD risk is the lack of reliable information about the isoflavones present in the foods or diet used. In some cases, isoflavone concentration is not determined [107,108]; or indirect literature values are used [109]. The use of literature and database isoflavone values in epidemiological studies may supply useful information to research determining the intake of phytoestrogens and its relation to potential biological activities. However, the use of literature values should be avoided in intervention studies since there is a large variation depending of variety, crop year, growth location, environmental and soil variables, agricultural practices, pathogen and insect attacks, processing and storage conditions [53, 110-118]. Differences in total isoflavone content in soybeans, for instance, can change by more than 20 times fold depending of these factors [53]. Moreover, even when isoflavone concentration is determined, not all chemical forms were measured and the total amount is expressed as total isoflavones, aglycone equivalents (by the use of conversion factors) [119-124] or as aglycones and respective glycosides [125].

It is noteworthy that there is a fundamental difference between providing total individual isoflavone concentration and aglycone equivalents. Total individual isoflavones refers to the sum of all chemical forms, including glucosides (malonyl, acetyl and derived glucosides) and their respective aglycone forms, while aglycone equivalents refers to the amount of aglycones present in all these chemical forms. Values expressed as total individual isoflavones can be between 57 and 97% higher than those expressed as aglycone equivalents due to differences in molecular weight of isoflavone derivatives. Illustratively, 1 gram of malonyl genistin is equivalent to only 0.521g of aglycones (genistein). The importance of these different expressed amounts is the final concentration in the food and the amount ingested by those subjects taking part on such intervention studies. On the other hand, not only the final concentration but also the profile of isoflavones occuring in the food or diet is important, since not all chemical forms have the same biological effects, mechanisms of action, bioavailability or metabolism.

In this context, the effects of soy foods processing are also important to be considered since it can substantially affect isoflavone concentration and profile. This aspect is particularly remarkable because some isoflavone chemical forms are relatively unstable. The application of high temperatures can promote decarboxilation of malonyl glucosides to acetyl glucosides and ester hydrolysis of malonyl and acetyl glucosides to underivatized glucoside. It is also possible for the different conjugated forms to generate the aglycone forms by cleavage of the glucosidic bond [53, 118,126-129]. An overview of the most common possible degradation paths of soy isoflavones are depicted in Figure 3 [53].There also reports indicating that some isoflavone aglycones are not entirely stable under acidic conditions [130].

The type and intensity of processing are also relevant regarding isoflavone concentration and profile in the final product. Soy protein concentrates, for example, when produced by alcohol extraction looses more than 95% of isoflavones initially present, while soy protein produced by aqueous washing preserves most isoflavones present [131,132]. This aspect is particularly relevant regarding intervention studies with soy protein where isoflavone content was not measured, since the lack of isoflavones may, at least in part, explain some of the controversial results reported.

Not only processing may affect isoflavone concentration and profile, but also storage conditions. The few reports available provide evidence that storage of soybeans and soy foods for prolonged times at room temperature can affect isoflavone concentration and profile. The usual effects of storage for long periods are a significant decrease or increase of individual isoflavone concentration. Concentration of malonyl glucosides tends to decrease leading to the respective glucoside or aglycones, which concentration tend to increase. It has been reported storage for 2 years reduced malonyl glucoside concentration by about 2 times, whereas glucosides and aglycone concentration increased up to 3–4 [133]. However, storage may also reduce concentratrion of glucosides and aglycone isoflavones [134]. Moreover, not only the isoflavone profile may be affected by the course of time, but also total isoflavone concentration, especially in the first year of storage. Nevertheless, changes on isoflavone profile and concentration caused by storage may be dependent of temperature, relative humidity and the soybean cultivar among other factors [133,134].

Therefore, the variability on isoflavone profile and concentration as produced by the previously mentioned factors may, to some extent, explain the lack of effects on CVDs risk markers of soy foods and associated isoflavones of certain studies. On the other hand, foods are complex matrices and other components are also present, and may as well be affected by variety, agricultural aspects, processing and storage [131, 135-138].

Although it is unknown exactly which are the active components of soy responsible for the potential lipid lowering effect of soy foods, it plausible to assume that may exist synergy between the different fractions of soybean and isoflavones [139]. The synergism between isoflavones and other components also explain the lack of effect of isolated isoflavones (extracts) alone observed in some meta-analysis. Evidence of synergism is also provided by different meta-analyses where soy protein with low isoflavone concentrations was not as efficient as soy protein with high isoflavone concentration.

In this context, it is of particular relevance one of the most recent studies in search of the mechanism and components responsible for the effect of soybean protein in the prevention of CVDs. In this study, an evaluation of the influence of di-and tri-peptides of digestible protein in combination with a soy isoflavone (genistein) on inhibition of platelet aggregation was conducted "in vitro". The results of this study indicated that only some peptides present in soybean protein (acid-type glutamate-aspartate and glut-asp-asp peptides) combined with genistein reduced the activity of platelets aggregation, indicating that there is a synergism between peptides and isoflavones present [140]. However, it is noteworthy that only one isoflavone was evaluated (genistein) and that other interactions and synergisms may affect differently other isoflavones and their derivatives. Therefore, more research is still needed in this direction in order to elucidate the potential role of isoflavones on reducing CVDs risk.

On the other hand, there are also trials indicating part of the lipid lowering effect of soy protein, and consequently soybeans and derived foods, may be due to amino acid composition, protein subunits or composite peptides. In fact, potential lipid lowering effects of soy products may be directly dependent of the type and profile of protein present. In one recent study, for example, supplementation with isolated soy protein that maintains the native protein structure lead to higher significant reductions in plasma concentrations of total and low-density lipoprotein cholesterol than a commercially available reference isolated soy protein [141].

However, not only the potential lipid lowering effect of soy foods and isoflavones may depend of the associated amino acid profile of the food. It has been reported that soy protein and associated isoflavones effects on reducing hyperinsulinemia depends on the interaction between the amino acid pattern and isoflavones [142]. This also puts in evidence the importance of processing on not only on isoflavones but also on other soybean components. In this context, it has been reported that different processing procedures used for the preparation of soy protein isolates can affect the intactness of protein subunits, and therefore affect biological functions of soy proteins [143]. Thus, it is important an adequate characterization of the soy protein used for intervention studies, including assessing the intactness and measuring the relative abundance of each protein subunit [144]. It may also explain the differences observed on some meta-analysis regarding isoflavones and soy protein. The use of soy protein produced by alcohol washing may not affect cholesterol concentrations, due to either the low isoflavone content of the soy protein, or disruption of the protein, the protein–isoflavone complex, or some other component of isolated soy protein that is important for its effect on plasma cholesterol concentrations [145,146].

4.2. Metabolism

Metabolism of individuals as well as metabolites resulting from the digestion are potentially relevant for the potential effects of soy foods and isoflavones on reducing CVD risk. The metabolism of isoflavones in animals and humans is complex mechanism and is a combination of mammalian and gut microbial processes [78, 147,148]. In soy products, isoflavones are present mainly as glucosides. After ingestion, intestinal glucosidases promote hydrolysis these isoflavone chemical forms, releasing the respective aglycone form (genistein, daidzein, and Glycitein) (Figure 4). Two of these aglycones, genistein and Daidzein, may also be derived from other precursors such as biochanin A and formononetin, respectively, after their breakdown by intestinal glucosidases [78]. Isoflavone aglycones may be absorbed or further metabolized to many specific metabolites such as equol, *O*-desmetillangolesin or ethylphenol [149-153].

However, absorption of isoflavones and their metabolites will depend of the subject. Since isoflavones are present mainly as glycosides that must be hydrolyzed to be absorbed, microbial population in the intestine of each individual is a key aspect of the metabolism of these compounds [154-156]. Aglycone isoflavones, however, may be absorbed to a greater extent than glycosides. Excretion of isoflavones has been reported to be different depending if subjects ingested glycosides or aglycones [157]. Moreover, Izumi et al. [158] found higher levels of isoflavones in plasma after aglycone consumption than after glycoside ingested in pill form. By contrast, Setchell et al. [159] reported greater bioavailability from glycosides than from aglycones on the basis of plasma concentrations. Furthermore, there are also differences on the bioavailability between isoflavone aglycones. Daidzein, for example, have been reported to be more bioavailable than genistein [160]. The combination of isoflavones present may also influence absorption, metabolism or both because single-dose pharmacokinetics suggests that the bioavailability of genistein and gaidzein differs in different mixes of the two isoflavones [161].

Furthermore, specific metabolites from the metabolism of certain isoflavones may also be of critical importance. Among isoflavone metabolite, equol has received much attention recently because of its biological activity is different from its precursors, displaying a higher antioxidant capacity, and estrogenic and anti-androgenic properties. It have been recently reported that equol may have vasorelaxant and antioxidant activity in carotid and cerebral arteries [162]. There are also studies indicating that equol rapidly activates eNOS in human aortic and umbilical vein endothelial cells, causing acute endothelium- and nitric oxide (NO)-dependent relaxation of aortic rings [163]

However, only between 30 and 50% of the population of Western countries excretes significant quantities of equol, which may explain in part the contradictory results observed in different intervention studies using isoflavones [155,164]. For example, it has been reported that, in some cases, positive effects of isoflavones on reducing some CVD risk markers was observed only on equol producers [105]. Therefore, absorption, metabolism and more importantly, bioavailability should be taken into account to improve the experimental design and the interpretation of the observed effects [165]. In this context, equol-production status is likely to be an important factor to be controlled on intervention studies.

4.3. Other Factors

There are several other factors that may have a major impact on the response of a possible effect from soy products in general and isoflavones in particular, to reduce the risk of CVDs. Pan et al., for example, reported gender-specific effects of soy protein intake on the risk of metabolic syndrome in middle-aged and elderly Chinese [166]. They reported that habitual soy protein intake was associated with moderately reduced risk of metabolic syndrome in middle-aged and elderly women, but elevated the risk in men. In middle-aged and elderly men, some contradictory effects were observed: it significantly increased risk for hyperglycemia but reduced risk for hypertension. It has also been reported that genistein affects adipose tissue deposition in a dose-dependent and gender-specific manner in mice [167]. It was suggested that the antiadipogenic action of genistein and down-regulation of adipogenic genes required the expression of ERβ. In this context, genetics make up one of the most important aspects in the reduction of CVD risk by soy foods and isoflavones. It has been reported that isoflavones affects the expression of a large number of genes, including induction of genes associated with cyclic adenosine 3V,5V-monophosphate (cAMP) signaling and cell differentiation and decreased expression of genes associated with cyclin-dependent kinase activity and cell division [168]. Interestingly, isoflavones affects gene expression of equol producers differently from nonproducers, supporting the role of this metabolite on the potential reduction of CVD risk by isoflavones at the gene level. Isoflavones also have been reported to affect gene expression of different antioxidant enzymes [85,169-173]. It has also been suggested that isoflavones may promote activation of estrogen receptor membrane, which in turn trigger signaling cascades, increasing activity of eNOS and the induction of antioxidant defense genes [85]. Since there are also evidences that age influences antioxidant gene expression [174,175], it plausible to assume that it will be an important factor influencing the possible effects of isoflavones on reducing CVD risk. Additionally, there is evidence that the age at which individuals are exposed to isoflavones could influence the body's response in relation to dietary intake of these compounds and the effects on cardiovascular prevention in the future life. The exposure of individuals, even before birth, could induce epigenetic modifications of DNA in heart cells, resulting in a reprogramming of the expression of certain genes related to mechanisms of action of isoflavones [176,177]. Moreover, recent research supports the theory that the potential chemoprotective property of isoflavones is dependent on the duration of exposure (lifelong exposure from prenatal through neonatal into adolescent and adult ages) [178,179]. It has been reported that chronic exposure to isoflavones during fetal and postnatal development differently affects gene expression in juvenile and adult rats [179]. Fetal epigenomic modification by isoflavones may also be responsible for the reduction of other risk factors associated with an increased incidence of CVD, such as obesity [180]. Therefore, both the time point on which phytoestrogen exposure starts together with the composition of the ingested phytoestrogen containing diet are of great importance for the biological response of subjects [179].

However, it is important to notice that other foods and phytochemicals are present on the habitual diet of individuals and are ingested together with soy foods and associated isoflavones. Overlapping effects may arise from different interactions among dietary bioactive compounds and/or food combinations. These interactions may either enhance or inhibit the response to specific foods [181]. Combination of wine, fish, dark chocolate, fruits, vegetables, garlic, and almonds, for instance, has been suggested to offer additive benefits and to reduce cardiovascular disease by more than 75% [182]. On a similar basis, combination of cholesterol-lowering foods or ingredients (plant sterols, soy protein, viscous fibers, and almonds) was recently evaluated and was found to be as efficacious as certain statins in lowering LDL cholesterol concentrations [183]. In contrast, the vascular effects of tea have been reported to be suppressed by soy milk intake. The authors suggested that structurally similar soy proteins in soy milk might mimic the effects of caseins in bovine milk in terms of blunting beneficial vascular effects of tea [184]. Curiously, combination of soy and tea bioactive components synergistically inhibits breast carcinoma and androgen-sensitive human prostate tumors in mice [185,186]. These studies provide supporting evidence about the interaction between phytochemicals present in different foods and potential health effects, and that cardiovascular protective effects of an individual's diet may reflect the combined effects of various bioactive components rather than from the effect of a single ingredient or compound. However, since multiple cellular and metabolic processes associated with the development of CVDs can be modulated simultaneously by the diet, it is a challenging task to determine the influence of individual components on the reduction of overall risk [181].

5. Conclusions

The available information provides sufficient evidence to suggest a plausible association between the consumption of soy protein and a modest improvement of lipid profiles, including reduction of total and LDL cholesterols and triglycerides and increase of HDL cholesterol. However, contradictory results have been reported and current researches are being directed to identify the components responsible for the beneficial health effects of soy protein and their mechanisms of action. Isoflavones are one of soy protein components that are being extensively studied because of "in vitro" and "in vivo" biological activity consistent with reduction of cardiovascular disease risk factors. There is indication that isoflavones, at least in part and depending of several factors, may play a role on the effects of soy protein on improvement of lipid profiles and therefore on the reduction of CVD risk. However, the specific components of soy protein and mechanisms of action have not yet been completely identified. It seems that several factors are interrelated and further studies are necessary to identify the interactions between different factors, including isoflavone concentration and profile of soy protein, soy protein intake, frequency and duration, synergism of associated isoflavones with other components, individual`s metabolism and genetics, age, gender, existence of other risk factors, and indirect effects of isoflavones on the development of the cardiovascular disease. These studies will allow identifying which segments of the population can more efficiently exploit the beneficial effects of soy protein and associated isoflavones, which will consequently lead to individualized dietary recommendations depending on the predisposition to respond to a specific diet.

Acknowledgments

The authors acknowledge funding from the Instituto Nacional de Investigación y Tecnología Agraria y Alimentaria (INIA) projects AT 07-003 and RTA 2009-00049-00-00.

References

[1] World Health Organization (http://www.who.int/topics/cardiovascular_diseases/en/index.html)

[2] Hackam, D. G. & Anand, S. S. (2003). Emerging Risk Factors for Atherosclerotic Vascular Disease: A Critical Review of the Evidence, *JAMA*, *290*, 932-940.
[3] Wilson, P. W. F., D'Agostino, R. B. & Levy, D. (1998). et al., Prediction of Coronary Heart Disease Using Risk Factor Categories, *Circulation*, *97*, 1837-1847.
[4] Pearson, T. A., Blair, S. N. & Daniels, S. R. (2002). et al., AHA Guidelines for Primary Prevention of Cardiovascular Disease and Stroke: 2002 Update, *Circulation*, *106*, 388-391.
[5] Rimbach, G., Boesch-Saadatmandi, C. & Frank, J. (2008). et al. Dietary isoflavones in the prevention of cardiovascular disease–A molecular perspective, *Food Chem. Toxicol.*, *46*, 1308-1319.
[6] Garcıa-Lafuente, A., Guillamón, E., Villares, A., Rostagno, M. A. & Martínez, J. A. (2009). Flavonoids as anti-inflammatory agents: implications in cancer and cardiovascular disease, *Inflamm. Res.*, *58*, 537-552.
[7] Blake, G., Ridker, P. M. (2002). Inflammatory bio-markers and cardiovascular risk prediction. *J. Intern Med.*, *252*, 283-294.
[8] Jenkins, D. C. (1988). Epidemiology of cardiovascular diseases. *J. Consult Clin. Psych.*, *56*, 324-332.
[9] Grundy, S. M. (1998). The role of cholesterol management in coronary disease risk reduction in elderly patients, *Endocrin. Metab Clin.*, *27*, 655-675.
[10] Li, T. S. C. (2008). Vegetables and fruits : nutritional and therapeutic values, CRC Press, *Taylor & Francis Group, Boca Raton*, FL, *286*, pages.
[11] Wildman, R. E. C. (2006). (Editor), *Handbook of Nutraceuticals and Functional Foods,* Second Edition, CRC; 560, pages
[12] Slavin, J. (2004). Whole grains and human health, *Nutr. Res. Rev.*, *17*, 99-110.
[13] Beaglehole, R. (1990). International trends in coronary heart disease mortality, morbidity, and risk factors. *Epidemiol Rev.*, *12*, 1-15.
[14] Keys, A., Menotti, A. & Aravanis, C. (1984). et al. The seven countries study: 2,289 deaths in 15 years, *Prev Med.*, *13*, 141-154.
[15] Tham, D. M., Gardner, C. D. & Haskell, W. L. (1998). Potential Health Benefits of Dietary Phytoestrogens: A Review of the Clinical, Epidemiological, and Mechanistic Evidence, *J. Clin. Endocr Metab.*, *83*, 2223-2235.
[16] Adlercreutz, H. & Mazur, W. (1997). Phyto-oestrogens and western diseases, *Ann. Med.*, *29*, 95-120.
[17] Moriguchi, E. H., Moriguchi, Y. & Yamori, Y. (2004). Impact of diet on the cardiovascular risk profile of Japanese immigrants living in Brazil: contributions of World Health Organization CARDIAC and MONALISA studies. *Clin Exp Pharmacol Physiol*, *31*, S5-7.
[18] Birk, Y. (1985). The Bowman-Birk protease inhibitor, *Int. J. Pept. Protein Res.*, *25*, 113-131.
[19] Anderson, R. L. & Wolf, W. J. (1995). Compositional changes in trypsin inhibitors, phytic acid, saponins and isoflavones related to soybean processing, *J. Nutr.*, *125*, 581S-588S.
[20] Phillips, K. M., Ruggio, D. M., Toivo, J. I., Swank, M. A. & Simpkins, A. H. (2002). Free and esterified sterol composition of edible oils and fats, *J. Food Compos. Anal. 15*, 123-142.

[21] Romani, A., Vignolini, P., Galardi, C., Aroldi, C., Vazzana, C. & Heimler, D. (2003). Polyphenolic content in different plant parts of soy cultivars grown under natural conditions, *J. Agric. Food Chem.*, *51*, 5301-5306.

[22] Cheng, E., Story, C. D., Yoder, L., Hale, W. H. & Burroughs, W. (1953). Estrogenic activity of isoflavone derivatives extracted and prepared from soybean oil meal. *Science*, *118*, 164-165.

[23] Walz, E. (1931). Isoflavon- und saponin-glucoside in Soja-Hispida, *Justus Liebigs Ann. Chem.*, *489*, 118-155.

[24] Naim, M., Gestetner, B., Kirson, I., Birk, Y. & Bondi, A. (1973). A new isofla- vone from soya beans, *Phytochemistry*, *12*, 169-170.

[25] Ohta, N., Kuwata, G., Akahori, H. & Watanabe, T. (1980). Isolation of a new isoflavone acely glucoside, 6"-O-acetylgenistin, from soybeans. *Agric. Biol.* Chem., *44*, 469-470.

[26] Ohta, N., Kuwata, G., Akahori, H. & Watanabe, T. (1979). Isoflavonoid constituents of soybeans and isolation of a new acetyldaidzin, *Agric. Biol. Chem.*, *43*, 1415-1419.

[27] Farmakalidis, E. & Murphy, P. A. (1985). Isolation of 6-O-acetylgenistin and 6"-O-acetyldaidzin from toasted defatted soyflakes, *J. Agric. Food Chem.*, *33*, 385-389.

[28] Kudou, S., Shimoyamada, M., Imura, T., Uchida, T. & Okubo, K. (1991). A new isoflavone glycoside in soybean seeds (Glycine max Merrill), glycitein 7-O-j3-D-(6"-O-acetyl)-glucopyranoside, *Agric. Biol. Chem.*, *55*, 859-860.

[29] Kudou, S., Fleury, Y., Welti, D., Magnolato, D., Uchida, T., Kitamura, K. & Okubo, K. (1991). Malonyl isoflavone glycosides in soybean seeds (Glycine max Merrill), *Agric. Biol. Chem.*, *55*, 2227-2233.

[30] Bingham, S. A., Atkinson, C., Liggins, J., Bluck, L. & Coward, A. (1998). Phytoestrogens: where are we now?, Brit. *J. Nutr.*, *79*, 393-406.

[31] Murphy, P. A., Song, T., Buseman, G. & Barua, K. (1997). Isoflavones in soy-based infant formulas, *J. Agric. Food Chem.*, *45*, 4635-4638.

[32] Murphy, P. A., Song, T., Buseman, G. & Barua, K. (1999). Isoflavones in Retail and Institutional Soy Foods, *J. Agric. Food Chem.*, *47*, 2697–2704.

[33] Franke, A. A., Hankin, J. H., Yu, M. C. & Maskarinec, G. (1999). Isoflavone levels in soy foods consumed by multiethnic populations in Singapore and Hawaii, *J. Agric. Food Chem.*, *47*, 977-986.

[34] Umphress, S. T., Murphy, S. P., Franke, A. A., Custer, L. J. & Blitz, C. L. (2005). Isoflavone content of foods with soy additives, *J. Food Compos. Anal.*, *18*, 533-550.

[35] Morton, M., Arisaka, O., Miyake, A. & Evans, B. (1999). Analysis of phyto- estrogens by gas chromatography-mass spectrometry, Environ. *Toxicol. Pharmacol.*, *7*, 221-225.

[36] Fukutake, M., Takahashi, M., Ishida, K., Kawamura, H., Sugimura, T. & Wakabayashi, K. (1996). Quantification of genistein and genistin in soybeans and soybean products, Food Chem. *Toxicol.*, *34*, 457-461.

[37] Morandi, S., D'Agostina, A., Ferrario, F. & Arnoldi, A. (2005). Isoflavone content of Italian soy food products and daily intakes of some specific classes of consumers, Eur. Food Res. *Technol.*, *221*, 84-91.

[38] Wiseman, H., Casey, K., Clarke, D. B., Barnes, K. A. & Bowey, E. (2002). Isoflavone aglycon and glucoconjugate content of high- and low-soy U.K. foods used in nutritional studies, *J. Agric. Food Chem.*, *50*, 1404-1410.

[39] Liggins, J., Mulligan, A., Runswick, S., Bingham, S. A. (2002). Daidzein and genistein content of cereals, *Eur. J. Clin. Nutr.*, *56*, 961-966.
[40] Prabhakaran, M. P., Hui, L. S. & Perera, C. O. (2006). Evaluation of the composition and concentration of isoflavones in soy based supplements, health products and infant formulas, *Food Res. Int.*, *39*, 730-738.
[41] Hutabarat, L. S., Greenfield, H. & Mulholland, M. (2001). Isoflavones and Coumestrol in Soybeans and Soybean Products from Australia and Indonesia, *J. Food Compos. Anal.*, *14*, 43-58.
[42] Genovese, M. I. & Lajolo, M. F. (2002). Isoflavones in soy-based foods consumed in Brazil: levels, distribution, and estimated intake, *J. Agric. Food Chem.*, *50*, 5987-5993.
[43] Thompson, L. U., Boucher, B. A., Liu, Z., Cotterchio, M. & Kreiger, N. (2006). Phytoestrogen Content of Foods Consumed in Canada, Including Isoflavones, Lignans, and Coumestan, *Nutr. Cancer*, *54*, 184-201.
[44] U.S. Department of Agriculture, Agricultural Research Service, USDA-Iowa State University Database on the Isoflavone Content of Foods, Release 1.4- 2007. Nutrient Data Laboratory, 2007, Web site: http://www.ars.usda.gov/nutrientdata.
[45] Venus database, Vegetal Estrogens in Nutrition and the Skeleton (Venus) project, http://www.venus-ca.org.
[46] Kiely, M., Faughnan, M., Wähälä, K., Brants, H. & Mulligan, A. (2003). Phyto-oestrogen levels in foods: the design and construction of the VENUS database, *Br. J. Nutr.* *89*, S19-S23.
[47] Finland National Food Composition Database (Fineli®), http://www.ktl.fi/fineli.
[48] Valsta, L. M., Kilkkinen, A., Mazur, W., Nurmi, T., Lampi, A. M., Ovaskainen, M. L., Korhonen, T., Adlercreutz, H. & Pietinen, P. (2003). Phyto-oestrogen database of foods and average intake in Finland, *Br. J. Nutr.*, *89*, S31-S38.
[49] Ritchie, M. R., Cummings, J. H., Morton, M. S., Michael Steel, C., Bolton-Smith, C., Riches, A. C. (2006). A newly constructed and vali- dated isoflavone database for the assessment of total genistein and daidzein intake, *Br. J. Nutr.*, *95*, 204-213.
[50] Reinli, K. & Block, G. (1996). Isoflavone content of soy foods, Nutr. *Cancer*, *26*, 123-148.
[51] Pillow, P. C., Duphorne, C. M., Chang, S., Contois, J. H., Strom, S. S., Spitz, M. R. & Hursting, S. D. (1999). Development of a database for assessing dietary phytoestrogen intake, Nutr. *Cancer*, *33*, 3-19.
[52] Mazur, W. (1998). Phytoestrogen content in foods, Baillieres Clin. Endocrinol. *Metab.*, *12*, 729-742.
[53] Rostagno, M. A., Villares, A., Guillamón, E., García-Lafuente, A., Martínez, J. A. (2009). Sample preparation for the analysis of isoflavones from soybeans and soy foods, *J Chromatogr.*, A, *1216*, 2-29.
[54] Stein, K. (2000). FDA Approves Health Claim Labeling for Foods Containing Soy Protein, *J. of the Am. Diet. Assoc.*, *100*, 292.
[55] Anderson, J. W., Johnstone, B. M., Cook-Newell, M. E. (1995). Meta-analysis of the effects of soy protein intake on serum lipids, *New Engl. J. Med.*, *333*, 276-282.
[56] Weggemans, R. M. & Trautwein, E. A. (2003). Relation between soy-associated isoflavones and LDL and HDL cholesterol concentrations in humans: a meta-analysis. *Eur. J. Clin. Nutr.*, *57*, 940–946.

[57] Zhuo, X. G., Melby, M. K. & Watanabe, S. (2004). Soy isoflavone intake lowers serum LDL cholesterol: a meta-analysis of 8 randomized controlled trials in humans. *J. Nutr.*, *134*, 2395-2400.

[58] Zhan, S. & Ho, S. C. (2005). Meta-analysis of the effects of soy protein containing isoflavones on the lipid profile. *Am. J. Clin. Nutr.*, *81*, 397-408

[59] Yeung, J., Yu, T. F. (2003). Effects of isoflavones (soy phyto-estrogens) on serum lipids: a meta-analysis of randomized controlled trials, *Nutr. J.*, *19*, 15.

[60] Taku, K., Umegaki, K., Ishimi, Y., Watanabe, S. (2008). Effects of extracted soy isoflavones alone on blood total and LDL cholesterol: Meta-analysis of randomized controlled trials, *Ther Clin Risk Manag.*, *4*, 1097–1103.

[61] Reynolds, K., Chin, A., Lees, K. A., Nguyen, A., Bujnowski, D., He, J. (2006). A meta-analysis of the effect of soy protein supplementation on serum lipids. *Am. J. Cardiol.*, *98*, 633-640.

[62] Jenkins, D. J. A., Kendall, C. W. C., Jackson, C. C. (2002). et al., Effects of high- and low-isoflavone soyfoods on blood lipids, oxidized LDL, homocysteine, and blood pressure in hyperlipidemic men and women, *Am. J. Clin. Nutr.,*, *76*, 365-372.

[63] Harland, J. I. & Haffner, T. A. (2008). Systematic review, meta-analysis and regression of randomised controlled trials reporting an association between an intake of circa 25 g soya protein per day and blood cholesterol, *Atherosclerosis*, *200*, 13-27.

[64] Steinberg, F. M. (2007). Soybeans or soymilk: does it make a difference for cardiovascular protection? Does it even matter? *Am. J. Clin. Nutr.*, *85*, 927-928.

[65] Dewell, A. A, Hollenbeck, P. L. W. & Hollenbeck, C. B. (2006). Clinical review: a critical evaluation of the role of soy protein and isoflavones supplementation in the control of plasma cholesterol concentrations, *J. Endocrinol. Metab.*, *91*, 772-780.

[66] Law, M. R., Wald, N. J. & Thompson, S. G. (1994). By how much and how quickly does reduction in serum cholesterol concentration lower risk of ischaemic heart disease? *Brit. Med. J.*, *308*, 367-372.

[67] Schaefer, E. J. (2002). Lipoproteins, nutrition, and heart disease, *Am. J. Clin. Nutr.*, *75*, 191-212.

[68] Lakka; H. M., Laaksonen; D. E. & Lakka, T. A. (2002). et al., The Metabolic Syndrome and Total and Cardiovascular Disease Mortality in Middle-aged Men, *JAMA*, *288*, 2709-2716

[69] Juturu, V. & Gormley, J. J. (2005). Nutritional Supplements Modulating Metabolic Syndrome Risk Factors and the Prevention of Cardiovascular Disease, *Curr. Nutr. Food Sci.*, *1*, 1-11.

[70] Azadbakht, L., Kimiagar, M. & Mehrabi, Y. (2007). et al., Soy inclusion in the diet improves features of the metabolic syndrome: a randomized crossover study in postmenopausal women, *Am. J. Clin. Nutr.*, *85*, 735-741.

[71] Lee, J. S. (2006). Effects of soy protein and genistein on blood glucose, antioxidant enzyme activities, and lipid profile in streptozotocin-induced diabetic rats, *Life Sci.*, *79*, 1578-1584.

[72] Mezei, O., Banz, W. J., Steger, R. W., Peluso, M. R., Winters, T. A. & Shay, N. (2003). Soy Isoflavones Exert Antidiabetic and Hypolipidemic Effects through the PPAR Pathways in Obese Zucker Rats and Murine RAW 264.7 *Cells, J. Nutr.*, *133*, 1238-1243.

[73] Villa, P., Costantini, B. & Suriano, R. (2009). et al., The differential effect of the phytoestrogen genistein on cardiovascular risk factors in postmenopausal women: relationship with the metabolic status, *J Clin Endocrinol Metab, 94*, 552–558.

[74] Noriega-López, L., Tovar, A. R. & Gonzalez-Granillo, M. (2007). et al., Pancreatic Insulin Secretion in Rats Fed a Soy Protein High Fat Diet Depends on the Interaction between the Amino Acid Pattern and Isoflavones, *J. Biol. Chem*., *282*, 20657-20666.

[75] Penza, M., Montani, C. & Romani, A. (2006). et al., Genistein affects adipose tissue deposition in a dose-dependent and gender-specific manner. *Endocrinology*, *147*, 5740-5751.

[76] Michael, M. R., Wolz, E. & Davidovich, A. (2006). et al.,, Acute, subchronic and chronic safety studies with genistein in rats. Food Chem. *Toxicol*., *44*, 56-80.

[77] Davis, J., Higginbotham, A. & O'Connor, T. (2007). et al., Soy protein and isoflavones influence adiposity and development of metabolic syndrome in the obese male ZDF rat. *Ann. Nutr. Metab*., *51*, 42-52.

[78] Ørgaard, A. & Jensen, L. (2008). The Effects of Soy Isoflavones on Obesity, *Exp. Biol. Med*., *233*, 1066-1080.

[79] Miksicek, R. J. (1994). Interaction of naturally occurring nonsteroidal estrogen with expressed recombinant human estrogen receptor. *J. Steroid. Biochem. Mol. Biol*., *49*, 153-160.

[80] DiSalvo, J., Nelson, S. R. & Kaplan, N. (1997). Protein tyrosine phosphorylation in smooth muscle: a potential coupling mechanism between receptor activation and intracellular calcium. *Proc. Soc. Exp. Biol. Med*., *214*, 285-301.

[81] Sauro, M. D. & Thomas, B. (1993). Decreased sensitivity of aorta from hypertensive rats to vasorelaxation by tyrphostin. *Life Sci*., *53*, 371–376.

[82] Washburn, S., Burke, G. L., Morgan, T. & Anthony, M. (1999. Effect of soy protein supplementation on serum lipoproteins, blood pressure, and me)nopausal symptoms in perimenopausal women. *Menopause*, *6*, 7-13.

[83] Teede, H. J., Dalais, F. S., Kotsopoulos, D., Liang, Y. L., Davis, S. & McGrath, B. P. (2001). Dietary soy has both beneficial and potentially adverse cardiovascular effects: a placebo-controlled study in men and postmenopausal women. *J. Clin. Endocrinol. Metab*., *86*, 3053-3060.

[84] Rivas, M., Garay, R. P., Escanero, J. F., Jr. Cia, P., Cia, P. & lda, J. O. (2002). A Soy milk lowers blood pressure in men and women with mild to moderate essential hypertension. *J. Nutr*., *132*, 1900-1902.

[85] Sagara, M., Kanda, T., Jelekera, M. N. & Teramoto, T. (2003). Effects of Dietary Intake of Soy Protein and Isoflavones on Cardiovascular Disease Risk Factors in High Risk, Middle-Aged Men in Scotland, *J. Am. Coll. Nutr*., *23*, 85-91.

[86] Raines, L. W. & Ross, R. (1995). Biology of atherosclerotic plaque formation: possible role of growth factors in lesion development and the potential impact of soy, *J. Nutr*., *125*, 624- 630.

[87] Ridker, P. M. (1988). Inflammation, Infection, and Cardiovascular Risk : How Good Is the Clinical Evidence? *Circulation, 97*, 1671-1674.

[88] Entman, M. L. & Ballanlyne, C. M. (1993). Inflammation in acute coronary syndromes, *Circulation*, *88*, 800-803.

[89] Murkies, A. L., Wilcox, G., Davis, S. R. (1998). Clinical review 92: *Phytoestrogens. J Clin Endocrinol. Metab., 83*, 297-303.

[90] Tham, D. M., Gardner, C. D. & Haskell, W. L. (1998). Clinical review 97: Potential health benefits of dietary phytoestrogens: a review of the clinical, epidemiological, and mechanistic evidence, *J. Clin. Endocrinol. Metab.*, *83*, 2223-2235.

[91] Akiyama, T., Ishida, J., Nakagawa, S. (1987). et al., Genistein, a specific inhibitor of tyrosine-specific protein kinases, *J. Biol. Chem.*, *262*, 5592-5595.

[92] Ozaki, Y., Yatomi, Y., Jinnai, Y. & Kume, S. (1993). Effects of genistein, a tyrosine kinase inhibitor, on platelet functions. Genistein attenuates thrombin-induced Ca2+ mobilization on human platelets by affecting polyphosphoinositide turnover. *Biochem. Pharmacol.*, *46*, 395-403.

[93] Mahn, K., Borrás, C. & Knock, G. A. (2005). Dietary soy isoflavone induced increases in antioxidant and eNOS gene expression lead to improved endothelial function and reduced blood pressure in vivo, *FASEB J.*, *19*, 1755-1757.

[94] Azadbakht, L., Kimiagar, M. & Mehrabi, Y. (2007). et al., Soy Consumption, Markers of Inflammation, and Endothelial Function: A cross-over study in postmenopausal women with the metabolic syndrome, *Diabetes Care*, *30*, 967-973.

[95] Colacurci, N., Chiàntera, A. & Fornaro, F. (2005). et al. Effects of soy isoflavones on endotelial function in healthy postmenopausal women. *Menopause*, *12*, 299-307.

[96] Hall, W. L., Vafeiadou, K. & Hallund, J. (2005). J., et al. Soy-isoflavone-enriched foods and inflammatory biomarkers of cardiovascular disease risk in postmenopausal women: interactions with genotype and equol production, *Am. J. Clin. Nutr.*, *82*, 1260-1268.

[97] Vafeiadou, K., Hall, W. L. & Williams, C. M. (2006). Does genotype and equol-production status affect response to isoflavones? Data from a pan-European study on the effects of isoflavones on cardiovascular risk markers in post-menopausal women, *Proc. Nutr. Soc.*, *65*, 106-115.

[98] Blum, A., Lang, N. & Peleg, A. (2003). et al., Effects of oral soy protein on markers of inflammation in postmenopausal women with mild hypercholesterolemia, *Am. Heart J.*, *145*, N1-N4

[99] Li, H. & Forstermann, U. (2000). Nitric oxide in the pathogenesis of vascular disease. *J. Pathol.*, *190*, 244-254.

[100] Lissin, L.W. & Cooke, J. P. (2000). Phytoestrogens and cardiovascular health, *J. Am. Coll. Cardiol.*, *35*, 1403-1410

[101] Liu, D., Homan, D. L.L. & Dillon, J. S. (2004). Genistein acutely stimulates nitric oxide synthesis in vascular endothelial cells by a cyclic adenosine 5 -monophosphate-dependent mechanism, *Endocrinology*, *145*, 5532-5539.

[102] Squadrito, F., Altavilla, D. & Morabito, N. (2002). et al., The effect of the phytoestrogen genistein on plasma nitric oxide concentrations, endothelin-1 levels and endothelium dependent vasodilation in postmenopausal women, *Atherosclerosis*, *163*, 339-347.

[103] Sheu, F., Lai, H. H. & Yen, G. C. (2001). Suppression Effect of Soy Isoflavones on Nitric Oxide Production in RAW 264.7 Macrophages, *J. Agric. Food Chem.*, *49*, 1767-1772.

[104] Clerici, C., Setchell, K. D. & Battezzati, P. M. (2007). et al., Pasta naturally enriched with isoflavone aglycons from soy germ reduces serum lipids and improves markers of cardiovascular risk, *J Nutr.*, *137*, 2270-2278.

[105] Blake, G. J., Ridker, P. M. (2002). Inflammatory bio-markers and cardiovascular risk prediction. *J. Intern. Med.*, *252*, 283-294.

[106] Vigna, G. B., Pansini, F. & Bonaccorsi, G. (2000). et al. Plasma lipoproteins in soy-treated postmenopausal women: a double-blind, placebo-controlled trial. Nutr. Metab. *Cardiovasc. Dis.*, *10*, 315-322.

[107] Van Horn, L., Liu, K. & Gerber, J. (2001). et al. Oats and soy in lipidlowering diets for women with hypercholesterolemia: is there synergy? *J Am Diet Assoc*, *101*, 1319-1325.

[108] Ashton, E. L., Dalais, F. S. & Ball, M. J. (2000). Effect of meat replacement by tofu on CHD risk factors including copper induced LDL oxidation, *J Am Coll Nutr*, *19*, 761-767.

[109] Wang, H., Murphy, P. A. (1994). Isoflavone composition of American and Japanese soybeans in Iowa: Effects of variety, crop year and location. *J. Agric. Food Chem*. *42*, 1674-1677.

[110] Riedl, K. M., Lee, J. H., Renita, M. (2007). et al., Isoflavone profiles, phenol content, and antioxidant activity of soybean seeds as influenced by cultivar and growing location in Ohio. *J. Sci. Food Agric*., *87*, 1197-1206.

[111] Lee, S. J., Yan, W., Ahn, J. K. & Chung, I. M. (2003). Effects of year, site, genotype and their interactions on various soybean isoflavones, *Field Crop Res*., *81*, 181-192.

[112] Hoeck, J. A., Fehr, W. R., Murphy, P. A. & Welke, G. A. (2000). Influence of genotype and environment on isoflavone contents of soybean. *Crop Sci*., *40*, 48-51.

[113] Piubelli, G. C., Hoffmann-Campo, C. B., De Arruda, I. C., Franchini, J. C. & Lara, F. M. (2003). Flavonoid Increase in Soybean as a Response to Nezara viridula Injury and Its Effect on Insect-Feeding Preference, *J. Chem. Ecol*., *29*, 1223-1233.

[114] Graham, T. L., Graham, M. Y. (1996). Signaling in soybean phenylpropanoid responses dissection of primary, secondary, and conditioning effects of light, wounding, and elicitor treatments. *Plant Physiol*., *110*, 1123-1133.

[115] Vyn, T. J., Yin, X. & Bruulsema, T. W. (2002). et al., Potassium Fertilization Effects on Isoflavone Concentrations in Soybean [Glycine max (L.) Merr.], *J. Agric. Food Chem*., *50*, 3501-3506.

[116] Rochfort, S. & Panozzo, J. (2007). Phytochemicals for Health, the Role of Pulses, *J. Agric. Food Chem*., *55*, 7981-7994

[117] Coward, L., Smith, M., Kirk, M. & Barnes, S. (1998). Chemical modification of isoflavones in soyfoods during cooking and processing, *Am. J. Clin. Nutr*., *68*, 1486S-1491S.

[118] Baum, J. A., Teng, H., Erdman, J. W. (1998). et al. Long-term intake of soy protein improves blood lipid profiles and increases mononuclear cell low-density-lipoprotein receptor messenger RNA in hypercholesterolemic, postmenopausal women. *Am. J. Clin. Nutr*., *68*, 545-551.

[119] Crouse, J. R., Morgan, T. & Terry, J. G. (1999). et al. A randomized trial comparing the effect of casein with that of soy protein containing varying amounts of isoflavones on plasma concentrations of lipids and lipoproteins. *Arch. Intern. Med*., *159*, 2070-2076.

[120] Teixeira, S. R., Potter, S. M. & Weigel, R. (2000). et al. Effects of feeding 4 levels of soy protein for 3 and 6 wk on blood lipids and apolipoproteins in moderately hypercholesterolemic men. *Am. J. Clin. Nutr*., *71*, 1077-1084.

[121] Dent, S. B., Peterson, C. T., Brace, L. D. (2001). et al. Soy protein intake by perimenopausal women does not affect circulating lipids and lipoproteins or coagulation and fibrinolytic factors. *J. Nutr.*, *131*, 2280-2287.

[122] Gardner, C. D., Newell, K. A. & Cherin, R. (2001). et al. The effect of soy protein with or without isoflavones relative to milk protein on plasma lipids in hypercholesterolemic postmenopausal women. *Am. J. Clin. Nutr.*, *73*, 728-735.

[123] Jenkins, D. J., Kendall, C. W. & Jackson, C. J. (2002). et al. Effects of highand low-isoflavone soyfoods on blood lipids, oxidized LDL, homocysteine, and blood pressure in hyperlipidemic men and women (UI 12145008). *Am. J. Clin. Nutr.*, *76*, 365-372.

[124] Teede, H. J., Dalais, F. S. & Kotsopoulos, D. (2001). et al. Dietary soy has both beneficial and potentially adverse cardiovascular effects: a placebo-controlled study in men and postmenopausal women. *J Clin Endocrinol Metab.*, *86*, 3053-3060.

[125] Jackson, C. J. C., Dini, J. P. & Lavandier, C. (2002). et al., of processing on the content and composition of isoflavones during manufacturing of soy beverage and tofu, *Process Biochem.*, *37*, 1117-1123.

[126] Simonne, A. H., Smith, M., Weaver, D. B., Vail, T., Barnes, S. & Wei, C. I. (2000). Retention and changes of soy isoflavones and carotenoids in immature soybean seeds (Edamame) during processing, *J. Agric. Food Chem.*, *48*, 6061-6069.

[127] Wang, H. J., Murphy, P. A. (1996). Mass balance study of isoflavones during soybean processing, *J. Agric. Food Chem.*, *44*, 2377-2383.

[128] Grun, I. U., Adhikari, K., Li, C., Li, Y., Lin, B., Zhang, J. & Fernando, L. N. (2001). Changes in the profile of genistein, daidzein, and their conjugates during thermal processing of Tofu, *J. Agric. Food Chem.*, *49*, 2839-2843.

[129] Franke, A. A., Custer, L. J., Cerna, C. M., Narala, K. K. (1994). Quantitation of phytoestrogens in legumes by HPLC, *J. Agric. Food Chem.*, *42*, 1905-1913

[130] Anderson, R. L. & Wolf, W. J. (1995). Compositional Changes in Trypsin Inhibitors, Phytic Acid, Saponins and Isoflavones Related to Soybean *Processing, J. Nutr.*, *125*, 581S-588S.

[131] Coward, L., Barnes, N., Setchell, K. D. R., Barnes, S. (1993). Genistein and daidzein and their b-glucoside conjugates: anti-tumor isoflavones in soybean foods from American and Asian diets. *J. Agric. Food Chem.*, *41*, 1961-1967.

[132] Lee, S. J., Chung, I. M. & Ahn, J. K. (2003). et al., Variation in isoflavone of soybean cultivars with location and storage duration, *J. Agric. Food Chem.*, *51*, 3382-3389.

[133] Kim, J. J., Kim, S. H., Hahn, S. J. & Chung, I. M. (2005). Changing soybean isoflavone composition and concentrations under two different storage conditions over three years, *Food Res. Int.*, *38*, 435-444.

[134] Liu, K. S. (1999). *Soybeans: chemistry, technology, and utilization*, Aspen Publ. Inc.: Gaithersburg,Maryland, USA, 532 pages

[135] Chitra, U., Singh, U. & Rao, R. V. (1996). Phytic acid, in vitro protein digestibility, dietary fiber, and minerals of pulses as influenced by processing methods, *Plant Food Hum Nutr.*, *49*, 307-316.

[136] Lusas, E. W. & Riaz, M. N. (1995). Soy Protein Products: Processing and use *J. Nutr.*, *125*, 573S-580S.

[137] Grieshop, C. M., Kadzere, C. T. & Clapper, G. M. (2003). et al., Chemical and Nutritional Characteristics of United States Soybeans and Soybean Meals, *J. Agric. Food Chem.*, *51*, 7684-7691

[138] Clair, R., Anthony, M. (2005). Soy, isoflavones and atherosclerosis. *Handb. experim. pharmacol., 170*, 301-323.
[139] Borgwardt, K., Bonifatius, S. & Gardemann, A. (2008). A. Acidic peptides enhanced genistein-dependent inhibition of human platelet aggregation: potential protective effect of digestible peptides plus genistein against atherosclerosis. *Nutr. Res.*, *28*, 523-531.
[140] Hoie, L. H., Guldstrand, M. & Sjoholm, A. (2007). et al., Cholesterol-lowering effects of a new isolated soy protein with high levels of nondenaturated protein in hypercholesterolemic patients, *Adv. Ther.*, *24*, 439-447
[141] Noriega-López, L., Tovar, A. R. & Gonzalez-Granillo, M. (2007). et al., Pancreatic Insulin Secretion in Rats Fed a Soy Protein High Fat Diet Depends on the Interaction between the Amino Acid Pattern and Isoflavones, *J. Biol. Chem.*, *282*, 20657-20666.
[142] Gianazza, E., Eberini, I., Arnoldi, A., Wait, R. & Sirtori, C. R. (2003). A proteomic investigation of isolated soy proteins with variable effects in experimental and clinical studies. *J Nutr.*, *133*, 9-14.
[143] Xiao, C. W. (2008). Health Effects of Soy Protein and Isoflavones in Humans, *J. Nutr.*, *138*, 1244S-1249S.
[144] Anthony, M. S., Blair, R. M. & Clarkson, T. B. (2002). Neither isoflavones nor the alcohol-extracted fraction added to alcohol-washed soy protein isolate restores the lipoprotein effects of soy protein isolate, *J. Nutr.*, *132*, 583S.
[145] Matthan, N. R., Jalbert, S. M. & Ausman, L. M. (2007). et al., Effect of soy protein from differently processed products on cardiovascular disease risk factors and vascular endothelial function in hypercholesterolemic subjects, *Am. J. Clin. Nutr.*, *85*, 960-966.
[146] Rowland, I., Faughnan, M. & Hoey, L. (2003). et al., Bioavailability of phyto-oestrogens. *Br. J. Nutr.*, *89*, S45-S58.
[147] Setchell, K. D. R. (2000). Absorption and metabolism of soy isoflavones - from food to dietary supplements and adults to infants. *J. Nutr.*, *130*, 654S-655S.
[148] Mortensen, A., Kulling, S. E. & Schwartz, H. (2009). et al., Analytical and compositional aspects of isoflavones in food and their biological effects, *Mol. Nutr. Food Res.*, *53*, S266-S309.
[149] Kelly, G. E., Nelson, C., Waring, M. A., Joannou, G. E. & Reeder, A. Y. (1993). Metabolites of dietary (soya) isoflavones in human urine. *Clin. Chim. Acta*, *223*, 9-22.
[150] Joannou, G. E., Kelly, G. E., Reeder, A. Y., Waring, M. & Nelson, C. (1995). A urinary profile study of dietary phytoestrogens – the identification and mode of metabolism of new isoflavonoids. J. Steroid Biochem. *Mol. Biol.*, *54*, 167-184.
[151] Faughnan, M. S., Hawdon, A. & Ah-Singh, E. (2004). et al., Urinary isoflavone kinetics: the effect of age, gender, food matrix and chemical composition, *Brit. J. Nutr.*, *91*, 567-574
[152] Chang, Y. C., Nair, M. G. M. G.(1995). Metabolism of daidzein and genistein by intestinal, *J. Nat. Prod.*, *58*, 1892-1896.
[153] Tsunoda, N., Pomeroy, S. & Nestel, P. (2002). Absorption in Humans of Isoflavones from Soy and Red Clover Is Similar, *J. Nutr.*, *132*, 2199-2201.
[154] Kelly, G. E., Joannou, G. E., Reeder, A. Y., Nelson, C. & Waring, M. A. (1995). The variable metabolic response to dietary isoflavones in humans. *Proc. Soc. Exp. Biol. Med.*, *208*, 40-43.
[155] Setchell, K. D. (1998). Phytoestrogens: the biochemistry, physiology, and implications for human health of soy isoflavones. *Am. J. Clin. Nutr.*, *68*, 1333S-1346S.

[156] Hutchins, A. M., Slavin, J. L., Lampe, J. W. (1995). Urinary isoflavanoid phytoestrogen and lignan excretion after consumption of fermented and unfermented soy products. *J. Am. Diet. Assoc., 95*, 545-551.

[157] Izumi, T., Piskula, M. K. & Osawa, S. (2000). et al., Soy isoflavone aglycones are absorbed faster and in higher amounts than their glucosides in humans. *J. Nutr., 130*, 1695-1699.

[158] Setchell, K. D. R., Brown, N. M. & Desai, P. (2001). et al., Bioavailability of pure isoflavones in healthy humans and analysis of commercial soy isoflavone supplements. *J. Nutr., 131*, 1362S-1375S.

[159] Xu, X., Wang, H. J., Murphy, P. A., Cook, L. & Hendrich, S. (1994). Daidzein is a more bioavailable soybean isoflavone than is genistein in adult women. *J. Nutr., 124*, 825-832.

[160] Busby, M. G., Jeffcoat, A. R. & Bloedon, L. T. (2002). et al., Clinical characteristics and pharmacokinetics of purified soy isoflavones: single-dose administration to healthy men. *Am. J. Clin. Nutr., 75*, 126-136.

[161] Jackmann, K. A., Woodman, O. L., Chrissobolis, S. & Sobey, C. G. (2007). Vasorelaxant and antioxidant activity of the isoflavone metabolite equol in carotid and cerebral arteries. *Brain Res., 1141*, 99-107.

[162] Setchell, K. D. R., Brown, N. M., Lydeking-Olsen, E. (2002). The Clinical Importance of the Metabolite Equol—A Clue to the Effectiveness of Soy and Its Isoflavones. *J. Nutr., 132*, 3577-3584.

[163] Joy, S., Siow, R. C. M. & Rowlands, D. J. (2006). et al., The Isoflavone Equol Mediates Rapid Vascular Relaxation Ca2 –Independent activation of endothelial nitric-oxide synthase/Hsp90 involving ERK1/2 and Akt phosphorylation in human endothelial cells, *J. Biol. Chem., 281*, 27335-27345

[164] Jackmann, K. A., Woodman, O. L., Sobey, C. G. (2007). Isoflavones, equol and cardiovascular disease: pharmacological and therapeutic insights. *Curr. Med. Chem., 14*, 2824-2830.

[165] Pan, A., Franco, O. H., Ye, J. (2008). et al., Soy Protein Intake Has Sex-Specific Effects on the Risk of Metabolic Syndrome in Middle-Aged and Elderly Chinese, *J. Nutr.,. 138*, 2413-2421.

[166] Penza, M., Montani, C., Romani, A. (2006). et al., Genistein affects adipose tissue deposition in a dose-dependent and gender-specific manner. *Endocrinology, 147*, 5740-5751.

[167] Borras, C., Gambini, J., Gomez-Cabera, M. C. (2006). et al., Genistein, a soy isoflavone, upregulates expression of antioxidant genes: involvement of estrogen receptors ERK1/2, and NFκB. *FASEB J., 20*, 2136-2138.

[168] Niculescu, M. D., Pop, E. A., Fischer, L. M. & Zeisel, S. H. (2007). Dietary isoflavones differentially induce gene expression changes in lymphocytes from postmenopausal women who form equol as compared with those who do not, *J. Nutr. Biochem., 18*, 380-390.

[169] Chung, M. J., Kang, A. Y. & Lee, K. M. (2006). et al., Water-soluble genistein glycoside isoflavones up-regulate antioxidant metallothionein expression and scavenge free radicals. *J. Agric. Food Chem., 54*, 3819-3826.

[170] Kampkotter, A., Wiegnand, C. & Timpel, C. (2008). et al., Increased expression of catalase in human hepatoma cells by soy isoflavone, daidzein. *Basic Clin. Pharmacol. Toxicol.*, *102*, 437-442.
[171] Mann, G. E., Rowlands, D. J., Li, F. Y. L. (2007). et al., Activation of endothelial nitric oxide synthase by dietary isoflavones: Role of NO in Nrf2-mediated antioxidant gene expression, *Cardiovasc. Res.*, *75*, 261-274.
[172] Blaya, M., Espinel, A. E. & Delgado, M. A. (2010). et al., Isoflavone effect on gene expression profile and biomarkers of inflammation, *J. Pharmaceut. Biomed.*, *51*, 382-390.
[173] Collins, A. R., Lyon, C. J. & Xia, X. (2009). et al., Age-Accelerated Atherosclerosis Correlates With Failure to Upregulate *Antioxidant Genes, Circ Res.*, *104*, e42-54.
[174] Tatone, C., Carbone, M. C. & Falone, S. (2006). et al., Age-dependent changes in the expression of superoxide dismutases and catalase are associated with ultrastructural modifications in human granulosa cells, *Mol. Hum. Reprod.*, *12*, 655-660.
[175] Jaenisch, R. & Bird, A. (2003). Epigenetic regulation of gene expression: how the genome integrates intrinsic and environmental signals, *Nat. Genet.*, *33*, S245-S254.
[176] Souzeau, E., Bélanger, S. & Picard, S. (2005). et al., Dietary isoflavones during pregnancy and lactation provide cardioprotection to offspring rats in adulthood, *Am. J. Physiol. Heart C.*, *289*, H715-H721.
[177] Warri, A., Saarinen, N. M., Makela, S. & Hilakivi-Clarke, L. (2008). The role of early life genistein exposures in modifying breast cancer risk, *Br. J. Cancer*, *98*, 1485-1493.
[178] Möller, F. J., Zieraua, O. & Hertrampf, T. (2009). et al., Long-term effects of dietary isoflavones on uterine gene expression profiles, *J. Steroid Biochem.*, *113*, 296-303
[179] Dolinoy, D. C., Weidman, J. R., Waterland, R. A. & Jirtle, R. L. (2006). Maternal Genistein Alters Coat Color and Protects Avy Mouse Offspring from Obesity by Modifying the Fetal Epigenome, *Environ Health Perspect*, *114*, 567-572.
[180] Davis, C. D. (2007). Nutritional Interactions: Credentialing of Molecular Targets for Cancer Prevention, *Exp Biol Med.*, *232*, 176-183.
[181] Franco, O. H., Bonneux, L. & de Laet, C. (2004). et al., The Polymeal: a more natural, safer, and probably tastier (than the Polypill) strategy to reduce cardiovascular disease by more than 75%. *BMJ, 329*, 1447-1450.
[182] Jenkins, D. J. A., Kendall, C. W. C. & Falkner, D. A. (2006). et al., Assessment of the longer-term effects of a dietary portfolio of cholesterol-lowering foods in hypercholesterolemia. *Am. J. Clin. Nutr.*, *83*, 582-591.
[183] Lorenz, M., Stangl, K. & Stangl, V. (2009). Vascular effects of tea are suppressed by soy milk, *Atherosclerosis*, *206*, 31-32.
[184] Zhou, J. R., Yu, L., Mai, Z. & Blackburn, G. L. (2004). Combined inhibition of estrogen-dependent human breast carcinoma by soy and tea bioactive components in mice. *Int J Cancer*, *108*, 8-14.
[185] Zhou, J. R., Yu, L., Zhong, Y., Blackburn, G. L. (2003). Soy phytochemicals and tea bioactive components synergistically inhibit androgen-sensitive human prostate tumors in mice. *J. Nutr.*, *133*, 516-521.

(Abstract is missing in this chapter)

In: Isoflavones Biosynthesis, Occurence and Health Effects ISBN: 978-1-61728-113-6
Editor: Melanie J. Thompson, pp. 111-135 © 2010 Nova Science Publishers, Inc.

Chapter III

Effects of Processing on Isoflavone Content and Profile in Foodstuffs: A Review

Antonietta Baiano*
Department of Food Science, University of Foggia, Via Napoli, 25-71100 Foggia, Italy, Istituto per la Ricerca e le Applicazioni Biotecnologiche per la Sicurezza e la Valorizzazione dei Prodotti Tipici e di Qualità, University of Foggia, Via Napoli, 25 - 71100 Foggia, Italy

Abstract

Isoflavones, together with lignans, and coumestans represent the three classes of phytoestrogens, compounds that have an estrogen-like and antioxidant activity and are structurally similar to estradiol. Isoflavones are widespread in a variety of leguminous plants such as red clover, kudzu root and soy and are subjected to increasing attention due the possibility of use as potential therapeutic agents. In fact, several studies suggest that consumption of isoflavone-containing foods is associated with the lowest level of cholesterol in blood, and prevention of cardiovascular disease, prostate, colon and breast cancers, osteoporosis, and menopausal symptoms in people whose soy-rich diet allow to increase of 100 times the concentration of endogenous estrogens. Isoflavones are associated with the protein fraction of soybeans and are present under several chemical forms: the β-glucosides genistin, glycitin, daidzin, their 6"-*O*-malonyl-glucosides and a 6"-*O*-acetyl-glucosides and the aglycone forms genistein, daidzein and glycitein. Processing greatly affect the content of total isoflavones and the concentrations of each of their chemical forms. Among soy foods, the highest concentration of isoflavones is detectable in the whole soybeans whereas soy milk, tofu, tempeh, miso, fermented bean curd, bean paste and the so called "second-generation soyfoods" (soy hot dog, tempeh burger, soy noodles) show isoflavone contents ranging from 2 to 46% with respect to the

* Corresponding author: Tel.: +39 (0)881 589117; fax: +39 (0)881 589308., e-mail address: a.baiano@unifg.it

whole soybeans. These reductions are mainly due to dilution of soy protein with other ingredients and heat treatments that also result in changes of the isoflavone profiles. In fact, the predominant isoflavones in soybeans and non-fermented soy products (soy protein, soymilk derivates, etc.) are malonyl glucosides, whereas fermented soy foods (miso, tempeh) contain mainly aglycones. These changes affect the metabolic activity of isoflavone-containing foods since the glycoside forms are biologically inactive and need to be cleaved by the β-glucosidases from intestine in the glucosyl residue and the biologically active aglycones.

This review gives an insight of the effects of processing and processing parameters (extraction, fermentation, pasta-making, addition of coagulants, addition of protease, temperature, pH) on isoflavone content, profile and antioxidant activity of soy foods.

Introduction

Isoflavones belong to the family of flavonoids and together with lignans and coumestans represent the three classes of the so-called phytoestrogens, naturally occurring polycyclic phenols found in the plant kingdom that, when ingested and metabolized, may compete with endogenous estrogens for receptor binding and act both as estrogen agonists and antagonists (Kris-Etherton & Hecker, 2002). The similar chemical structures of 17ß-estradiol and equol, a phytoestrogen metabolite (both the molecules contain the phenolic ring and an identical distance between the hydroxyl groups) are considered prerequisites for estrogen binding (Tham et al., 1998). Although the estrogenic activity of the main isoflavones (genistein and daidzein), is from 100 to 1000 times lower than that of 17ß-estradiol (Miksicek et al., 1994), their concentration in the plasma of individuals consuming the traditional Japanese diet (50-80 mg/d soy-derived phytoestrogens) can be 100 times higher than the concentration of endogenous estrogens and thus sufficient to give a physiological response (Setchell & Cassidy, 1999). Isoflavones are widespread in a variety of leguminous plants included red clover and kudzu root but the largest contribution to the dietary intake comes from soybean and soy-based foods (Kirk et al., 1999; Morandi et al., 2005; Umphress et al., 2005). In fact, kudzu root and red clover are generally used in some dietary supplements whereas soybeans and soy flour are involved in the production of a variety of foodstuffs including soy milk, tofu, tempeh, miso, fermented bean curd, bean paste, soy hot dog, tempeh burger, soy noodles, breads, and meat.

Isoflavones are present under 12 chemical forms: three aglycones (daidzein, genistein, glycitein), and the corresponding β-glucosides (daidzin, genistin, glycitin), 6"-*O*-malonyl-glucosides and 6"-*O*-acetyl-glucosides (Murphy et al., 1999).

Daidzein and genistein are the main isoflavones in soy whereas kudzu root mainly contains daidzein. Naringenin is a precursor to genistein present in some citrus fruits whereas formononetin and biochanin A are methylated isoflavones found in clovers and metabolized to daidzein and genistein, respectively.

Several studies have shown that there is a large variability (up to 3-fold within the same cultivar growth in different geographical areas and years) in isoflavone content and profile of soybeans as a consequence of environmental conditions, changes in temperature, light exposure nutritional status, and biotic stresses (Wang & Murphy, 1994; Caldwell et al., 2005). Generally, the synthesis of phytoestrogens represents a response to environmental stresses. Flavonoids are also involved in legume-rhizobial nodulation (Subramanian et al.,

2007) since this process begins with signal exchange between the symbiotic partners. Flavonoids and isoflavonoids exuded through the plant roots whereas the rhizobial symbionts respond by producing a lipo-chitooligosaccharide signal (Peters et al., 1986). The health benefits provided by stress-induced phytochemicals have been recently related to a phenomenon called xenohormesis. According to this theory, such chemical compounds from plants provide a warning about the deterioration of the environment, allowing mammals to activate defense mechanisms while conditions are still favourable (Howitz & Sinclair, 2008).

Isoflavonoids are formed by a branch of the flavonoid biosynthetic pathway, and originate from a flavanone intermediate (naringenin or liquiritigenin, in the case of genistein and daidzein, respectively). The flavanone first undergoes abstraction of hydrogen radical at C-3, then a B-ring migration from C-2 to C-3 and subsequently the hydroxylation of the resulting C-2 radical. This reaction needs NADPH and molecular oxygen, and is catalyzed by a microsomal stereoselective cytochrome P450 enzyme (isoflavone synthase or IFS). The resulting unstable 2-hydroxyisoflavanone undergoes dehydration to yield genistein or daidzein (Dixon & Ferreira, 2002).

The different chemical structures of isoflavones are responsible for their different bioavailability in humans (Chun et al., 2008). Isoflavone aglycones are absorbed faster and in larger amounts than their glucosides (Izumi et al., 2002) that require a previous hydrolysis by bacterial β-glucosidases in the intestinal wall, resulting in the conversion to their corresponding bioactive aglycones (Setchell, 1998; Setchell et al., 2002) that are then absorbed by the intestinal tract and linked in the intestinal wall and liver with glucuronic acid to make them more water-soluble (glucuronidation). According to Cederroth & Nef (2009), after ingestion of the aglycones, the average time requested to reach peak plasma concentrations is 4-7 hours, versus the 8-11 necessary for the corresponding β-glycosides. However, a study performed by Richelle et al. (2002) showed that hydrolysis of isoflavone glycosides does not alter plasma and urine isoflavone pharmacokinetics in women. Daidzein and genistein can be further metabolized to equol and *O*-demethyangolensin (the first) and p-ethyl phenol (the second) although the ability to operate the conversion of daidzein to equol has been found only in 30% of humans (Atkinson et al., 2005). In fact, biotransformations of isoflavones from food intake in human appear to be highly dependent on the nature of the individual intestinal microflora (Omoni & Aluko, 2005; Xu et al., 1995) and are also affected by dietary habits. In fact, according to Setchell & Cole 2006), the frequency of equol producers in the vegetarians is 59%, similar to the frequency in Japanese adults consuming soy, whereas in nonvegetarian adults is 25%. Several studies have investigated the plasma concentration of phytoestrogens and their metabolites in subjects consuming a diet with or without soy. In humans having a soy-free diet, plasma concentration of isoflavones is usually in the nanomolar range (van Erp-Baart et al., 2003) whereas acute ingestion of dietary soy leads to a fast increase up to the micromolar range (King & Bursill, 1998; Watanabe et al., 1998).

Isoflavones are excreted within about 24 hours after ingestion, mainly in urine and, to a lesser extent, in feces (Setchell et al., 2001). Urinary concentrations of daidzein and genistein may not correlate well with the ingested amounts, perhaps because of the limited absorption of these isoflavones at higher doses (Setchell et al., 2003). Urinary isoflavone concentrations show small variations by demographic variables such as age and race/ethnicity. In fact, urinary isoflavone concentration is generally lower in adults than in children and adolescents whereas non-Hispanic whites have higher equol concentration than non-Hispanic blacks and Mexican Americans.

Isoflavone receive increasing attention due to the health benefits associated with their consumption and, consequently, soy food consumption has markedly increased due to the extended availability of soy products and isoflavone supplements used as pharmaceutical products or food fortificants. The hypothesis that soyfoods may lower the risk of coronary artery disease by lowering serum total cholesterol and low-density lipoprotein (LDL) cholesterol concentrations gave rise from epidemiological observations that Asian populations, whose diets include soyfoods have a lower incidence of this disease than West populations (Beaglehole, 1990). Some works have also suggested that the higher isoflavone intake in Asian diets may account for the lower incidence of menopause-related symptoms, the reduced risk for breast, prostate, and colon cancer and the modulation of osteoporosis. Estrogens have been reported to affect adiposity either directly, by modulating lipogenesis and lipolysis, or indirectly, by modulating appetite or energy consumption (Cooke & Naaz, 2004). Estrogens also play an important role in glucose homeostasis and modulate insulin sensitivity (Godsland, 2005). According to a recent report (Balk et al., 2005), there is no conclusive evidence of a dose-response effect of isoflavone on cardiovascular diseases, menopausal symptoms, endocrine function, cancer, bone health, reproductive health, kidney diseases, cognitive function, or glucose metabolism. A significant number of other studies reported an absence of beneficial effects of soy on bodyweight, serum lipid profile, fatmass, blood glucose and insulin profiles (Anderson & Hoie, 2005; Li et al., 2005; Hall et al., 2006; Ikeda et al., 2006; Anderson et al., 2007).

Some of the healthy effects of isoflavones are mainly related to their antioxidant activity that is equal to or greater than that of the other polyphenols. Among isoflavones, the most potent antioxidants are represented by genistein and daidzein and both aglycone and glycoside forms are effective (Xu et al., 2000). For example, genistin protects against the oxidative DNA damage induced by hydroxyl radicals, has superoxide anion scavenging capacity (Russo et al., 2006), and reduces the oxidation of low-density lipoprotein (Lee et al., 2005) whereas genistein exerts an inhibitory effect against Cu^{++}-induced lipid peroxidation of HDL (Ferretti et al., 2004).

Some studies concern with the potential adverse effects of soy components (Fort et al., 1990). Over the well known detrimental effects of antinutrients such as hemagglutinins and enzyme inhibitors, the ingestion of isoflavones has been shown to inhibit thyroxine synthesis thus inducing goiter and hypothyroidism in infants fed with soy-based formulas. Soy isoflavones inhibit the iodination of tyrosine by competing with tyrosine. Furthermore, high concentrations of isoflavones have been reported to result in genetic abnormalities in human lymphocytes, oviduct cells, and testis cells. Since isoflavones may act both like estrogen and antiestrogen, they could cause direct and indirect toxicological effects on the reproductive organs. According to Nagata et al. (1998), high intake of soy milk can alter the mestrual cycle length.

The concentrations of total isoflavones and their different forms in foods depend on how the food is prepared (Wang & Murphy, 1994; Wang & Murphy, 1996). Heat processing, enzymatic hydrolysis, defoaming and fermentation can alter the isoflavone content and profile depending on the process severity. Isoflavones are associated with the protein fraction of soybeans that, among the soy foods, show the highest concentration of these compounds. The isoflavone concentration progressively decreases as the degree of processing increases thus, the most common soy products (soy milk, tofu, and second-generation soy foods) contain from 2 to 23% of the isoflavone content of whole soybeans due to aqueous processing and dilution (Dewell & Hollenbeck, 2003).

This work represents an overview of the extent of changes induced by processing on isoflavone content, profile and antioxidant activity of soy foods. An insight of the isoflavone contents of different foods is also given.

Daily Intake of Isoflavones and Isoflavone Content of Foods

The estimation of the isoflavone daily intake is not easy. It can be based on the total content of isoflavones per serving and the estimated consumption frequency of specific classes of people having soy based diets or can be inferred by the results of administered questionnaire.

In order to develop a dietary assessment instrument to measure the soy food consumption and the consequent isoflavone intake, Kirk et al. (1999) designed a frequency questionnaire including 40 soy food products classified into 5 groups (main dishes, breakfast foods, soy dairy products, beverages, and others). People were asked to indicate the frequency of consumption of each food and the serving size. In more recent years (2007), French et al. tested a phytoestrogen food frequency questionnaire including 53 soy foods founding a good correlation between isoflavone intakes and urinary concentrations of isoflavone metabolites in 26 premenopausal, Caucasian women aged 25 to 42 years.

The isoflavone intake in the traditional Asian diet is estimated to be about 20-50 mg/day (Adlercreutz et al., 1991; Nagata et al., 1997; Chen et al., 1999). A progressive increase in soy derivative consumption has also been observed in Western countries, since soybean is a good source of protein having a high content of essential amino acids and can be used in the preparation of products directed to vegetarians and milkintolerant people (Morandi et al., 2005). A further great impulse to the consumption of soy foods has been given by the American Food and Drug Administration (FDA) that, in 1999, authorized the labelling of health claims on food containing soy proteins, referring to the beneficial role of soy protein in reducing the risk of coronary heart disease. Nevertheless, the isoflavone intake in USA and European Countries is still low (0.15μg to 3mg per day and 0.49 to 1mg per day, respectively) (De Kleijn et al., 2001; Horn-Ross et al., 2001; Van Erp-Baart et al., 2003). Morandi et al. (2005) estimated the isoflavone intake of specific classes of Italian consumers since the Italian Health Authority recommended the limit for adults of 80 mg/day. The found that hypercholesterolemic patients fed a soy-based diet containing from 25 to 113mg/day of isoflavone and that infants fed soy formulas and 5-6-year-old celiac children largely exceeds the theoretical maximum daily intake.

Concerning the isoflavone content of foods, the first USDA-Iowa State University Isoflavones Database for 128 foods was released in 1999 by the Nutrient Data Laboratory (NDL) of the Agricultural Research Service (ARS) of the United State Department of Agriculture (USDA). The updated version of this database, created on the basis of literature published in peer-reviewed journal, was released in 2008 and can be downloaded from the ARS site (www.ars.usda.gov/SP2UserFiles/Place/12354500/Data/isoflav/Isoflav_R2.pdf). It contains values for 557 foods. Isoflavones included in the database are daidzein, genistein, glycitein, and total isoflavones. Two hundred and sixty eight of the 557 foods do not contain any isoflavones whereas 200 selected foods containing other plant estrogens such as coumestrol, formononetin, and biochanin A are listed in a separate table.

Umphress et al. (2005) analyzed 167 processed and fast foods containing soy additives such as isolated soy protein, soy protein concentrate, soy flour, and hydrolyzed soy protein) and 12 traditional soybean products for the isoflavone content and profile. Concentrations of total isoflavones, expressed as aglycones, ranged from not detected (49 foods) to 149.9mg/100g. The highest concentrations of total isoflavones were found in imitation bacon bits. Concerning each specific food groups, the ranges were the following: bread and grain products (not detected - 93.9mg/100g), gravies and sauces (not detected - 2.7mg/100g), meat and poultry products (not detected - 15.9mg/100g), meat substitutes (0.3 - 149.9mg/100g), nutritional products (0.4 - 36.5mg/100g), reduced fat peanut butters (0.1 - 4.0mg/100g), seafood products (not detected - 1.6mg/100g), and soybean products (2.9 - 109.3mg/100g).

In their study on food products purchased in Italian supermarkets, Morandi et al. (2005) found that the total isoflavone contents were in the range 21-803 mg/g of dry weight and that the highest values were detectable in imitation dairy and meat products. Infant formulas and gluten-free products contained from 121 to 427 mg/g and from 77–220 mg/g of dry weight, respectively.

Haron et al. (2009) made a survey on the isoflavone content of some selected soy foods (raw and fried tempeh, boxed and homemade soybean drink, egg tofu, firm tofu, soft tofu, and tofu skin). They found the highest isoflavone content in raw tempeh (26±6mg of daidzein, 28±11mg of genistein, and 206±56mg of total isoflavones per 100g of dry weight. The deep frying decreases of about 45% the total isoflavone content. The lowest total isoflavone contents were detected in in processed soy foods like egg tofu and home made soybean drink (less than 20mg/100g).

Akitha Devi et al. (2009) performed a survey on 4 soybean cultivars and 26 soy products founding that the total isoflavone contents of the soybean cultivars was in the range 52.5–98.6mg/100g per kg, and those of soy products was from 3.3 to 79.5mg/100g. The isoflavone content decreased in the order: soy sprouts, soy seeds, soy flour, soy milk, soy meals, and soy sauce.

Ribeiro et al. (2009) studied the soflavone content and β-glucosidase (the popular name of glucan *endo*-1,6- β-glucosidase) activity in 18 soybean cultivars of different maturity groups under the same cultivation conditions. The results showed that the predominant isoflavones were malonyl daidzin and malonyl genistin in all cultivars, followed by β–glucosides. Aglycones corresponded to 2.0% of the total forms whereas the acetylated forms were not detected. Regression analysis also showed the absence of correlation between number of days of maturity and the distribution of the isoflavone forms.

Since the phytoestrogen content in plants is variable and depends on genetic and environmental factors, growth, harvesting and processing, Kuhnle et al. (2009) analysed nine different types of foods from different sources and countries of origin to compare phytoestrogen content and assess variability. Isoflavones were detected in apple, baked beans, carrots, cauliflower, frozen peas, lettuce, pears, red potatoes, and savoy cabbage. The highest contents were found in red potatoes, baked beans, and pears (23.83, 20.33, and 20.30μg/100g of dry weight, respectively) whereas the lowest were detected in savoy cabbage, cauliflowers, and carrots (3.09, 4.50, and 5.20μg/100g of dry weight, respectively). The coefficient of variation ranged from a minimum of 21% (pear) to a maximum of 60% (baked beans).

Baiano et al. (2009) determined the isoflavone content of two types of soy flour founding a minimum for the toasted flour and a maximum for the defatted flour (1234 and 2210.28mg/kg of dry weight, respectively). The main isoflavone forms in both types of soy flour were malonyl genistin, genistin, malonyl daidzin, and daidzin. The percentage of the aglycones forms ranged from 0.6% of the toasted flour to 7% of the defatted flour. According to Coward et al. (1998), defatted soy flour that had not been heat treated consisted mostly of 6"-*O*-malonyl-β-glucoside conjugates whereas toasted soy flour contained large amounts of 6"-*O*-acetyl-β-glucoside conjugates, as a consequence of the heat-induced decarboxylation of the malonate group to acetate.

Effects of Germination on Isoflavone Content and Profile of Soybeans

Soybean is a major source of proteins and isoflavones nevertheless some disadvantages are related to their consumption, including undesirable flavours and odours due to lipoxygenase activity and the presence of antinutritional factors such as trypsin inhibitors, phytates, and flatulent. It is known that germination process determine an improvement of nutritional quality by prevention of lipid oxidation, an increase of ascorbic acid and riboflavin contents, decrease of trypsin inhibitors and the hydrolysis of raffinose and stachyose that are the sugars responsible for flatulence (Ahmad & Pathak, 2000). Concerning the effects of germinartion on isoflavone content, Terrence et al. (1991) found that it increased after 1 day of germination and then concentration slowly decreased. Zhu et al. (2005) evaluated the changes in isoflavone content and profile at varying length of hypocotyls during germination and determined the germination stage that maximizes the isoflavone content. They studied two soybean varieties, *Hutcheson* and *Caviness*, selected for their high isoflavone contents and submitted them to soaking (25°C, 12 hours), germination at 40°C, and freeze-drying. The isoflavone content and profile of dry, soaked, germinated (at hypocotyl lengths of 0.5, 2.5, and 6.5 mm), and nongerminated seeds was determined. The results showed that the maximum amount of total isoflavone, genistein, and daidzein with their β-glucosides was reached when the length of hypocotyl of the germinated-seed was 0.5mm for *Hutcheson* (2.491, 1.500, and 0.671 mg/g, respectively), and 2.5mm for *Caviness* (2.78, 1.523, and 0.905 mg/g, respectively). At the same hypocotyls lengths, an increase in malonylgenistin and malonyldaidzin was registered (1.305 and 0.476mg/g in *Hutcheson*; 1.308 and 0.677mg/g in *Caviness*, respectively) followed by a decrease after this stage. The highest genistein and daidzein contents were observed just after soaking whereas glycitein and its β-glucosides remained stable during germination. Pauchar-Menaco et al. (2010) study the way to optimize time and temperature of germination of soybean BRS 133 in order to maximize the contents of bioactive compounds included isoflavones. They applied the response surface methodology (RSM), with a 2^2 central composite rotational design founding that the highest isoflavone concentration (453mg/100g) was obtained with low germination time (21h) and temperature (20°C) whereas the optimal increase in the concentrations of aglycones (90.3mg/100g) was observed when germination was carried out at 30°C for 63h. The total isoflavone concentration of the non-germinated freeze-dried soybean flour was 390mg/100g (aglycones amounted to 26.7mg/100g). Concerning genistein, germination for 63h at 30°C contributed to an increase of about 212%.

Effects of Processing on Isoflavone Content and Profile of Soy Protein Isolates

Soy protein isolates are largely used as ingredients in several food products since, beside the nutritional and functional properties, there are some possible beneficial effects associated with their isoflavone contents. Nevertheless, isoflavone content and profile can appreciably changes depending on processing conditions. Lopes Barbosa et al. (2006) study the influence of pH (range between 2 and 10), ionic strength (0.01, 0.1, 0.2, 0.5, 1.0 M), water-to-soy flour ratio (10:1, 20:2, 30:1), time (1, 3 hours), and temperature (4, 25, 50°C) in order to obtain soy protein isolates rich in isoflavones. They found that the last three parameters did not affect in a significant manner the total amount of isoflavones extracted but temperatures below 50°C caused hydrolysis of β-glucosides catalyzed by endogenous β-glucosidase activity with increase of aglycones. Concerning ionic strengths, aqueous solutions 0.1 and 0.2 M caused a decrease in the extraction of isoflavones, and changes of the ionic strength beyond this range resulted in an increase of extraction, reaching a maximum at about 0.5 M. However, the highest yield of isoflavones was always obtained in a simple aqueous solution. Concerning pH, in the region close to the isoelectric point of soy proteins (pI 4.5) there was a drastic decrease in the efficiency of isoflavone extraction. The lower yield of isoflavones at the isoelectric pH of proteins is a result of both a lower solubility and a decreased extraction due to their associations with solubilized proteins.

Yu et al. (2007) was able to obtain soy protein isolate rich in isoflavones from soy hypocotyls defatted by supercritical CO_2. Hypocotyls represent a byproduct of soy protein industry but, thanks to their richness in isoflavones, they may become a functional food. Defatting was performed at 35MPa and 45°C, for 2 hours at a constant flow rate of CO_2 (5l/min) and allowed to obtain a product containing about 2.18g/100g of isoflavones. The soy protein isolate was extracted from defatted soy hypocotyls under optimized conditions (pH 6, ratio water solution-to-defatted soy flour 10:1, T=30°C, time 30min) and showed a concentrations of isoflavones equal to 0.64mg/100g with three major β-glycosides (glycitin, daidzin and genistin) accounting for about 83.7% of the total isoflavones.

Effects of Processing on Isoflavone Content and Profile During Soymilk and Tofu Manifacture

Soaking is one of the most important steps in the preparation of soybean foods (Toda et al., 2001) since it reduces the energy required for further processing (cooking time included). Furthermore, it greatly affects the isoflavone profiles due to the β-glucosidase activity and the consequent hydrolysis of isoflavone glucosides into aglycones. Wardhani et al. (2008) applied a modified first order kinetic equation to predict the transformation of isoflavones into their aglycones during soybean soaking at different temperatures (30, 50, 60, and 85°C). Authors described these kinetics in terms of daidzin and genistin disappearance and daidzein and genistein formation. The production of daidzein was not equimolar to the disappearance of daidzin due to the daidzin–daidzein degradation into unknown products. Temperature enhanced diadzin–genistin disappearance and diadzein–genistein formation. Concerning the effect of temperature on the β-glucosidase, the maximum activity was observed after 1h at 50-60°C and after 6h at 30°C whereas soaking at 85°C did not develop this activity. Enzyme inactivation, elution into soaking water or a combination of both could be responsible for this loss of activity.

Prabhakaran et al. (2006) investigated the effects of hot (95°C) and cold (45°C) grinding (used as an alternative to the traditional soaking) of *Harovinton* soybeans during the soymilk preparation but also of two types (direct and indirect) of UHT treatment process on soymilk isoflavones. The results showed that the grinding at a lower temperature determined an increase of the β-glucosidase activity and thus the increase in the concentration of aglycones. The isoflavone concentration of soymilk prepared by hot-grind process was significantly higher than that produced by a cold-grind process due to the higher solubility of isoflavones in hot water than in cold water. Concerning UHT treatment, there were no apparent differences in the loss of isoflavones between direct and indirect heating.

Ishihara et al. (2007) compared the isoflavone content and composition of commercial and homemade soymilk and tofu founding that homemade soymilk, made by the extended boiling method, yielded the highest total isoflavone and glucoside content.

Processing of soymilk and tofu includes soaking, heating and, in the case of tofu, the addition of coagulants. Kao et al. (2004) studied the effects of soaking (at 25°C for 4h, 35°C for 8h, or 45°C for 12h), heating of the soaked soybean slurry (at 100°C for 10, 20 or 30min), and addition of coagulants (calcium sulfate and calcium chloride) on the stability of isoflavone glucosides. The concentrations of the three aglycones, daidzein, glycitein and genistein were increased with increasing soaking temperature and time. After soaking at 45°C for 12h, the concentrations of daidzein, glycitein and genistein increased, probably because of conversion the relative glucosides. The other nine isoflavones decreased with soaking. The malonylglucosides can be leached into water or converted to acetylglucosides, which can be further converted to glucosides or aglycones as a function of soaking temperature and time. The concentrations of total isoflavones also decreased as the soaking temperature and time increased. Concerning the effects of heating, the aglycones did not show significant change whereas the contents of daidzin, glycitin and genistin increased. Acetyldaidzin and acetylglycitin were not detected while a slight increase was observed for acetylgenistin, probably formed from malonylgenistin. The malonyl-type isoflavones decreased, probably as a consequence of their conversion to glucosides by cleaving the malonyl ester groups. Concerning the addition of coagulants, the highest concentration of isoflavones was obtained in tofu with 0.3% calcium sulfate when compared to the higher percentage of the same coagulant. Comparing calcium sulfate and calcium chloride at the same level, a greater yield of isoflavones was obtained with calcium sulfate probably because it is a better protein coagulator than calcium chloride. In comparison to soymilk, the aglycones are present in the highest content in tofu, followed by malonylglucosides, glucosides and acetylglucosides.

According to Jackson et al. (2002), the recoveries of isoflavones in soy beverage and tofu in relation to their initial concentration in the soybeans were 54 and 36%, respectively whereas the isoflavones losses in water used to soak raw soybeans, okara (waste from heat-treated slurry), and whey were 4, 31, and 18%, respectively thus indicating the need to develop suitable techniques to recover and utilize these functional compounds from by-products.

Coagulation of soymilk consists in the cross-linking of protein molecules with the divalent cations. Different coagulants may have a different influence towards the retention of isoflavones in tofu and whey and, thus, Prabhakaran et al. (2006) investigated the effects of calcium sulphate $CaSO_4 \cdot 2H_2O$, calcium chloride $CaCl_2 \cdot 2H_2O$, calcium acetate $Ca(CH_3\text{-}COO)_2$, calcium lactate $(CH_3CHOHCOO)_2Ca \cdot 5H_2O$, magnesium chloride $MgCl_2 \cdot 6H_2O$, magnesium sulphate $MgSO_4 \cdot 7H_2O$ and acetic acid CH_3COOH starting from a soymilk having 182.19±0.36 μg/g of isoflavones on a wet weight basis (equivalent to 1570.64μg/g on a dry matter basis). Calcium sulphate was found to be the most suitable coagulant for tofu making in terms of yield, firmness, and retention of isoflavones (from 1317.88±6.77 to 1424.00±0.67μg/g on a dry matter basis). Calcium chloride was quicker than calcium sulphate in coagulating the soy protein but a quick coagulation resulted in more exclusion of isoflavones into the whey during the coagulation process. An average of 104.02μg/g of isoflavones were lost into the tofu wheys, on a wet weight basis thus indicating the possibility to use this by-product as a functional liquid.

Effects of Enzymic Treatments and Fermentation on Isoflavone Content and Profile of Soyfoods

There are very few reports on the effects of proteases on isoflavones associated with many soybean products and ingredients. Wu & Muir (2010) investigated the effects on the isofavone content and profile of a soy flour of the following enzymes: a protease from *Aspergillus oryzae*, a neutral protease from *Bacillus subtilis*, an alkaline protease from *Bacillus subtilis*, and a protease from *Bacillus licheniformis*. After hydrolysis performed at the optimum pH value of each enzymes, the resultant residues contained higher amount of isoflavones than the corresponding hydrolysates. The isoflavone profiles greatly changed as a function of pH. In fact, is is well known that malonylglucosidic and acetylglucosidic conjugates are strongly vulnerable to environmental factors and processing conditions. Hydrolysis at pH 10 of the alkaline protease resulted in complete loss of malonylglucosidic and acetylglucosidic conjugates both in the hydrolysate and residue whereas 6”-*O*-malonylgenistin and 6”-*O*-malonyldaidzin (57.8% in soybean flour) accounted for 66.2, 58.3, and 70.5% in proteases from *Aspergillus oryzae, Bacillus licheniformis*, and *Bacillus subtilis* (neutral protease), respectively. There was a large increase in aglycone content in the residue compared to that of the flour (4.7%). In fact, aglycones accounted for 62, 49, 44.1, and 44.7% of the total isoflavones in residues from *Aspergillus oryzae, Bacillus licheniformis, Bacillus subtilis* (alkaline), and *Bacillus subtilis* (neutral), proteases, respectively. The remarkable increase in aglycone content was attributed to the contaminated present in these protease preparations β-glucosidase present in the protease preparations.

The possibility to use microbial β-glucosidases to increase the aglycone content of soy products has been largely investigated. Choi et al. (2002) found that a glucoside hydrolyzing activity was induced in some lactic acid bacteria only when cultured in soymilk medium whereas *Lactobacillus delbrueckii* subsp. *delbrueckii* KCTC 1047, independently on the medium, completely hydrolyzed genistin and daidzin at 50μg/ml into their respective aglycones within 30 min.

Yang et al. (2009) published the first report on the possibility to use a thermostable fungal β-glucosidase isolated from the *Paecilomyces thermophila* J18 in order to increase the isoflavone aglycones in soy products. This enzyme was able to retain more than 95% of its initial activity after 8h at 50°C and converted isoflavone glycosides of in soybean flour extract and soybean embryo extract to their aglycones, resulting in more than 93% of hydrolysis after 4h of incubation.

The production of β-glucosidase in recombinant *Aspergillus oryzae* was investigated by Horii et al. (2009). Their results showed that the yield of recombinant β-glucosidase was 960mg/l in liquid culture (20-fold higher than the yield of β-glucosidase produced by *Saccharomyces cerevisiae*. The recombinant β-glucosidase converted isoflavone glycosides into their aglycones more efficiently than β-glucosidase from almond and produced isoflavone aglycones even in the presence of the insoluble form of isoflavone glucosides.

Chien et al. (2006) study the changes in isoflavones during the fermentation of soymilk. They fermented soymilk with inoculation of lactic acid bacteria (*Streptococcus thermophilus* BCRC 14085, *Lactobacillus acidophilus* BCRC 14079) and bifidobacteria (*Bifidobacterium infantis* BCRC 14633, *B. longum* B6) individually and in combination. They observed that fermented soymilk showed a lower total isoflavone content (81.94–86.61mg/ml) than non-fermented soymilk (87.61mg/ml). Fermentation determined a decrease in the contents of glucosides, malonylglucosides, and acetylglucosides along with a significant increase of aglycone content. Soymilk fermented with *S. thermophilus* showed the highest β-glucosidase activity and the greatest increase in the contents of daidzein, genistein and glycitein that from the initial 14.24%, 6.89% and 2.45%, respectively, changed to 36.20%, 28.80% and 12.44% after 24 h of fermentation.

Chun et al. (2008) tried to enrich soymilk of isoflavone aglycones by fermentation with single and combined cultures of *Streptococcus infantarius* 12 and *Weissella* sp. 4 founding that the rate of hydrolysis of isoflavone glucosides was the least with the first and the highest with the latter, resulting in hydrolysis of about 23–33% and 98–99% of the glucosides, respectively. Some mixed cultures (*Streptococcus infantarius* 12:*Weissella* sp. 4 ratios 1:3, 1:5, and 1:10) seem to be more effective.

Yang & Zhang (2009) used the basidiomycete *Ganoderma lucidum* WZ02 in soymilk fermentation. *Ganoderma lucidum* is one of the most famous traditional Chinese medicines and is called ‘‘Lingzhi” in China and ‘‘Reishi” in Japan. The soymilk was inoculated with 10% (v/v) of the seed culture and fermentated under the following conditions: temperature, 30°C; aeration rate, ratio volume air/volume fermentation broth per minute, 1vvm; agitation speed, 150 rpm. The contents of daidzein and genistein were increased from 1.89 and 2.34μg/ml at the beginning to 16.7 and 17.1μg/ml after 96h of fermentation at 30°C.

Hubert et al. (2008) studied the modifications occurring in the phytochemical profile of soy germ finely ground, suspended in ultra-pure distilled water and inoculated with three standardized mixtures of lactic acid bacteria. The raw soy germs contained 39.0 ± 2.5μmol/g of total isoflavones, with β-glucosides, malonylglucosides, aglycones and acetylglucosides accounting for 46.9%, 44.5%, 7.3%, and 1.3%, respectively. After 48h incubation at 37°C, the aglycones ranged from 62.7 to 75.3% of total isoflavones. The fermented soy germ extracts contained less than 22% malonylglucosides, 15% β-glucosides, and 2% acetylglucosides.

A popular Chinese soy product is represented by sufu that is a fermented tofu. Lin-Jun et al. (2004) analysed the changes in isoflavone content and profile occurring during sufu preparation. It included production of tofu by salt precipitation from boiled soymilk, fermentation of diced tofu at 28°C for 48h by the starter (*Actinomucor elegans* As3.227), salting and ripening under a dressing mixture of kojic red rice, alcohol beverage, sugar, wheat-based miso, and spices. They found a loss of 83% of isoflavone from soybean to sufu mainly occurring during tofu preparation and salting. The levels of aglycones increased especially during tofu preparation and, in the final product, aglycones and glucosides accounted for 99.7% and 0.3% of the total, respectively.

An analogous work was performed by Wang et al. (2007) on douchi, a traditional Chinese fermented soybean food. Douchi preparation included the following steps: washing, soaking and steaming of soybeans; inoculation of soybeans with *Aspergillus oryzae* 3.951; addition of NaCl; ripening. The results indicate that 61% of the isoflavones were lost from raw soybeans to douchi and the losses mainly occurred as a consequence of fermentation (43%) and addition of 10% NaCl (18%), respectively. The pre-treatment did not generate differences in total isoflavone content altered the isoflavone composition. During processing, the levels of aglycones increased (exceeded 90% of total content following post-fermentation), whereas the corresponding levels of β-glucosides, malonylglucosides, and acetylglucosides decreased. Finally, the activity of β-glucosidase was affected by NaCl supplementation.

Effects of Heat Treatments, High-Pressure Processing and Gamma Irradiation on Isoflavone Content and Profile of Soyfoods

The production of several soy foods include a thermal treatment, which can modify sensory properties and acceptability but also the isoflavone content and distribution.

According to Yuan et al. (2009), the isoflavone profiles of soybean hypocotyls and cotyledons change during dry heating. In particular, they found that hypocotyls and cotyledons contained high amounts of malonylglycosides (69.1 and 69.4%, respectively) and β-glycosides (27.1 and 25.4%), and only a very small quantity of aglycones (3.8 and 5.2%) whereas no acetylglycosides were detected. As a consequence of dry heating, acetylglycosides and β-glycosides formed, respectively, as the thermal decarboxylation and deesterification products of malonylglycosides, which are thermally unstable isoflavones.

The stability of isoflavones during extrusion processing of corn/soy mixture were investigated by Mahungu et al. (1999). According to their results, the extrusion barrel temperature had the most influence on isoflavone profile, especially in terms of decarboxylation of the malonyl β-glucoside, followed by the moisture content. The amount of extractable isoflavones decreased after extrusion whereas remained approximately the same when first hydrated with water before extraction. Nevertheless, hydration with water produced enzymatic glycolysis in the unextruded samples, thus, increasing the aglycone fraction.

Chiarello et al. (2006) investigated the influence of heat treatment on the isoflavone profile of soymilk prepared from BRS 213 variety. Heat treatment by autoclaving for 5 min did not affect the total isoflavone content, but led to a 90% decrease in malonyl forms and a 70% increase of β-glycosides. Instead, autoclaving for 15 min led to a 20% loss of total isoflavones.

An investigation performed by Lee & Lee (2008) highlighted the effects of some thermal treatments on isoflavone distribution in soybeans. They found that during oven-drying at 100°C, concentration of malonyl derivatives of isoflavones decreased whereas that of β-glucosides increased and malonyl-β-genistin had higher slopes of decreases than malonyl-β-daidzin. Roasting and explosive puffing determined a significant decrease of malonyl derivatives and significant increases of acetyl-β-glucosides and β-glucosides.

According to Coward et al. (1998), baking or frying of textured vegetable protein at 190°C and baking of soy flour in cookies did not change the total isoflavone content but determined an increase in β-glucosides conjugates and the consequent reduction of 6"-O-malonyl-β-glucoside conjugates.

In contrast to the changes occurring during thermal processing, high-pressure treatments only act on non-covalent bonds such as hydrogen bonding, ionic, and hydrophobic interactions, which are involved in protein-isoflavone interactions. When submitted to high-pressure processing, unfolding and denaturation of the two main soy proteins, glycinin and β-conglycinin, can modify the affinity for isoflavones, thus inducing changes of the isoflavone extractability. Jung et al. (2008) studied the effects of high-pressure processing (HPP) of soymilk and soybeans on the isoflavone content, profile and water extractability of soymilk. They applied high hydrostatic pressure to hydrated soybeans (100–700 MPa, 25°C) and soymilk (400–750 MPa; 25 and 75°C). They found that the pressure alone did not affect the isoflavone content of pressurized soymilk, and that the isoflavone profile may be significantly modified due to the adiabatic heating occurring during pressurization combined with mild temperature treatment. For example, at 75°C, the isoflavone profile shifted from malonylglucosides toward β-glucosides. Results also indicated that high-pressure processing applied as a pre-treatment of the hydrated beans in soymilk production did not significantly change isoflavone content and profile of the obtained soymilk.

Aguiar et al. (2009) evaluated the effects of gamma irradiation on the isoflavone contents of defatted soybean flour and soybean molasses, a derivative from the soybean food production. They observed reduction in the concentration of isoflavones in soy molasses whereas isoflavones in defatted soy flour were not eliminated by gamma radiation at rates up to 50 kGy. According to de Oliveira et al. (2009), the total isoflavone content of gamma-irradiated soybeans remained almost unchanged with the increase of radiation dose up to 10kGy. Although a general correlation among total isoflavone content and radiation dose was not found, some data suggest that the increase in the radiation dose induced a decrease in the content of glucosyl glucosides and malonyl isoflavones, as well as an increase in their aglycones.

Changes in Isoflavone Content and Profile during Bread-and Pasta-Making

Zhang et al. (2004) investigated the effects of each step of bread-making on the isoflavone content and profile. They combined and mixed the ingredients (soymilk powder and soy flour included) to form a dough that was proofed at 48°C for 1–4 hr and baked at 165°C for 50 min. Soy ingredients and wheat flour were the major contributors of the β-glucosidase activity. Although no degradation of isoflavones was observed during breadmaking, the isoflavone distribution was largely modified. During proofing, β-glucosides were converted to aglycones by specific β-glucosidase activity whereas baking determined a significantly decrease of malonylglucosides and an increase of β-glucosides.

In order to enhance aglycone content in soy bread, Riedl et al. (2005) looked for the time and temperature conditions that allowed to obtain the highest β-glucosidase activity. Proofing was performed for 1, 2, 3, and 4h at 22, 32, and 48°C and the results showed that aglycones increased, whereas the simple glucosides decreased, malonyl-glucosides decreased of 3-15%, and acetylglucosides were fairly constant. In order to obtain a rapid and large glucoside conversion, the optimum time-temperature conditions were 2h at 48°C.

According to Shao et al. (2009), the bread making processes did not affect the total isoflavone content, but changed glucosides and acetylglucosides to aglycones and malonylglucosides prior to acetylglucosides and further to glucosides during baking.

Another way to enhance isoflavone aglycones in the soy bread formula was represented by the addition of 5% almonds (Zhang, 2004), since they are a rich source of β-glucosidase. The same author found that no significant changes occurred in the isoflavone content during a 14 day-storage at room temperature.

Baiano et al. (2009) studied the changes in isoflavone content and profile during pasta-making. They produced soy spaghetti in which 5, 10, 15, 25 and 50% of durum wheat semolina was replaced with a toasted soy flour or a defatted soy flour. Spaghetti were dried at 50°C for 16h at a relative humidity of the heating medium of about 65–75%. The pasta-making procedure determined an increase of the aglycone forms that was at least 37 and 21% in pasta made with toasted soy flour and defatted soy flour, respectively. Cooking at the optimal cooking time induced noticeable decreases of the isoflavone content, with losses ranging from 60.44 to104.91mg/kg in pasta made with toasted soy flour and from 85.41 to 469.33mg/kg in pasta made with defatted soy flour as a consequence of the leaching in the cooking water.

Effects of Processing on Isoflavone Content and Profile of Soy Foods

Storage is known to determine changes in the phenolic content and profile, but very few works have been performed on the changes of isoflavone concentration and distribution during food shelf-life.

Kim et al. (2005) investigated the variation of isoflavone concentrations in soybean seeds under different storage conditions during longer storage periods. They found that changes in isofavones were cultivar-dependent. The *Daweon* cultivar showed a greater decrease at room (-2039.0μg/g) than at low temperature (-1822.0μg/g) over three years, whereas only a slight decrease were found under the same conditions for *Hannam* cultivar. With storage at room temperature, the acetylglucoside group slightly increased the isoflavone concentration, while the other three groups, aglycones, glucosides, and malonyl glucosides, decreased it. With storage at low temperature, malonylglucosides decreased whereas aglycones, glucosides, and acetylglucoside increased the isoflavone concentration.

In a study on soy-milks enriched with isoflavone aglycones deriving from soy germ, soy protein isolate, and bifidobacteria, Tsangalis et al. (2004) found that content and composition of isoflavones were unchanged after 14 day refrigerated storage.

According to Otieno (2007), the degradation of each isoflavone compound in fermented soymilk at each storage temperature was found to follow a first order kinetic model. Aglycones as well as glucosides were stable during storage at the 4 storage temperatures whereas glycone forms had smaller degradation constants compared to glucosides at all storage temperature.

Antioxidant Activity of Soy Foods

Antioxidants are molecules able to counteract the damaging effects of oxygen and free radicals in tissues. Anticancer and anti-inflammatory effects, cardiovascular protectors and enzyme inhibitor functions (Arora et al., 1998) of isoflavones are mainly related to their antioxidant activity. Among isoflavones, the strongest antioxidants are represented by genistein and daidzein. Lee et al. (2005) measured the antioxidant activity of genistein, daidzein and their corresponding glycosides using three methods: human low-density lipoprotein (LDL) oxidation, the ferric reducing-antioxidant power (FRAP) and the anti-DPPH free radical assays. According to FRAP and the anti-DPPH assays, genistin, daidzein, glycitin, malonyl glycitin and malonyl genistein had similar antioxidant activities to their corresponding aglycones whereas the antioxidant power of these glycosides was weaker than their corresponding aglycones, when the LDL oxidation assay was applied.

Since skin care research demonstrated the ability of isoflavones in reducing damage to the skin from the UVB rays, Chiu et al. (2009) performed *in vitro* and *in vivo* studies on the anti-photoaging effects of an isoflavone extract from soybean cake. The *in vitro* effects were determined in Human immortalized keratinocytes (HaCaT cells) irradiated with UVB. In the *in vivo* study, six-week old male ICR-Foxn/nu mice were irradiated with UVB and epidermal thickness, catalase, expressions of cyclooxygenase-2, and proliferating cell nuclear antigen were measured in order to evaluate the antioxidant and anti-inflammatory effects of the isoflavone extract. The results of the *in vitro* studies showed that the UVB-induced HaCaT cell death decreased in presence of the isoflavone extract. The in vivo studies demonstrated that the topical application of isoflavone extract before UVB irradiation decreased the epidermal thickness and the expressions of cyclooxygenase-2, and proliferating cell nuclear antigen and increased catalase concentration.

According to Record et al. (1995), genistein is an effective hydrogen peroxide scavenger but exhibited a lower antioxidant capacity against other peroxidative systems.

Rimbach et al. (2003) measured the antioxidant activity, free radical-scavenging properties and selected cellular effects of the isoflavone metabolites equol, 8-hydroxydaidzein, O-desmethylangiolensin, and 1,3,5 trihydroxybenzene in comparison with their parent genistein and daidzein. They found that 8-hydroxydaidzein was the most potent scavenger of hydroxyl and superoxide anion radicals. Isoflavone metabolites exhibited higher antioxidant activity than parent compounds in FRAP and TEAC assays but genistein showed the highest potency, followed by equol and daidzein for the suppression of nitric oxide production by activated macrophages.

Rüfer & Kulling (2006) investigate the antioxidant activities of six naturally occurring isoflavones and their corresponding oxidative and bacterial metabolites by means of the oxygen radical absorbance capacity assay, the in vitro oxidation of low density lipoproteins with the conjugated diene and the thiobarbituric acid reacting substances formation. The oxidative metabolites of genistein and daidzein as well as equol exhibited the highest antioxidant activities in all three assays. Formononetin, the 4'-*O*-methylether of daidzein, showed the lowest antioxidant property.

According to Kampkötter et al. (2008), none of the main isoflavone compounds showed prominent antioxidant effects in cell-free assay systems (trolox equivalent antioxidant capacity assay and 2,2-diphenyl-1-picrylhydrazyl assay). Authors concluded that daidzein up-regulated the antioxidant enzyme catalase and, furthermore, the antioxidant effects of this isoflavonone was also due to formation of the antioxidant metabolites 6-OH-daidzein and 3′-OH-daidzein by intestinal bacteria.

Akitha Devi et al. (2009) measured the DPPH radical-scavenging capacity of isoflavone extracts from soybeans and soyproducts founding that it correlated well with the total isoflavon content. Antioxidant potential was more pronounced in soybean seed extracts than in processed products. Amongst the soy products, soy sauce showed high antioxidant activity, in spite of low levels of Total phenolics and total isoflavones, probably due to the presence of other antioxidant phytochemicals as preservatives.

Fritz et al. (2003) evaluated the *in vivo* antioxidant activity of soy isoflavones in human subjects determined by the urinary excretion of secondary lipid peroxidation products. A soy protein isolate was added to the diet of ten healty women provided 3 levels of isoflavones: 0.15, 1.01, and 2.01 mg/kg body weight. Results show that levels the individual urinary nonpolar compounds were significantly lower due to consumption of isoflavones, especially at the higher level.

Isoflavones (genistein and daidezin, in particular) act as noncompetitive inhibitors of lipoxygenase since they reduce active state iron to ferrous state and prevent the activation of the resting enzyme (Mahesha et al., 2007).

Germination of lupin seeds was found able to produced significant increases of flavonoids and non-flavonoid phenolic compounds and an increase in the antioxidant activity measured through the DPPH assay was also observed as a consequence of the process (Dueñas et al., 2009).

Ishihara et al. (2007) determined the antioxidant capacity of commercial and homemade soymilk and tofu using a modified 2,2-azino-bis-(3-ethylbenzthiazoline-6-sulfonic acid) method. The results showed that homemade soymilk made by the extended boiling method had the highest antioxidant activity and that there was a positive correlation between total isoflavone, aglycone conjugates and antioxidant capacity. Rekha & Vijayalakshmi (2008) investigated the antioxidant activity of soymilk fermented with probiotic yeast (Saccharomyces boulardii) and lactic acid bacteria (Lactobacillus acidophilus B4496, Lactobacillus bulgaricus CFR 2028, Lactobacillus casei B1922, Lactobacillus plantarum B4495, and Lactobacillus helviticus B4526) by means of 1,1-diphenyl-2-picrylhydrazyl free radical scavenging assay, inhibition of ascorbate autoxidation, and measurement of reducing activity. The activities varied with the starters used but, nevertheless, were significantly higher than those found in unfermented soymilk. The increase of antioxidant activity corresponded to increase in aglycone content and decreases in glucosides.

Kwak et al. (2007) measured the antioxidant activity of Chungkookjang (a fermented soybean paste) by in vitro (lipid peroxidation inhibition and 1,1-diphenyl-2-picrylhydrazyl radical scavenging abilities) and in vivo (hepatic superoxide dismutase activity, hepatic thiobarbituric acid-reactive substance, and catalase activity) assays. Authors found that Chungkookjang had a stronger antioxidant activity than unfermented steamed soybeans, probably due to its increased aglycones and malonylglycosides. According to Kim et al. (2008), the 100% ethanol extract of Chungkookjang inhibited the generation of 1,1-diphenyl-2-picryl hydrazin radicals, had an inhibitory effect on LDL oxidation, inhibited H_2O_2-induced DNA damage from NIH/3T3 fibroblasts, and showed the cytoprotective effects against H_2O_2-induced cell death. The in vivo trials also demonstrated that an oral administration of Chungkookjang inhibited the formation of malondialdehyde, the damage of DNA, and the formation of micronucleated reticulocytes in $KBrO_3$-treated mice.

According to Park et al. (2008), the position of hydroxyl substituents at the aromatic ring of isoflavone plays an important role in radical scavenging effect. They applied the DPPH assay in order to evaluate the antioxidant activity of three trihydroxyisoflavones, namely 7,8,4'-trihydroxyisoflavone, 7,3',4'-trihydroxyisoflavone, and 6,7,4'-trihydroxyisoflavone, isolated from 5-year-old Doenjang (a Korean fermented soypaste) in comparison with other known isoflavones (daidzein, glycitein, genistein). The results showed that the trihydroxyisoflavones showed stronger DPPH radical scavenging activity than daidzein, glycitein, and genistein. 7,8,4'-trihydroxyisoflavone and 7,3',4'-trihydroxyisoflavone ahowed inhibition of xanthine oxidase and superoxide scavenging effect higher than 6,7,4'-trihydroxyisoflavone, daidzein, glycitein, and genistein.

Wang et al. (2008) measured the in vitro and in vivo antioxidant activities of aqueous Douchi extracts. They applied assays such as the □□□-diphenyl-□-pricrylhydrazyl (DPPH) and 2,20-azino-bis-(3-ethylbenzthiazoline-6-sulfonic acid) (ABTS) in order to measure the radical scavenging activity and the chelating ability of ferrous ions. They also studied the antioxidant enzymatic activities of extracts in cholesterol-fed rats, an index of lipid peroxidation (thiobarbituric acid reactive substances (TBARS)), and observed the hepatic tissue ultramicrostructure under transmission electron microscope (TEM). Their results showed that superoxide dismutase activities in liver and kidney, catalase activity in liver, and glutathione peroxidase activity in kidney increased significantly compared with the negative control group. TBARS in liver and kidney of extracts groups decreased and less fatty degeneration in hepatocytes was found.

In their study on soy-pasta, Baiano et al. (2009) found that cooking induced significant decreases of the antioxidant activity in almost all soy-spaghetti. This result appears well correlated with the decrease of the total isoflavone content and the detection of antioxidant activity in the cooking water.

Conclusion

Current evidence from a great number of studies suggests that variety, operations such as germination, fermentation, cooking, pasta-making, bread-making, chemical hydrolysis, enzymatic hydrolysis and processing conditions (temperature, time, pH, ionic strength), storage induce changes in isoflavone content (due to dilution of soy protein with other ingredients, losses in by-products, and severe treatments) and profile (hydrolysis of β-glucosides, conversion of malonyl glucosides to acetyl glucosides and β-glucosides). These changes affect bioavalability and bioactivity of soy isoflavones and, thus, their beneficial quality. Food science needs to further investigate on the modulation of the processing parameters in order to increase the isoflavone retention, especially of the aglycone fraction and, consequently increase the isoflavone intake. Since the loss of noticeable amounts of isoflavone in soy by-products, food industry should think them as interesting source of these phytochemicals to be used as food supplements and additives.

References

Adlercreutz, H., Honjo, H., Higashi, A. Fotsis, T., Hamalainen, E., Hasegawa, T. & Okada, H. (1991). Urinary excretion of lignans and isoflavonoid phytoestrogens in Japanese men and women consuming a traditional Japanese diet. *American Journal of Clinical Nutrition*, *54*, 1093-1100.

Aguiar, C. L., Baptista, A. S., Walder, J. M. M., Tsai, S. M., Carrão-Panizzi, M. C. & Kitajima, E. W. (2009). Changes in isoflavone profiles of soybean treated with gamma irradiation. International *Journal of Food Sciences and Nutrition*, *60,* 387-394.

Ahmad, S. & Pathak, D. K. (2000). Nutritional changes in soybean during germination. *Journal of Food Science and Technology India*, *37,* 665-666.

Akitha Devi, M. K., Gondi, M., Sakthivelu, G., Giridhar, P., Rajasekaran, T. & Ravishankar, G. A. (2009). Functional attributes of soybean seeds and products, with reference to isoflavone content and antioxidant activity *Food Chemistry*, *114,* 771-776.

Anderson, J. W. & Hoie, L. H. (2005). Weight loss and lipid changes with low-energy diets: comparator study of milk-based versus soy-based liquid meal replacement interventions. *Journal of the American College of Nutrition*, *24,* 210-216.

Anderson, J. W., Fuller, J., Patterson, K., Blair, R. & Tabor, A. (2007). Soy compared to casein meal replacement shakes with energy-restricted diets for obese women: randomized controlled trial. *Metabolism*, *56,* 280-288.

Arora, A., Nair, M. G. & Strasburg, G. M. (1998). Structure-activity relationship for antioxidant activities of a series of flavonoids in a liposomal system. *Free Radical Biology & Medicine*, *24,* 1355-1363.

Atkinson, C., Frankenfeld, C. L., Lampe, J. W. (2005). Gut bacterial metabolism of the soy isoflavone daidzein: exploring the relevance to human health. *Experimental Biology and Medicine (Maywood)*, *230,* 155-170.

Baiano, A., Terracone, C., Gambacorta, G. & La Notte, E. (2009). Evaluation of isoflavone content and antioxidant activity of soy-wheat pasta. *International Journal of Food Science and Technology*, *44,* 1304-1313.

Balk, E., Chung, M., Chew, P., Ip, S., Raman, G., Kupelnick, B., Tatsioni, A., Sun, Y., Wolk, B., DeVine, D. & Lau, J. (2005). *Effects of Soy on Health Outcomes. Evidence Report/Technology Assessment No. 126.* (Prepared by Tufts-New England Medical Center Evidence-based Practice Center under Contract No. 290-02-0022.) AHRQ Publication No. 05-E024-2. Rockville, MD: Agency for Healthcare Research and Quality.

Beaglehole, R. (1990). International trends in coronary artery disease mortality, morbidity, and risk factors. *Epidemiological Review*, *12,* 1-15.

Caldwell, C. R., Britz, S. J. & Mirecki, R. M. (2005). Effect of temperature, elevated carbon dioxide, and drought during seed development on the isoflavone content of dwarf soybean [Glycine max (L.) Merrill] grown in controlled environments. Journal of *Agricultural and Food Chemistry*, *53,* 1125-1129.

Cederroth, C. R. & Nef, S. (2009). Soy, phytoestrogens and metabolism: A review. *Molecular and Cellular Endocrinology*, *304,* 30-42.

Chen, Z., Zheng, W., Custer, L. J., Dai, Q., Shu, X. O., Jin, F. & Franke, A. A. (1999). Usual dietary consumption of soy foods and its correlation with the excretion rate of isoflavonoids in overnight urine samples among Chinese women in Shanghai. *Nutrition and Cancer*, *33,* 82-87.

Chiarello, M. D., Le Guerroe, J. L., Chagas, C. M. S., Franco, O. L., Bianchini, E. & Joao, M. J. (2006). Influence of heat treatment and grain germination on the isoflavone profile of soy milk. *Journal of Food Biochemistry*, *30,* 234-247.

Chien, H. L., Huang, H. Y. & Chou, C. C. (2006). Transformation of isoflavone phytoestrogens during the fermentation of soymilk with lactic acid bacteria and bifidobacteria. *Food Microbiology*, *23,* 772-778.

Chiu, T. M., Huang, C. C., Lin, T. J., Fang, J. Y., Wu, N. L. & Hung, C. F. (2009). In vitro and in vivo anti-photoaging effects of an isoflavone extract from soybean cake. Journal of Ethnopharmacology, *126,* 108-113.

Choi, Y. B., Kim, K. S. & Rhee, J. S. (2002). Hydrolysis of soybean isoflavone glucosides by lactic acid bacteria. *Biotechnology Letters*, *24,* 2113-2116.

Chun, J., Kim, J. S. & Kim J. H. (2008). Enrichment of isoflavone aglycones in soymilk by fermentation with single and mixed cultures of *Streptococcus infantarius* 12 and *Weissella* sp. 4. *Food Chemistry*, *109,* 278-284.

Chun, J., Kim, J. S. & Kim, J., H. (2008). Enrichment of isoflavone aglycones in soymilk by fermentation with single and mixed cultures of *Streptococcus infantarius* 12 and *Weissella* sp. 4. *Food Chemistry*, *109,* 278-284.

Cooke, P. S. & Naaz, A. (2004). Role of estrogens in adipocyte development and function. *Experimental Biology and Medicine*, *229,* 1127-1135.

Coward, L., Smith, M., Kirk, M. & Barnes, S. (1998). Chemical modification of isoflavones in soyfoods during cooking and processing. *American Journal of Clinical Nutrition*, *68*, (suppl), 1486-1491.

De Kleijn, M. J. J., van der Schouw, Y. T., Wilson, P. W. F., Adlercreutz, H., Mazur, W., Grobbee, D. E. & Jacques, P. F. (2001). Intake of dietary phytoestrogens is low in postmenopausal women in the United States: The Framnigham Study. *Journal of Nutrition*, *131,* 1826-1832.

De Oliveira, M. R. R., Mandarino, J. M. G. & del Mastro, N. L. (2009). *The effects of gamma irradiation on soybean isoflavones contents.* International Nuclear Atlantic Conference - INAC 2009, Rio de Janeiro,RJ, Brazil, September 27 to October 2, 2009.

Dewell, A. & Hollenbeck, C. B. (2003). Soy and isoflavones: a review of the potential role of phytoestrogens in the management of hypercholesterolemia. *Menopause Management*, July/August, 10-14.

Dixon, R. A., & Ferreira, D. (2002). *Genistein. Phytochemistry*, *60,* 205-211.

Dueñas, M., Hernández, T., Estrella, I. & Fernández, D. (2009). Germination as a process to increase the polyphenol content and antioxidant activity of lupin seeds (*Lupinus angustifolius* L.). *Food Chemistry*, *117,* 599-607.evidence. *Journal of Clinical Endocrinology and Metabolism*, *83,* 2223-2235.

Ferretti, G., Bacchetti, T., Menanno, F. & Curatola, G. (2004). Effect of genistein against copper-induced lipid peroxidation of human high density lipoproteins (HDL). *Atherosclerosis*, *172,* 55-61.

Fort, P., Moses, N. & Fasano, M. (1990). Breast and soy-formula feedings in early infancy and the prevalence of autoimmune thyroid disease in children. *Journal of American College of Nutrition*, *9,* 164-67.

French, M. R., Thompson, L. U. & Hawker, G. A. (2007). Validation of a phytoestrogen food frequency quesrionnaire with urinary concentrations of isoflavones and lignan metabolites in premenopausal women. *Journal of the American College of Nutrition*, *26,* 76-82.

Fritz, K. L., Seppanen, C. M., Kurzer, M. S. & Csallany, A. S. (2003). The in vivo antioxidant activity of soybean isoflavones in human subjects. Nutrition Research, 479-487.

Godsland, I. F. (2005). Oestrogens and insulin secretion. *Diabetologia*, *48,* 2213-2220.

Hall, W. L., Vafeiadou, K., Hallund, J., Bugel, S., Reimann, M., Koebnick, C., Zunft, H. J., Ferrari, M., Branca, F., Dadd, T., Talbot, D., Powell, J., Minihane, A. M., Cassidy, A., Nilsson, M., Dahlman-Wright, K., Gustafsson, J. A. & Williams, C.M. (2006). Soy-isoflavone-enriched foods and markers of lipid and glucose metabolism in postmenopausalwomen: interactions with genotype and equol production. American *Journal of Clinical Nutrition*, *83,* 592-600.

Haron, H., Ismail, A., Azlan, A., Shahar, S. & Su Peng, L. (2009). Daidzein and genestein contents in tempeh and selected soy products. *Food Chemistry*, *115,* 1350-1356.

Horn-Ross, P. L., John, E. M., Lee, M., Stewart, S. L., Koo, J., Sakoda, L. C., Shiau, A. C., Goldstein, J., Davis, P. & Perez-Stable, E. J. (2001). Phytoestrogen consumption and breast cancer risk in a multiethnic population: the Bay Area Breast Cancer Study. *American Journal of Epidemiology*, *154,* 434-441.

Horri, K., Adachi, T., Matsuda, T., Tanaka, T., Sahara, H., Shibasaki, S., Ogino, C., Hata, Y., Ueda, M. & Kondo, A. (2009). Improvement of isoflavone aglycones production using β-glucosidase secretory produced in recombinant Aspergillus oryzae. *Journal of Molecular Catalysis B. Enzimatic*, *59,* 297-301.

Howitz, K. T. & Sinclair, D. A. (2008). Xenohormesis: sensing the chemical cues of other species. *Cell*, *133,* 387-391.

Hubert, J., Berger, M., Nepveu, F. Paul, F. & Daydé, J. (2008). Effects of fermentation on the phytochemical composition and antioxidant properties of soy germ. *Food Chemistry*, *109,* 709-721.

Ikeda, Y., Iki, M., Morita, A., Kajita, E., Kagamimori, S., Kagawa, Y. & Yoneshima, H. (2006). Intake of fermented soybeans, natto, is associated with reduced bone loss in postmenopausal women: Japanese Population-BasedOsteoporosis (JPOS) Study. *Journal of Nutrition, 136,* 1323-1328.

Ishihara, M., Singh, H., Chung, G. & Tam, C. (2007) Content, composition and antioxidant activity of isoflavones in commercial and homemade soymilk and tofu. *Journal of the Science of Food and Agriculture, 87,* 2844-2852.

Izumi, T., Piskula, M. K., Osawa, S., Obata, A., Tobe, K., Saito, M., Kataoba, S., Kubota, Y. & Kikuchi, M. (2000). Soy isoflavone aglycones are absorbed faster and in higher amounts than their glucosides in humans. *Journal of Nutrition, 130,* 1695-1699.

Jackson, C. J. C., Dini, J. P., Lavandier, C., Rupasinghe, H. P. V., Faulkner, H., Poysa, V., Buzzell, D. & DeGrandis, S. (2002). Effects of processing on the content and composition of isoflavones during manufacturing of soy beverage and tofu. *Process Biochemistry, 37,* 1117-1123.

Jung, S., Murphy, P. A. & Sala, I. (2008). Isoflavone profiles of soymilk as affected by high-pressure treatments of soymilk and soybeans. *Food Chemistry, 111,* 592-598.

Kampkötter, A., Chovolou, Y., Kulawik, A., Röhrdanz, E., Weber, N., Proksch, P. & Wätjen, W. (2008). Isoflavone daidzein possesses no antioxidant activities in cell-free assays but induces the antioxidant enzyme catalase. *Nutrition Research, 28,* 620-628.

Kao, T. H., Lu, Y. F., Hsieh, H. C. & Chen, B. H. (2004). Stability of isoflavone glucosides during processing of soymilk and tofu. *Food Research International, 37,* 891-900.

Kim, J. J., Kim, S. H., Hahn, S. J. & Chung, I. M. (2005). Changing soybean isoflavone composition and concentrations under two different storage conditions over three years. *Food Research Intenational, 38,* 435-444.

Kim, N. Y., Song, E. J., Kwon, D. Y., Kim, H. P., Heo, M. Y. (2008). Antioxidant and antigenotoxic activities of Korean fermented soybean. *Food and Chemical Toxicology, 46,* 1184-1189.

King, R. A. & Bursill, D. B. (1998). Plasma and urinary kinetics of the isoflavones daidzein and genistein after a single soy meal in humans. *American Journal of Clinical Nutrition, 67,* 867-872.

Kirk P., Patterson, R. E. & Lampe, J. (1999). Development of a soy food frequency questionnaire to estimate isoflavone consumption in US adults. *Journal of American Dietetic Association, 99,* 558-563.

Kirk, P., Patterson, R. E. & Lampe, J. (1999). Development of a soy food frequency questionnaire to estimate isoflavone consumption in US adults. *Journal of the American Dietetic Association, 99,* 558-563.

Kris-Etherton, P. M. & Hecker, K. D. (2002). Bioactive compounds in foods: Their role in the prevention of cardiovascular disease and cancer. *American Journal of Medicine, 113,* 71-88.

Kuhnle, G. G. C., Dell'Aquila, C., Runswick, S. A., Bingham, S. A. (2009). Variability of phytoestrogen content in foods from different sources. Food Chemistry, *113,* 1184-1187.

Kwak, C. S., Lee, M. S. & Park, S. C. (2007). Higher antioxidant properties of Chungkookjang, a fermented soybean paste, may be due to increased aglycone and malonylglycoside isoflavone during fermentation. *Nutrition Research, 27,* 719-727.

Lee, C. H., Yang, L., Xu, J. Z., Yeung, S. Y. V., Huang, Y. & Chen, Z. Y. (2005). Relative antioxidant activity of soybean isoflavones and their glycosides. *Food Chemistry, 90,* 735-741.

Lee, C. H., Yang, L., Xu, J. Z., Yeung, S. Y., V., Huang, Y. & Chen, Z. Y. (2005). Relative antioxidant activity of soybean isoflavones and their glycosides. *Food Chemistry, 90,* 735-741.

Lee, S. W. & Lee, J. H. (2008). Effects of oven-drying, roasting, and explosive puffing process on isoflavone distributions in soybeans. *Food Chemistry, 112,* 316-320.

Li, Z., Hong, K., Saltsman, P., DeShields, S., Bellman, M., Thames, G., Liu, Y.,Wang, H. J., Elashoff, R. & Heber, D. (2005). Long-term efficacy of soy-based meal replacements vs an individualized diet plan in obese type II DM patients: relative effects on weight loss, metabolic parameters, and C-reactive protein. *European Journal of Clinical Nutrition, 59,* 411-418.

Lin-Jun, Y., Li-Te, L., Zai-Gui, L., Tatsumi, E. & Saito, M. (2004). Changes is isoflavone contents and composition of sufu (fermented tofu) during manufacturing. *Food Chemistry, 87,* 587-592.

Lopes Barbosa, A. C., Lajolo, F. M. & Genovese, M. I. (2006). Influence of temperature, pH and ionic strength on the production of isoflavone-rich soy protein isolates *Food Chemistry, 98,* 757-766.

Mahesha, H. G., Singh, S. A. & Rao, A. G. A. (2007). Inhibition of lipoxygenase by soy isoflavones: Evidence of isoflavones as redox inhibitors. *Archives of Biochemistry and Biophysics, 461,* 176-185.

Mahungu, S. M., Diaz-Mercado, S., Li, J., Schwenk, M., Singletary, K. & Faller, J. (1999). Stability of isoflavones during extrusion processing of corn/soy mixture. *Food Chemistry, 47,* 279-284.

Miksicek, R. J. (1994). Interaction of naturally occurring nonsteroidal estrogens with expressed recombinant human estrogen receptor. *Journal of Steroid Biochemistry and Molecular Biology, 49,* 153-60.

Morandi, S., D'Agostina, A., Ferrario, F. & Arnoldi, A. (2005). Isoflavone content of Italian soy food products and daily intakes of some specific classes of consumers. *European Food Research and Technology, 221,* 84-91.

Morandi, S., D'Agostina, A., Ferrario, F. & Arnoldi, A. (2005). Isoflavone content of Italian soy food products and daily intakes of some specific classes of consumers. *European Food Researches and Technology, 221,* 84-91.

Murphy, P. A., Song, T. T., Buseman, G., Barua, K., Beecher, G. R., Trainer, D. & Holden, J. (1999). Isoflavones in retail and institutional soy foods. *Journal of Agricultural and Food Chemistry, 47,* 2697-2704.

Nagata, C., Kabuto, M., Kurisu, Y. & Shimizu, H. (1997). Decreased serum estradiol concentration associated with high dietary intake of soy products in premenopausal *Japanese women. Nutr Cancer, 29,* 228-233

Nagata, C., Takatsuka, N., Inaba, S., Kawakami, N. & Shimizu, H. (1998). Effect of soy milk consumption on serum estrogen concentrations in premenopausal Japanese women. *Journal of National Cancer Institute, 90,* 1830-1835.

Omoni, A. O. & Aluko, R. E. (2005). Soybean foods and their benefits: *Potential mechanisms of action. Nutrition Reviews, 63,* 272-283.

Otieno, D. O., Ashton, J. F. & Shah, N. P. (2007). Role of microbial strain and storage temperatures in the degradation of isoflavone phytoestrogens in fermented soymilk with selected β-glucosidase producing Lactobacillus casei strains. *Food Research International*, *40,* 371-380.

Park, J. S., Park, H. Y., Kim, D. H., & Kim, H. K. (2008). ortho-Dihydroxyisoflavone derivatives from aged Doenjang (Korean fermented soypaste) and its radical scavenging activity Bioorganic & Medicinal Chemistry Letters, *18,* 5006-5009.

Pauchar-Menacho, L. M., Berhow, M. A., Mandarino, J. M. G., de Mejia, E. G. & Chang, Y. K. (2010). Optimisation of germination time and temperature on the concentration of bioactive compounds in Brazilian soybean cultivar BRS 133 using response surface methodology. *Food Chemistry*, *119,* 636-642.

Peters, N. K., Frost, J. W. & Long, S. R. (1986). A plant flavone, luteolin, induces expression of *Rhizobium meliloti* nodulation genes. *Science*, *233,* 977-980.

Prabhakaran, M. P. & Perera, C. O. (2006). Effect of extraction methods and UHT treatment conditions on the level of isoflavones during soymilk manufacture. *Food Chemistry*, *99,* 231-237.

Prabhakaran, M. P., Perera, C. O. & Valiyaveetil, S. (2006). Effect of different coagulants on the isoflavone levels and physical properties of prepared firm tofu. *Food Chemistry*, *99,* 492-499.

Record, I. R., Dreosti, I. E. & McInerney, J. K. (1995). The antioxidant activity of genistein in vitro. *Journal of Nutritional Biochemistry*, *6,* 481-485.

Rekha, C. R. & Vijayalakshmi, G. (2008). Biomolecules and nutritional quality of soymilk fermented with probiotic yeast and bacteria. *Applied Biochemistry and Biotechnology*, *151,* 452-463.

Ribeiro, M. L. L., Mandarino, J. M. G., Carrão-Panizzi, M. C. & de Oliveira, M. C. N. (2007). Isoflavone content and b-glucosidase activity in soybean cultivars of different maturity groups. *Journal of Food Composition and Analysis*, *20,* 19-24.

Richelle, M., Pridmore-Merten, S., Bodenstab, S., Enslen, M. & Offord, E. A. (2002). Hydrolysis of isoflavone glycosides to aglycones by β-glycosidase does not alter plasma and urine isoflavone pharmacokinetics in postmenopausal women. *Journal of Nutrition*, *132,* 2587-2592.

Riedl, K. M., Zhang, Y. C., Schwartz, S. J. & Vodovotz, Y. (2005). Optimizing dough proofing conditions to enhance isoflavone aglycones in soy bread. *Journal of Agricultural and Food Chemistry*, *53,* 8253-8258.

Rimbach, G., De Pascual, T. S., Ewins, B. A., Matsugo, S., Uchida, Y., Minihane, A. M., Turner, R., VafeiAdou, K. & Weinberg, P. D. (2003). Antioxidant and free radical scavenging activity of isoflavone metabolites. *Xenobiotica*, *33,* 913-925.

Rüfer, C. E. & Kulling, S. E. (2006). Antioxidant activity of isoflavones and their major metabolites using different in vitro assays. *Journal of Agricultural and Food Chemistry*, *54,* 2926-2931.

Russo, A., Cardile, V., Lombardo, L., Vanella, L. & Acquaviva, R. (2006). Genistin inhibits UV light-induced plasmid DNA damage and cell growth in human melanoma cells. The *Journal of Nutritional Biochemistry*, *17,* 103-108.

Setchell K. D. R. & Sidney J. C. (2006). Method of defining equol-producer status and its frequency among vegetarians. *Journal of Nutrition*, *136,* 2188-2193.

Setchell, K. D. R. & Cassidy, A. (1999). Dietary isoflavones: biological effects and relevance to human health. *Journal of Nutrition*, *129*, 758-677.

Setchell, K. D. R. (1998). Phytoestrogens: the biochemistry, physiology, and implications for human health of soy isoflavones. *American Journal of Clinical Nutrition*, *68,* 1333-1346.

Setchell, K. D. R., Brown, N. M., Zimmer-Nechemias, L., Brashear, W. T., Wolfe, B., Kirschner, A. S. & Heubi, J. E. (2002). Evidence for lack of absorption of soy isoflavone glycosides in humans, supporting the crucial role of intestinal metabolism for bioavailability. *American Journal of Clinical Nutrition*, *76,* 447-453.

Setchell, K. D., Brown, N. M., Desai, P., Zimmer-Nechemias, L., Wolfe, B. E., Brashear, W. T., Kirschner, A. S., Cassidy, A. & Heubi, J. E.. (2001). Bioavailability of pure isoflavones in healthy humans and analysis of commercial soy isoflavone supplements. *Journal of Nutrition*, *131*(4 Suppl), 1362-1375.

Setchell, K. D., Faughnan, M. S., Avades, T., Zimmer-Nechemias, L., Brown, N. M., Wolfe, B. E., Brashear, W. T., Desai, P., Oldfield, M. F., Botting, N. P. & Cassidy, A. (2003). Comparing the pharmacokinetics of daidzein and genistein with the use of 13C-labeled tracers in premenopausal women. *American Journal of Clinical Nutrition*, *77,* 411-419.

Shao, S., Duncan, A. M., Yang, R., Marcone, M. F., Rajcan, I. & Tsao, R. (2009). Tracking isoflavones: from soybean to soy flour, soy protein isolates to functional soy bread. *Journal fo the Functional Foods*, *1,* 119-127.

Subramanian, S., Stacey, G. & Yu, O. (2007). Distinct, crucial roles of flavonoids during legume nodulation. *Trends in Plant Science*, *12,* 282-285.

Terrence, L. G. (1991). Flavonoid and isoflavonoid distribution in developing soybean seedling tissues and in seed and root exudates. *Plant Physiology*, *95,* 594-603.

Tham, D. M., Gardner, C. D. & Haskell, W. L. (2008) *Clinical Review 97*, Potential health benefits of dietary phytoestrogens: A review of the clinical, epidemiological, and mechanistic

Toda, T., Sakamoto, A., Takayanagi, T. & Yokotsuka, K. (2001). Changes in isoflavone composition of soybean during soaking in water. *Food Science Technology Research*, *6,* 314-319.

Tsangalis, D., Ashton, J. F., Stojanovska, L., Wilcox, G. & Shah, N. P. (2004). Development of an isoflavone aglycone-enriched soymilk using soy germ, soy protein isolate and bifidobacteria. *Food Research International*, *37,* 301-312.

Umphress, S. T., Murphy, S. P., Franke, A. A., Custer, L. J. & Blitz, C. L. (2005). Isoflavone content of foods with soy additives. *Journal of Food Composition and Analysis*, *18,* 533-550.

Umphress, S. T., Murphy, S. P., Franke, A. A., Custer, L. J. & Blitz, C. L. (2005). Isoflavone content of foods with soy additives. *Journal of Food Composition and Analysis*, *18*, 533-550.

van Erp-Baart, M. A., Brants, H. A., Kiely, M., Mulligan, A., Turrini, A., Sermoneta, C., Kilkkinen, A. & Valsta, L. M. (2003). Isoflavone intake in four different European countries: the VENUS approach. *British Journal of Nutrition*, *89* (Suppl 1), 25-30.

Wang, D., Wang, L. J., Zhu, F., Zhu, J., Chen, X. D., Zou, L., Saito, M. & Li, L. (2008). In vitro and in vivo studies on the antioxidant activities of the aqueous extracts of Douchi (a traditional Chinese salt-fermented soybean food). *Food Chemistry*, *107*, 1421-1428.

Wang, H. & Murphy, P. A. (1994). Isoflavone content in commercial soybean foods. *Journal of Agricultural Food Chemistry*, *42*, 1666-1673.

Wang, H. & Murphy, P. A. (1996). Mass balance study of isoflavones during soybean processing. *Journal of Agricultural Food Chemistry, 44*, 2377-2383.

Wang, H. J. & Murphy, P. A. (1994). Isoflavone composition of American and Japanese soybeans in Iowa: effects of variety, crop year, and location. *Journal of Agricultural and Food Chemistry, 42*, 1674-1677.

Wang, L. J., Yin, L. J, Li, D., Zou, L. & Saito M. (2007). Influences of processing and NaCl supplementation on isoflavone contents and composition during douchi manufacturing. *Food Chemistry, 101*, 1247-1253.

Wardhani, D. H., Vázquez, J. A. & Pandiella, S. S. (2008). Kinetics of daidzin and genistin transformations and water absorption during soybean soaking at different temperatures. *Food Chemistry, 111*, 13-19.

Watanabe, S., Yamaguchi, M., Sobue, T., Takahashi, T., Miura, T., Arai, Y., Mazur, W., Wahala, K. & Adlercreutz, H. (1998). Pharmacokinetics of soybean isoflavones in plasma, urine and feces of men after ingestion of 60 g baked soybean powder (kinako). *Journal of Nutrition, 128*, 1710-1715.

Wu, J. & Muir, A. D. (2010). Isoflavone during protease hydrolysis of defatted soybean meal. *Food Chemistry, 118*, 328-332.

Xu, X., Keecha, S. H., Wang, H. J., Murphy, P. A. & Hendrich, S. (1995). Bioavailability of soybean isoflavones depends upon gut microflora in women. *Journal of Nutrition, 125*, 2307-2315.

Xu, X., Wang, H. J., Murphy, P. A. & Hendrich, S. (2000). Neither background diet nor type of soy food affects short-term isoflavone bioavailability in women. *Journal of Nutrition, 130*, 798-801.

Yang, H. & Zhang, L. (2009). Changes in some components of soymilk during fermentation with the basidiomycete Ganoderma lucidum. *Food Chemistry, 112,* 1-5.

Yang, S., Wang, L., Yan, Q., Jiang, Z. & Li, L. (2009). Hydrolysis of soybean isoflavone glycosides by a thermostable β-glucosidase from Paecilomyces thermophila. *Food Chemistry, 115*, 1247-1252.

Yu, J., Liu, A. Y & Wang, X. G. (2007). Preparation of isoflavones enriched soy protein isolate from defatted soy hypocotyls by supercritical CO_2. LWT-Food *Science and Technology, 40*, 800-806.

Yuan, J. P., Liu, Y. B., Peng, J., Wang, J. H. & Liu, X. (2009). Changes of isoflavone profile in the hypocotyls and cotyledons of soybeans during dry heating and germination. *Journal of Agricultural and Food Chemistry, 57*, 9002-9010.

Zhang, Y. C. (2004). *Physicochemcal properties and isoflavon content of bread made with soy.* Dissertation presented in partial fulfillment of the requirements for the Degree Doctor of Philosophy in the Graduate School of the Ohio State University, available at http://etd.ohiolink.edu/etd/send-pdf.cgi/Zhang%20Yu%20Chu.pdf?acc_num=osu 108744-4553.

Zhang, Y. C., Lee, J. H., Vodovotz, Y. & Schwartz, S. (2004). Changes in distribution of isoflavones and β-glucosidase activity during soy bread proofing and baking. *Cereal Chemistry, 81*, 741-745.

Zhu, D., Hettiarachchy, N. S., Horax, R. & Chen, P. (2005). *Plant Foods for Human Nutrition, 60*, 147-151.

In: Isoflavones Biosynthesis, Occurence and Health Effects ISBN: 978-1-61728-113-6
Editor: Melanie J. Thompson, pp. 137-161 © 2010 Nova Science Publishers, Inc.

Chapter IV

Dietary Isoflavones and Intestinal Microbiota: Metabolism and Transformation into Bioactive Compounds

Maddalena Rossi, Alberto Amaretti, Lucia Roncaglia, Alan Leonardi and Stefano Raimondi*
Dept. Chemistry, University of Modena and Reggio Emilia, Italy

Abstract

Edible plants provide the human with hundreds of non-nutritional phytochemicals which are recognized as beneficial, such as isoflavones. Likewise other polyphenols, isoflavones may undergo extensive transformations during passage through human digestive tract, especially in the colon, where members of the complex commensal microbiota are capable to carry out synergistically a broad range of metabolic transformations affecting the fate and the biological activity of phytochemicals. Diverse bacterial species occurring in the large intestine hydrolyze the glucose conjugated forms of isoflavones, releasing the corresponding aglycones, which may undergo further microbial conversions (especially reductions) giving rise to a wide spectrum of isoflavone-derived compounds. The end-products of microbial transformations are subjected to substantial person-to-person variation, reflecting the impact of the colonic microbiota, since intestinal bacteria may greatly increase or compromise the biological activity of dietary isoflavones. In fact, certain reduction products (e.g. S-equol) possess superior antioxidant and estrogen-like activities, while others are precursors for C-ring cleavage and lead to isoflavone degradation. Extensive research has been performed to characterize the intestinal bacteria which are responsible for isoflavones transformations. In particular, many efforts are being carried out to identify single bacteria that convert the

* Corresponding author: Dept. of Chemistry, University of Modena and Reggio Emilia, via Campi 183, 41125, Modena, Italy, tel. +39-059-2055567, fax. +39-059-373543, E-mail: maddalena.rossi@unimore.it

major soy isoflavone daidzein into S-equol, in the perspective to enrich soy products with such a valuable isoflavone-derived metabolite. Besides, diverse bacteria which are capable of deglycosylation, demethylation, and various reduction reactions are increasingly being identified. Nonetheless, only bifidobacteria and lactobacilli, which produce β-glucosidase that hydrolyze isoflavone glycosides into the corresponding aglycones, are currently exploited to improve the biological activity of soymilk.

Introduction

Edible plants are dietary sources of hundreds of non-nutritional phytochemicals. These molecules are typically secondary metabolites and occur in small quantities in plant tissues and plant-derived foods. The major classes of bioactive phytochemicals are polyphenols (which include phenolic acids, flavonoids, stilbenes, lignans, and tannins), organosulphur compounds (which include glucosinolates and isothiocynates), carotenoids, plant sterols, monoterpenes, and alkaloids (Kris-Etherton et al. 2002). A number of beneficial activities, including antioxidant, antitumoral, and estrogenic-like properties, have been demonstrated for thousands of diverse phytochemicals through numerous epidemiologic, clinical, and experimental studies (Cederroth and Nef 2009, Crozier et al. 2009).

The level of bioactive phytochemicals within the body is largely determined by diverse phenomena, such as the digestive transformation of native compounds, the absorption in the intestine, the hepatic activity, and the biliary or urinary excretion (Scalbert and Williamson 2000, Manach et al. 2004). The phytochemicals which are not absorbed in the small intestine reach the colon, where they may undergo extensive biotransformation by the resident microbiota (Manach et al. 2004, Aura 2008, Selma et al 2009). The bacterial transformation may lead to the inactivation and/or degradation of phytochemicals or may cause the production of compounds with enhanced biological activity or bioavailability. Examples of the specific conversion of diverse molecules into bioactive metabolites accomplished by the microbiota are the conversion of lignans into enterolactone and enterodiol and the conversion of soy isoflavones into S-equol (Clavel et al. 2006a, Clavel et al. 2006b, Tsangalis et al 2007, Wang et al. 2007). During the course of absorption, phytochemicals are conjugated in the intestine and later in the liver, being subject to methylation, sulfation, and β-glucuronidation. Then, they can be secreted in the duodenum with the bile as hydrophilic conjugates. The colonic microbiota is involved in the enterohepatic recycling of phytochemicals. In fact, the microbial enzymes, especially β-glucuronidase, can deconjugate the excreted phytochemicals in the colon, where they can be reabsorbed, leading to a longer presence in the body (Manach et al. 2004, Zhang et al. 2007).

Likewise many other phytochemicals, isoflavones are converted by host enzymes and mostly by bacteria of the human intestinal microbiota. Bacterial transformations are essential for their absorption, bioavailability, and estrogenic properties. *In vivo* studies demonstrated that variations in the activity of isoflavones occur among different subjects, this interindividual variation being mainly attributed to differences in the composition of the gut microbiota (Xu et al. 1995, Zhang et al. 1999, Hur et al. 2002).

The mammalian gut is one of the most densely populated ecosystems on Earth. In particular, the human large intestine is colonized by a complex community composed largely of anaerobic bacteria, whose cell numbers exceed 10^{11} per gram of intestinal content. The adult human gastrointestinal tract contains all three domains of life (bacteria, archaea, and eukarya), but it is predominated by bacteria belonging to *Firmicutes*, *Bacteroidetes* and *Actinobacteria* (Eckburg et al. 2005) (Figure 1). *Firmicutes* are by far the most abundant and diverse group, and include the *Clostridia* and *Bacilli* class; *Bacteroidetes* are also present in high numbers. This vast microbiota fulfils its energy needs using, through hydrolytic and fermentative pathways, various types of substrates left undigested and unabsorbed in the small intestine. Oligo- and polysaccharides, proteins and peptides are broken down by bacterial enzymes to their oligomeric and/or monomeric components. Then, sugars and amino acids are fermented, yielding organic acids (such as lactic, propionic, and butyric acids), branched chain fatty acids (such as isobutyric, isovaleric, and 2-methylbutyric acids), H_2, CO_2, ammonia, amines and several other end-products. Short chain fatty acids (SCFA) are, from a nutritional point of view, the major fermentation products. Butyrate is of particular importance because it is the major energy source for colonocytes, the epithelial cells that line the colon. SCFA affect the metabolism, growth, and differentiation of colonocytes, influence the hepatic control of lipids and carbohydrates, and provide muscle, kidney, heart and brain with energy (O'Keefe 2008).

In the mutualistic relationship between mammals and commensal intestinal bacteria, the bacterial metabolism and the cross-talk between bacterial and animal cells have a major impact on the nutrition and the overall health status of the host. Microorganisms are provided with a broad spectrum of fermentable compounds and a temperature controlled anoxic environment, while the host gains the protection against infections, the modulation of the immune system and the supply of carbon, energy, vitamins, and bacterial-activated dietary metabolites, including isoflavones (Leser and Mølbak 2009, O'Keefe et al. 2009).

In this chapter, it is discussed the role of intestinal bacteria in transforming the major isoflavones occurring in diet. Attention is focused on the microbial hydrolysis of isoflavone glycoconjugates, on the reactions which transform the aglycones into molecules with enhanced biological activity or, conversely, into inactive compounds, and on the bacterial groups which are responsible for such transformations. Most information is presented for soybean isoflavones (daidzein, genistein, and glycitein and their corresponding glycoconjugates), which have received the greatest interest and for which plenty of scientific literature is available. Furthermore, what is known about the bacterial transformations of other less common isoflavones, such as puerarin and the red clover isoflavones formononetin and biochanin A is described.

Hydrolysis of Isoflavone Glycoconjugates

Even though the diet can provide the human with beneficial isoflavones, their bioavailability and biological activity are largely determined by the form in which they occur. In food, and especially in soy-based products, the aglycone forms of isoflavones (e.g. daidzein, genistein and glycitein) are generally less abundant than the respective glycoconjugates, which include glucosides (daidzin, genistin and glycitin), acetylglucosides (acetyldaidzin, acetylgenistin and acetylglycitin), and malonylglucosides (malonyldaidzin, malonylgenistin and malonylglycitin).

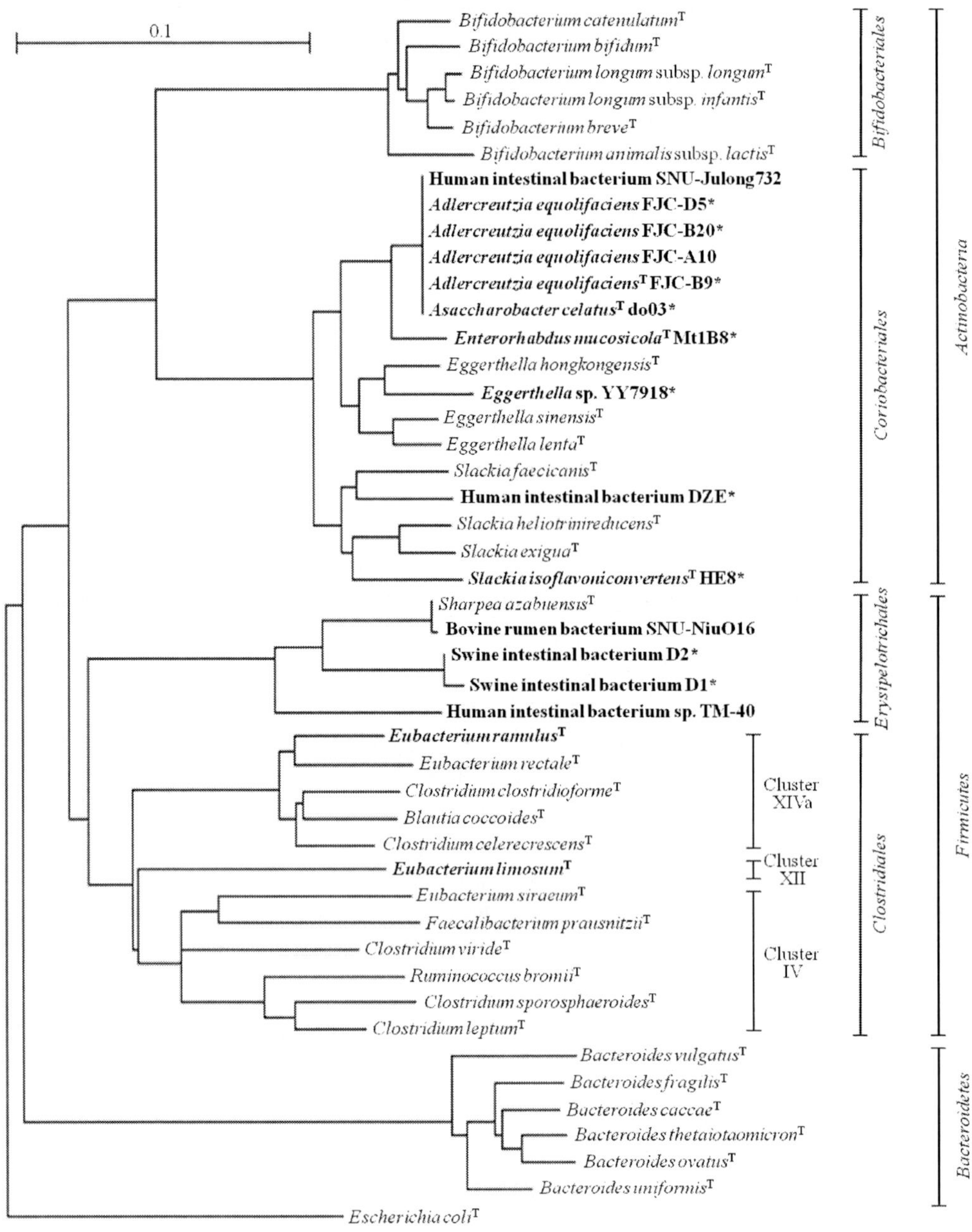

Figure 1. Phylogenetic tree showing representative species of the major taxa of the intestinal microbiota and, in bold, the bacteria which are known to be involved in isoflavones metabolism. The superscript * indicates bacteria which are capable of daidzein conversion into equol. The comparison is based on the 16S rRNA gene sequences, available from the Ribosomal Database Project. The superscript ^{T} indicates type strain bacteria.

The absorption efficiency of the conjugated forms is generally poor, causing the bioavailability of glucosides, acetylglucosides, and malonylglucosides to be low. Thus, activation of naturally occurring conjugated isoflavones, with the release of the bioactive aglycone, is necessary to promote the absorption and to enable further biotransformation by colonic microbiota.

Upon ingestion, small intestinal brush-border membrane enzymes and microbial β-glucosidases are involved in the removal of the glycosyl group. The hydrolysis releases the aglycones, which are readily absorbed and become bioactive. Glucosidases are common and widely distributed in nature. A major cytosolic β-glucosidase has been detected in a variety of mammalian tissues (McMahon et al. 1997). It consists of an enzyme with broad specificity β-glucosidase activity, capable of hydrolyzing a variety of aryl β-D-glycosides. However, the major contribution to activation of phytochemicals seems to be carried out by bacterial enzymes. Potentially, bacterial enzymes of the small intestine can hydrolyze all glycoconjugates. On the contrary, the enzymes of colonic bacterial exert their activity only on molecules which escape hydrolysis and absorption in the small intestine. In the colon, the microbiota produces a number of glycosyl hydrolases which act on a variety of glycosidic bonds, including those of the glycoconjugates of plant secondary metabolites.

A major part of gut anaerobic bacteria is charged with the task of breaking down undigested polysaccharides and fermenting the resulting monosaccharides to short-chain fatty acids (SCFA). To perform this metabolism, a vast number of enzymes such as β-glucuronidase, β-glucosidase, β-xylosidase, α-arabinosidase, β-galactosidase, is produced by intestinal bacteria (Scheline 1973). These enzymes can be located extracellularly, cell-bound or released into the lumen environment by cell lysis. Several studies investigated bulk enzyme activities in faecal samples. A high interindividual variation in enzymatic hydrolytic activities, including β-glucosidase, has been demonstrated in human faeces (McBain and Macfarlane 1998). These differences have been attributed to factors which are affected by the diet, such as the composition of the gut microbiota and the availability of carbon and energy sources. For instance, β-glucosidase activity increased in faeces following soy consumption by human volunteers (Weiserman et al. 2004).

In decreasing order of concentration, the bacteria belonging to the groups of *Bacteroidetes*, *Clostridium* cluster XIVa, *Clostridium* cluster IV, and *Actinobacteria* (mostly *Atopobium* and *Bifidobacterium*) dominate the colonic microbiota (Harmsen et al. 2002, Aminov et al., 2006, Mueller et al. 2006, Mariat et al. 2009). β-glucosidase activity has been found against the model substrate *p*-nitrophenyl β-D-glucopyranoside within members of all these major bacterial groups, mostly as a cell-bound activity (McBain and Macfarlane 1998, Nakamura et al. 2002, Dabek et al. 2008). Thus, it is conceivable that diverse bacteria may be involved in the hydrolysis of isoflavones glycoconjugates. Even though β-glucosidase seems to be widespread among the above bacterial groups, marked differences have been observed among the members of the same group.

Only few strains of *Bacteroides* have been screened for β-glucosidase and some of them exhibited high β-glucosidase activity, while other were negative to the assay. In a study, three strains of the genus *Bacteroides* (*B. ovatus*, *B. fragilis*, and *B. vulgatus*) were all positive for β-glucosidase, the highest activity being exhibited by *B. ovatus* (McBain and Macfarlane 1998). Diversely, β-glucosidase was not found in other two strains of *B. ovatus* and *B. vulgatus*, but only in one of *B. thetaiotaomicron* (Dabek et al. 2008). However, data concerning hydrolysis of glycosylated isoflavones by *Bacteroides* are lacking. It is possible that this bacterial group, which possess a major array of enzymes for sugar hydrolysis and can break down a large variety of glycosidic bonds, is involved in the release of isoflavone aglycones from glycoconjugates as well.

Within *Clostridiales*, β-glucosidase activity have been examined in 22 strains of cluster XIVa and 6 of cluster IV, including both human faecal isolates and reference strains (Dabek et al. 2008). Among the strains of cluster XIVa, β-glucosidase was present in 8 strains of *Roseburia* (*R. intestinalis*, *R. fecis*, *R. hominis*, *R. inulinivorans*), 3 of *Eubacterium rectale*, 2 of *Coprococcus*, 1 of *Ruminococcus obeum*, and 1 of *Butyrivibrio fibrisolvens*, while it was not found in 2 strains of *Eubacterium hallii*, 2 of *Coprococcus comes*, 1 of *Anaerostipes caccae*, and 2 unidentified isolates. Among the strains of cluster IV, only 1 strains of *Eubacterium siraeum* and 1 of *Ruminococcus* sp. produced β-glucosidase, while 1 strain of *Ruminococcus bromii* and 3 of *Faecalibacterium prausnitzii* did not (Dabek et al. 2008).

Most information about β-glucosidase production by human intestinal bacteria is available for the genus *Bifidobacterium*, since considerable interest has focused on the utilization of bifidobacteria to hydrolyze isoflavone glycoconjugates in fermented soy products. Likewise the other saccharolytic intestinal bacteria, bifidobacteria can produce a number of diverse glycosyl hydrolases which are necessary to hydrolyze oligo and polysaccharide into fermentable carbohydrates in the colon (Schell et al 2002). Many strains belonging to diverse bifidobacterial species have been selected for their ability to hydrolyze the glycosilated forms of soy isoflavones (Bae et al. 2000, Bae et al. 2002, Tsangalis et al. 2002, Bae et al. 2004, Cheng et al. 2004, Marotti et al. 2007, Dabek et al. 2008, Raimondi et al. 2009,). Coherently, several strains of *Bifidobacterium* have been demonstrated to produce β-glucosidase (Marotti et al. 2007, Raimondi et al. 2009) and many of them exhibited an enzymatic activity that was higher or at least comparable than that of most strains of *Bacteroidetes* and *Clostridiales* (Dabek et al 2008). Among bifidobacteria, low production or lack of β-glucosidase was observed only within the species *B. longum* (Marotti et al. 2007, Raimondi et al. 2009). The absence of β-glucosidase activity in a few strains was in agreement with the incapability of hydrolyzing daidzin (Raimondi et al. 2009).

In conclusion, currently available information do not allow to establish which bacterial group is the major responsible for deconjugation of isoflavones *in vivo*. *Bacteroidetes* and *Clostridiales* are numerically predominant by approximately one magnitude order over bifidobacteria (Harmsen et al. 2002). Conversely, β-glucosidase seems to be more widespread within the genus *Bifidobacterium*, which have been also reported to produce higher activity albeit they are less numerous.

Bacterial Transformations of Isoflavone Aglycones

Isoflavone aglycones are extensively transformed by intestinal bacteria into a variety of metabolites which may exert enhanced or decreased biological activity (Aura 2008, Selma et al. 2009, Crozier at al. 2009). The production of different end-products by intestinal bacteria is affected by differences in the structure of the parent aglycone. In fact isoflavones differ in the number and/or position of hydroxyl and methoxyl substituents of A and B rings which may promote certain types of reactions (Figure 2).

The major transformations occurring to all classes of isoflavones generally start with the hydrogenation of the activated double bond between C-2 and C-3 of the C-ring. This reduction generates dihydroisoflavones, such as dihydrodaidzein, dihydrogenistein, and dihydroglycitein (Figure 2). The involvement of the gut microbiota in the production of all these dihydroisoflavones have been extensively documented and several isolates of intestinal bacteria have been discovered to carry out this reaction (Hur et al. 2000, Wang et al. 2005a, Wang et al. 2005b, Tamura et al. 2007). Since germ-free rats excreted dihydrodaidzein and dihydrogenistein with urine, it is likely that the production of dihydroisoflavones can occur in the liver without any contribution of intestinal bacteria, presumably through a cytochrome P450 mediated reaction (Bowey et al. 2003). However, it is necessary that the hydrogenation is carried out by intestinal bacteria to make dihydroisoflavones available in the colon for further microbial transformation. Moreover, the hydrogenation of the double bond causes the appearance of a chiral center on C3 position of dihydroisoflavone, but a tautomerization reaction likely occurs to convert the R-dihydroisoflavone into the S-isomer (Wang et al. 2005b, Kim et al. 2009). Dihydroisoflavones constitute branch point of two divergent routes of bacterial transformation. They may undergo a reductive pathway yielding equol and equol-related compounds or, alternatively, the cleavage of C ring which, in certain cases, may lead to the fission of the molecules into two moieties (Figure 3, Figure 4, Figure 5).

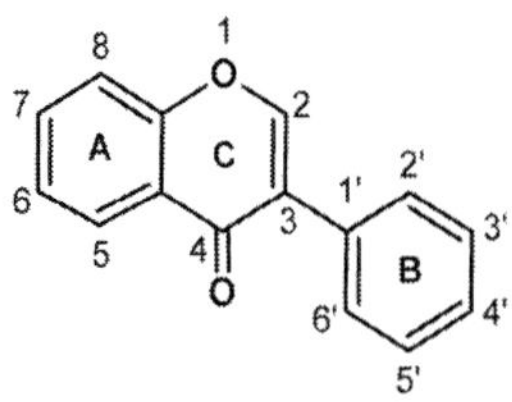

Isoflavones

7,4'-$(OH)_2$	Daidzein
5,7,4'-$(OH)_3$	Genistein
7-OH, 4'-OCH_3	Formononetin
5,7-$(OH)_2$, 4'-OCH_3	Biochanin A
5,4'-$(OH)_2$, 7-OCH_3	Prunetin
5,7-$(OH)_2$, 6,4'-$(OCH_3)_2$	Pratensein
7,3'-$(OH)_2$, 4'-OCH_3	Calycosin
7-OH, 3'-O-CH_2-O-4'	Pseudobaptigenin
7,4'-$(OH)_2$, 6-OCH_3	Glycitein

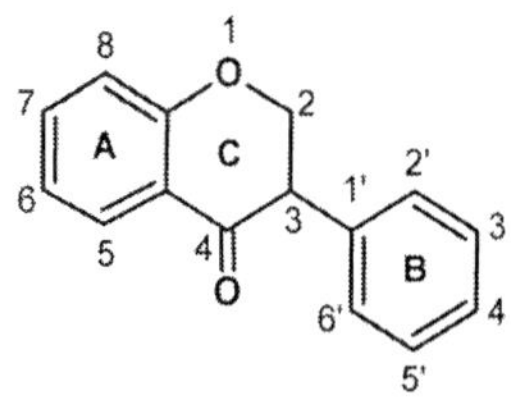

Dihydroisoflavones

7,4'-$(OH)_2$	Dihydrodaidzein (DHD)
5,7,4'-$(OH)_3$	Dihydrogenistein (DHG)
7-OH, 4'-OCH_3	Dihydroformononetin
5,7-$(OH)_2$, 4'-OCH_3	Dihydrobiochanin A

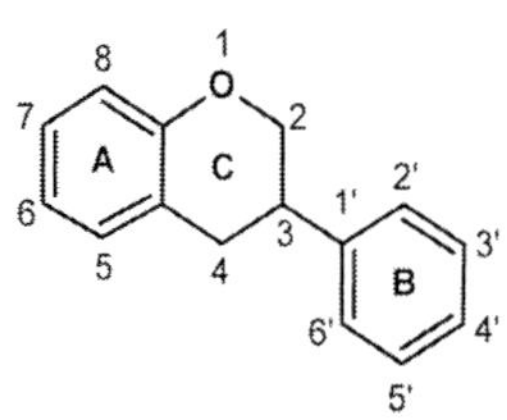

Isoflavans

7,4'-$(OH)_2$	Equol
7-OH, 4'-OCH_3	4'-*O*-Me-equol
4,7,4'-$(OH)_3$	4-OH-equol

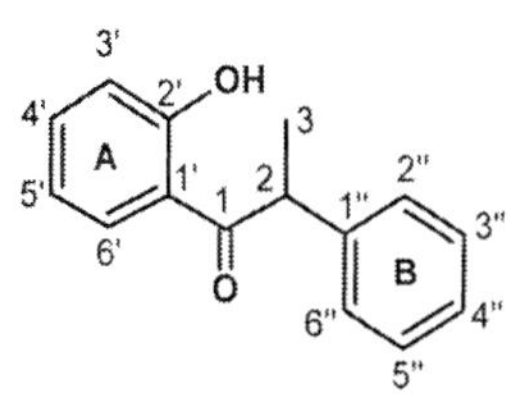

2'-OH-α-methyldeoxybenzoin

4',4''-$(OH)_2$	*O*-desmethylangolensin (*O*-DMA)
4',6',4''-$(OH)_3$	6'-OH-*O*-DMA
4',4''-$(OH)_2$,5'-OCH_3	5'-OCH_3-*O*-DMA

Figure 2. Chemical structures of the major dietary isoflavones and their bacterial metabolites

Production of equol-related compounds requires the reduction of the cheto-group in position 4 of the C-ring of isoflavones, yielding a compound related to 4-hydroxy-equol (Figure 3). This molecule possesses two chiral centers on positions C3 and C4, but only the *cis*-diastereoisomer of 4-hydroxy-equol, and specifically 3S, 4S-configuration, is selectively produced by the gut microbiota. (Setchell et al. 2005, Rüfer et al. 2006). The hydroxyl group on position C4 can be removed by intestinal bacteria, producing a reduced compound similar to equol. Therefore, S-equol is the sole enantiomer arising from daidzein (Setchell et al. 2005). The configuration of the chiral centers formed within the compounds which are produced from isoflavones other than daidzein (e.g. genistein or glycitein) has never been investigated, but it is conceivable that it is the same of daidzein metabolites.

The bacterial cleavage of C-ring of dihydroisoflavones always starts off with the break of the bond between the oxygen in the pyrone ring and the C-2 to yield a compound related to *O*-desmethylangolensin (*O*-DMA) (Figure 2). If this compound does not present a hydroxyl group on position 6' (corresponding to position 5 of the A-ring of the isoflavone), depending on structure of the parent aglycone, no further reactions occur on C-ring. Therefore C-ring cleavage of dihydrodaidzein does not progress beyond *O*-DMA.

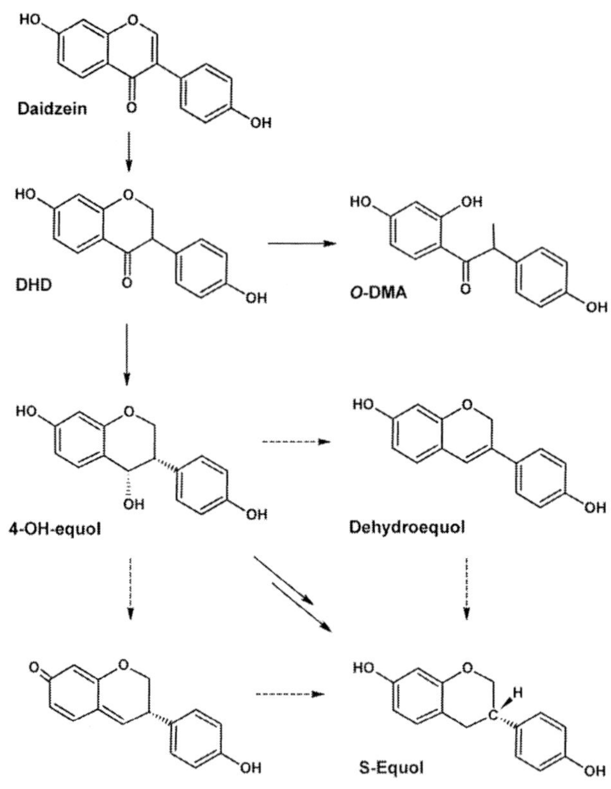

Figure 3. Intestinal bacterial metabolism of the isoflavone daidzein to O-DMA and equol. The dashed arrows indicate hypothesized pathways

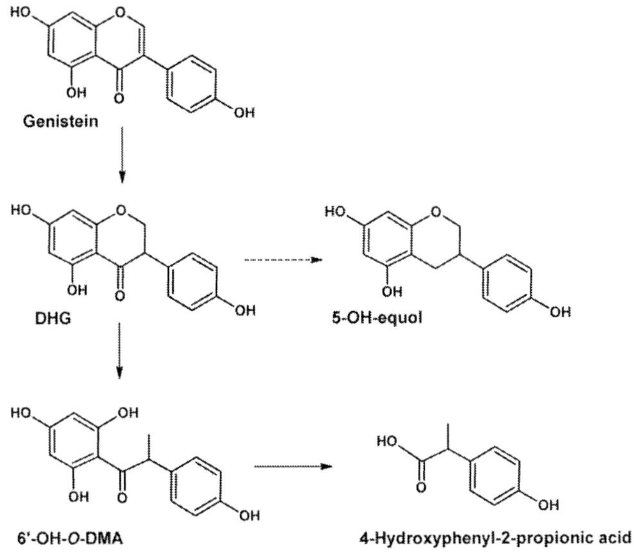

Figure 4. Intestinal bacterial metabolism of the isoflavone genistein: C-ring fission and production of 5-OH-equol. The dashed arrow indicates a reaction which is performed by intestinal isolates which have never been demonstrated *in vivo*.

On the contrary, if the compound related to *O*-DMA bears a hydroxyl group in position 6', the bond between position 1 and 1' is broken by intestinal bacteria and the molecule is cleaved into two moieties (Figure 4): phloroglucinol, originating from isoflavone A-ring, and a compond related to 2-(4-hydroxyphenyl)-propionic acid (HPPA), originating from B-ring. Similar ring cleavage occurs in the bacterial metabolism of other flavonoids, such as the flavanol quercetin, the flavone apigenin, the flavanone hespertin, and the flavanols catechins. Likewise genistein, all these flavonoids possess a hydroxyl group in position 6' and can be similarly cleaved into phloroglucinol and a derivative of HPPA (Rechner et al. 2004). Furthermore, different schemes of ring cleavage occur in quercetin metabolism, giving rise to a variety of fission products (e.g. hydroxyphenylacetic and hydroxybenzoic acids), but they have never been reported for genistein and the other isoflavones (Rechner et al. 2004, Aura 2008, Selma et al. 2009).

Daidzein

Daidzein (4', 7-dihydroxyisoflavone) is the major isoflavone in soy germ (Figure 3). In soybeans and non-fermented food it almost totally occurs in the conjugated forms of glucoside (daidzin), acetylglucoside (acetyldaidzin) and malonylglucosides (malonyldaidzin). Only after the release of the sugar moiety by human and bacterial β-glucosidase, the aglycone can be absorbed through the gut epithelium or undergo a reductive metabolic pathway catalyzed by components of the gut microbiota. Among soy isoflavones, daidzein is the most studied because of the great biological activity of equol, one of its major metabolites. Equol exists in the two chiral forms R and S, where S-equol is the most effective in stimulating an estrogenic response among the isoflavone derivatives and has a longer half life and a superior antioxidant activity (Setchell and Cassidy 1999, Setchell and Cole 2003, Wang et al. 2007).

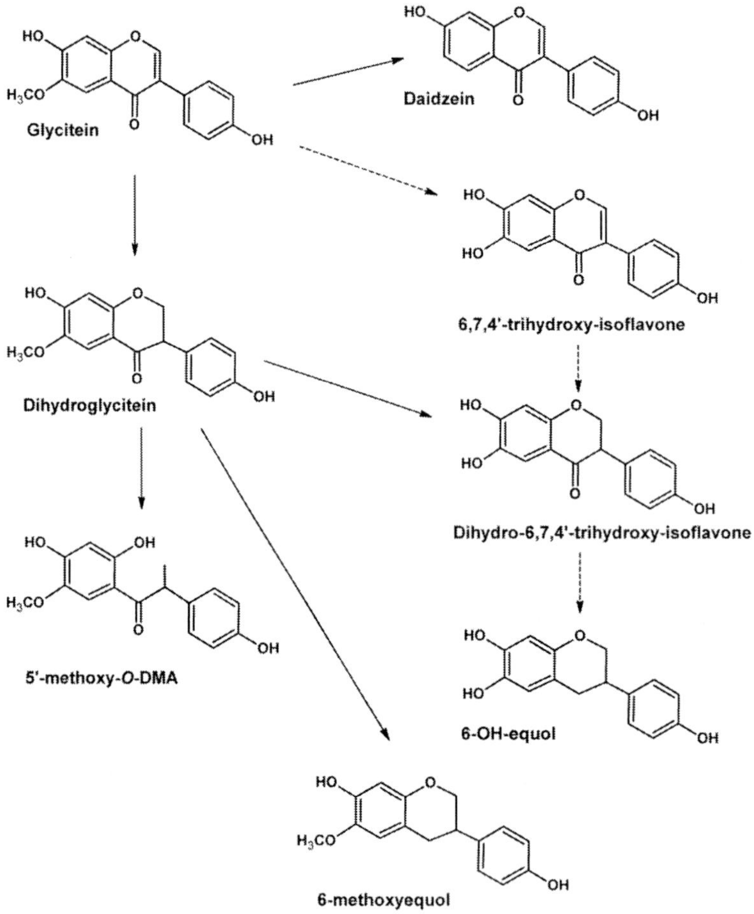

Figure 5. Intestinal bacterial metabolism of the isoflavone glycitein. Dashed arrow indicate hypothesized bacterial reactions.

The reductive pathway of daidzein starts with the production of dihydrodaidzein (DHD). DHD can undergo C-ring cleavage, leading to the production of *O*-desmethylangolensin (*O*-DMA) or, conversely, can be transformed into S-equol through reduction reactions (Atkinson et al. 2005, Wang et al. 2005, Wang et al. 2007, Yuan et al. 2007, Kim et al. 2009) (Figure 3).

The role of the gut microbiota in the reductive pathway of daidzein is well documented. Germ-free animals do not produce S-equol or *O*-DMA (Bowey et al. 2003), and *in vitro* incubation of human fecal samples with daidzein can give equol and/or *O*-DMA as products (Atkinson et al. 2004). Great interindividual variability affects the capacity to produce S-equol and *O*-DMA, as a consequence of the impact of the diverse composition of colonic microbiota (Zhang et al. 1999). In fact, the production of S-equol occurs only in 30-40% of people (Lampe et al. 1998), whereas approximately 80-90% can produce *O*-DMA, not all the S-equol producers being *O*-DMA producers and *vice versa*.

The central role of the gut microbiota in S-equol production was demonstrated through bioconversion experiments with fecal cultures. Fecal samples from equol producers successfully converted daidzein into S-equol, while fecal samples from non-producers did not, albeit some of the latter produced *O*-DMA and DHD.

Reduction to Dihydrodaidzein

The first step of daidzein metabolic pathway is the reduction to DHD, through the hydrogenation of the activated double bond between C-2 and C-3 of the C-ring (Figure 3). This step attracted great interest, since DHD represents the branch point of different fates of daidzein metabolism. DHD is a stable compound found in feces incubated with daidzein and in the urine of individuals consuming soy isoflavone, that can be easily extracted and revealed by HPLC coupled with UV (Chang and Nair 1995, Joannou et al. 1995).

Since the gut microbiota consists of a multitude of species, it is likely that several microbial groups can carry out the reduction of daidzein to DHD. Single strains performing this conversion have been isolated. Some of them can perform only this reaction of daidzein metabolism, such as the *Clostridium*-like strains SNU Niu-O16, TM-40, and HGH6, still lacking a final taxonomic classification, but exhibiting similarity to members of *Erysipelotrichales* (Figure 1). The strain Niu-O16 was isolated from bovine rumen (Wang et al. 2005a), whereas the strains TM-40 and HGH6 were obtained from human feces (Hur et al. 2000, Tamura et al. 2007). Transformation of daidzein by Niu-O16 resulted in (*R*)- and (*S*)-DHD enanthiomers with the same abundance, but it was not determined whether this was due to a no enanthioselective reduction or to a tautomerization reaction (Wang et al. 2005b).

O-DMA Production and Isoflavone Degradation

The intestinal microbiota acts by two distinctive pathways to convert daidzein to the more estrogenic metabolite equol or to the less estrogenic one *O*-desmethylangolensin (*O*-DMA) (Joannou et al. 1995) (Figure 3). In this latter case daidzein is first converted to dihydrodaidzein, then to *O*-DMA through 2-dehydro-*O*-DMA (Hur et al. 2002). Among several intermediates, *O*-DMA and S-equol were the first to be isolated and characterized, since they are easily extracted and detected. They occur in urine and blood of humans and animals and were found as final products of *in vitro* experiments with fecal cultures as well (Chang and Nair 1995, Joannou et al. 1995, Heinonen et al. 1999).

The C-ring cleavage of daidzein can be accomplished by the colonic species *Eubacterium ramulus* (Schoefer et al. 2002), the major product observed being *O*-DMA. *E. ramulus* is a member of *Clostridium* cluster XIVa and is regarded as one of the major isoflavones degrading bacteria in the human gastrointestinal tract (Schoefer et al. 2002) (Figure 1). Consistently, during a flavonoid-free diet, both total cell counts of intestinal bacteria and *E. ramulus* decreased significantly, while there was an increase in the *E. ramulus* amount of up to 10-fold during a flavonoid-rich feeding period (Schoefer et al. 2002). These results suggest that dietary secondary plant metabolites may influence the composition of intestinal microbiota (Blaut et al. 2003).

Among intestinal bacteria, also *Clostridium* sp. HGH 136, isolated from human feces, cleaved the C-ring of daidzein to *O*-DMA under anoxic conditions. Dihydrodaidzein and 2-dehydro-*O*-DMA were transient intermediates and did not accumulate in the culture (Hur et al. 2002). Also the human strain SY8519, belonging to a novel genus in the *Clostridium* cluster XIVa (Yokoyama et al. 2009), produced *O*-DMA from daidzein. Being phylogenetically related to *E. ramulus*, it is possible that it can metabolize a broad spectrum of polyphenols (Yokoyama et al. 2009).

Equol Production

Extensive search for intestinal bacteria capable of producing equol has been carried out and most reactions of the biotransformation have been unraveled. DHD is transformed by components of fecal microbiota into S-equol through *cis*-4-hydroxy-equol, the major diastereoisomer of 4-hydroxy-equol occurring in urine samples and in fecal cultures (Heinonen et al. 1999, Setchell et al. 2005, Rüfer et al. 2006). 4-hydroxy-equol is converted to equol by either reductive rearrangement or by dehydration and subsequent reduction (Joannou et al. 1995, Kim et al. 2009) (Figure 3).

Several fecal cultures and consortia of intestinal bacteria convert daidzein into equol (Chang and Nair 1995, Hur et al. 2000, Rafii et al. 2007) (Figure 1). A stable mixed culture transforming daidzein into equol was obtained from a fecal sample of an equol-producing human subject. It was composed of *Lactobacillus mucosae*, *Enterococcus faecium*, *Finegoldia magna* and *Veillonella* sp. The addition of this bacterial consortium to a fecal culture that could not metabolize daidzein provided the capacity to produce equol (Decroos et al. 2005). A diverse mixed culture, containing the *Clostridium*-like strain SNU Niu-O16 and *Eggerthella* sp. Julong 732, was able to produce S-equol (Wang et al. 2007). The former strain was capable to reduce daidzein to dihydrodaidzein, the latter produced S-equol, and together they completely and stereoselectively (100% e.e.) converted daidzein into S-equol. The strain named SNU Niu-O16 yielded a racemic mixture of R and S dihydrodaidzein, while *Eggerthella* sp. Julong 732 did not exhibit racemase activity. Thus, it is conceivable that the complete and stereoselective bioconversion resulted from enantioselective preference of *Eggerthella* sp. Julong 732 toward S-dihydrodaidzein, with a putative tautomerization reaction that converted R-dihydrodaidzein into the S-isomer (Wang et al. 2005a, Wang et al. 2005b, Wang et al. 2007, Kim et al. 2009).

Considerable efforts focused on the search for single bacterial strains capable to carry out the complete bioconversion have been unsuccessful for a long time. The general approach was to screen bacteria isolated from feces of equol-producing subjects. Most of research explored the ability of commensal lactobacilli and bifidobacteria to carry out the complete reaction, in the perspective to use positive strains as probiotic supplements or starters for the production of equol-enriched soy-based fermented food. These studies failed to identify positive strains and demonstrated that few lactobacilli strains, for instance *L. mucosae*, can participate to some steps of equol production (Decroos et al. 2004, Wang et al. 2005b). Up to now, the involvement of bifidobacteria in the reduction of daidzein toward S-equol seems to be excluded, even if they exert an important role in the hydrolysis of isoflavone glyco-conjugates to release the aglycone. Albeit bifidobacteria have never been reported among the single bacteria which perform the whole reductive bioconversion from daidzein to S-equol (Wang et al. 2005a, Minamida et al. 2006, Tamura et al. 2007, Wang et al. 2007, Minamida et al. 2008, Maruo et al. 2008, Yu et al. 2008, Matthies et al. 2008), one study described the formation of S-equol in soymilk fermented with some strains of *Bifidobacterium* (Tsangalis et al. 2002). Nonetheless, S-equol has never been claimed again as a bioconversion product of soymilk isoflavones by bifidobacteria and, in a further investigation, 22 bifidobacteria failed to transform daidzein into reduced metabolites under all the experimental conditions, excluding any role in the reductive pathway of daidzein toward the production of S-equol (Raimondi et al. 2009).

Some intestinal strains which carry out the bioconversion of daidzein to S-equol self-sufficiently have been recently isolated and characterized from the feces of humans and animals. In some cases, the phenotypic and biochemical features and the phylogenetic position of the new isolates impelled the proposal of new bacterial genera and/or species or caused the description of previously defined genera and/or species to be updated (Minamida et al. 2006, Maruo et al. 2008, Matthies et al. 2009). Since identification of an increasing number of equol producing bacteria is in progress, and many of them are the forerunner of novel species, the list of equol-producing bacteria is likely far to be conclusive.

Nearly all the equol-producing bacterial isolates have been classified within the high G+C% content gram-positives, in the class of *Actinobacteria*, and specifically in the family of *Coriobacteriaceae*, whose role in isoflavonoid metabolism within the mammalian intestine seems crucial. Within *Coriobacteriaceae*, equol-producers belong to the genus *Eggerthella* and *Slackia* (Jin et al. 2008, Yokoyama et al. 2008, Matthies et al. 2009), and to the species *Asaccharobacter celatus* and *Adlercreutzia equolifaciens* (Minamida et al. 2006, Minamida et al. 2008, Maruo et al. 2008, Matthies et al. 2008) (Figure 1).

The human strain *Eggerthella* sp. YY7918 converts daidzein into DHD and completes the transformation to S-equol with high stereoselectivity. Unlike other *Eggerthella* strains, it can't hydrolyze daidzin, due to the lack of β-glucosidase activity. Furthermore, it doesn't carry out any transformation of other isoflavone aglycones, such as glycitein, genistein, and formonetin (Yokoyama et al. 2008). Few strains within the family of *Coriobacteriaceae* metabolize only DHD to equol, but other strains do not present equol-producing activity (Minamida et al. 2006, Minamida et al. 2008, Maruo et al. 2008). Therefore, a reasonably wide diversity in equol-producing capability seems to exist within this family. Furthermore, whereas growing cells of human strain *Slackia isoflavoniconvertens* HE8 produced equol in high amount (62% of yield), they hardly converted daidzein when it was added during the stationary growth phase, suggesting that isoflavone conversion might be inducible (Matthies et al. 2009).

Few equol-producing intestinal isolates were not classified within the family of *Coriobacteriaceae*, such as *Eubacterium* sp. (Yu et al. 2008). Furthermore, three strains of *Bacteroides ovatus*, *Streptococcus intermedius*, and *Streptococcus constellatus* were patented for equol production (Uchiyama et al. 1999). However, the ability of *B. ovatus*, *S. intermedius*, and *S. constellatus* seems to be questionable, since these species have never been claimed again as equol producers since the patent registration.

It is important to mention also the strain *Lactococcus garviae* 20-92 that, albeit isolated from human fecal samples (Uchiyama et al. 2007), is typically found in fish and aquacultures and does not normally occur in the human gut microbiota, although it is not pathogenic or toxic. This strain converts daidzein to equol during the stationary phase of growth, and represents the first lactic acid bacterium which perform the complete transformation of daidzein into equol (Uchiyama et al. 2005, Yee et al. 2008, Setchell et al. 2009).

Effects of the Gut Environment on Equol Production

The pivotal role of the gut microbiota in daidzein metabolism and equol production is unquestionable. Therefore, it seems possible to improve equol production through changes in the environmental conditions in the hindgut, hence modulating the microbiota. In particular, the diet and other physiological factors (e.g. the intestinal transit time and redox potential), influence the microbiota composition and/or activity and affect equol production (Setchell et al. 1984).

Several studies established a comparison between the habitual diet of equol-producers and non-producers, but led to contrasting conclusions. A positive relationship between carbohydrate and fiber intake and equol production was identified in people consuming a western diet (Lampe et al. 1998, Rowland et al. 2000). Furthermore it was reported that subjects non-producers were converted to equol-producers with continual soy ingestion (Lu et al. 1996). However, it is hard to be conclusive about this argument and it seems that the question is still open, since a diverse study demonstrated that it was impossible to induce equol production in non-producers by supplementing diets with high-fiber wheat bran cereal or soy protein (Lampe et al. 2001).

Equol production by single strains or mixed microbial culture was affected, *in vitro*, by environmental conditions, such as the presence of hydrogen and the addition of FOS, arginine, butyric and propionic acids (Decroos et al. 2005, Minamida et al. 2006). Hydrogen gas and SCFA occur within the gut, since they are fermentation products of diverse intestinal bacteria. H_2 exerted a major impact on daidzein metabolism by fecal cultures: incubation with H_2 increased equol production, while equol was not detected in presence of N_2 or metallic palladium. Furhtermore, addition of SCFA influenced equol production *in vitro*: butyric and propionic acids enhanced equol production, while the opposite effect was attained by acetic acid (Decroos et al. 2005, Minamida et al. 2006). It is known that the production of SCFA and of H_2 are connected through complex syntrophic relationship among bacteria, involving some of the most represented bacterial groups within the gut microbiota. In fact, as a result of the mutualistic cross-feeding among bacteria, hydrogen production and consumption, acetate production and consumption, and production of butyrate and propionate are strictly related each other (Duncan et al. 2002; Flint et al. 2007). Thus, it was proposed that the influence of SCFA on equol production is somehow in relationship to bacterial H_2 consumption and production (Duncan et al. 2002, Yu et al. 2008).

The investigation of whether the prebiotic fructo-oligosaccharides (FOS) may improve equol production lead to contrasting conclusions. The addition of FOS to a fecal microbial culture inhibited equol production (Decroos at al. 2005), while rats fed a diet supplemented with FOS exhibited increased intestinal bioavailability of daidzein (Uehara et al. 2001). Since FOS fermentation increases hydrogen release by certain intestinal bacteria (Cummings et al. 2001), it was proposed that FOS may stimulate equol production (Decroos at al. 2005). However, FOS are known to modify the composition of the human colon microbiota, causing a significant improvement of the population of bifidobacteria (Gibson and Roberfroid 1995, Hopkins and Macfarlane 2003), a genus which has a major role in hydrolysis of isoflavone glycoconjugates, but which is incapable of daidzein metabolism.

Genistein

Genistein (4', 5, 7-trihydroxyisoflavone) is the second soy isoflavone for importance and occurrence after daidzein (Dixon et al. 2002) (Figure 2). It is found in soybean in high amounts (up to 1 g Kg-1) mostly in conjugated form within the cotyledons (Franke et al. 1995). Pharmacological effects of genistein include tyrosine kinase inhibition, chemoprotectant activities against cancer, and cardiovascular disease (Banerjee et al. 2008), but few information are still available about biological activity of its metabolites.

Likewise daidzein, genistein undergoes extensive biotransformation by the gut microbiota (Figure 4). It is first converted into dihydrogenistein (DHG) through the reduction of the activated double bond between C-2 and C-3 to a single bond. This reaction is the same occurring to daidzein and the above mentioned strains SNU sp. Niu-O16 and HGH6, converting daidzein to DHD, were capable of reducing genistein to DHG as well, but could not carry out any further transformation of both DHD and DHG (Hur et al. 2000, Wang et al. 2005b). Likewise (*R*)- and (*S*)-DHD, (*R*)- and (*S*)-DHG were produced by the strain SNU Niu-O16 with the same abundance, probably as a consequence of the tautomerization reaction (Wang et al. 2005b).

As well DHD, DHG may be subject to two different fates, such as the cleavage of C-ring or the transformation into 5-OH-equol through a reductive pathway similar to the one leading to equol. For a long time, it has been assumed that the equol-like transformation of DHG could not occur, because the putative intermediate (tetrahydrogenistein) and end-product (5-OH-equol) have never been observed in urine or feces nor identified when genistein or DHG were incubated with equol-producing fecal cultures (Griffiths and Smith 1972, Chang and Nair 1995). Albeit 5-OH-equol has been never found in urine, feces, or blood, and it is still unclear whether it is produced in the hindgut (Heinonen et al. 1999, Coldham et al. 1999, Coldham et al. 2002), few intestinal bacterial strains capable to transform genistein into 5-OH-equol have been identified among the ones which can transform daidzein into equol (Jin et al. 2008, Matthies et al. 2008, Matthies et al. 2009).

Genistein metabolites arising from the cleavage of C-ring, such as 4-ethylphenol, have been identified in fecal cultures and in animal studies (Griffiths and Smith 1972, Chang and Nair 1995). Other metabolites of genistein, such as dihydrogenistein (DHG) and 6'-hydroxyl-*O*-DMA (6'-OH-*O*-DMA), were found in human urine (Kelly et al. 1993, Joannou et al. 1995, Heinonen et al. 1999). Furthermore, a human intestinal strain of *Eubacterium ramulus* cleaving the C- ring and degrading genistein has been identified (Schoefer et al. 2002). This strain cleaved the C-ring of genistein producing 6'-OH-*O*-DMA, which was further cleaved into two moieties, phloroglucinol and 2-(4-hydroxyphenyl)-propionic acid (HPPA), originating from A-ring and B-ring, respectively. Phloroglucinol is utilized as a substrate by intestinal bacteria, thus it is quickly degraded and could therefore not be detected (Schoefer et al. 2002, Rechner et al. 2004). Any further conversion of HPPA has never been observed by means of human intestinal bacteria (Joannou et al. 1995), even if it is transformed into 4-ethylphenol in some animal models (Batterham et al. 1965, Chang and Nair 1995).

The microbial transformation of genistein into HPPA and 4-ethylphenol represents a substantial loss of biological activity, since estrogenic or other healthy properties have never been reported for these compounds (Steer et al. 2003). Therefore, the prevention of C-ring cleavage is of interest to maintain the biological activity of genistein. Albeit prediction of *in vivo* behaviors based on results from *in vitro* experiments should be carried out with caution, the addition of FOS to fecal cultures preserved genistein from degradation. It was proposed that FOS limited degradation of genistein through the selective stimulation of bifidobacteria. In fact they cannot perform reduction of complex molecules (McMullen et al. 2006, Chun et al. 2007, Raimondi et al. 2009, Iqbal and Zhu 2009) and their growth can hinder the growth of other bacteria responsible for genistein metabolism (Steer 2003).

Other Isoflavone Compounds

Other naturally occurring isoflavones include glycitein, formononetin, biochanin A, prunetin, pratensein, calycosin, and pseudobaptigenin (Figure 2). They differ in the number and in the presence of hydroxy and methoxy groups. Because of the limited occurrence in the human diet, their metabolism has not been studied in detail. Only some information about glycitein, formononetin, biochanin A and puerarin are present.

Glycitein

Glycitein (7, 4'-dihydroxy-6-methoxyisoflavone) (Figure 2) accounts for less than 10% of the total isoflavones in soybeans and soybean foods and for about 50% in soy germ (Song et al. 1998). A mouse uterine growth assay demonstrated that glycitein is more estrogenic than genistein, (Song et al. 1999), but, probably because of its scarcity in soy-based foods, its metabolism has not been explored yet.

Glycitein is structurally similar to daidzein, with the exception of the methoxyl group at the 6-position. Diverse molecules arising from glycitein reduction, including dihydroglycitein, dihydro-6, 7, 4'-trihydroxyisoflavone, 5'-O-methoxyl-*O*-desmethylangolensin, and 6-O-methoxyl-equol (Figure 2), were found in the urine of soy consuming humans (Heinonen et al. 2003), likely as a consequence of gut micriobiota transformation. In fact, the same metabolites were produced when glycitein was incubated in fecal cultures from the metabolite-producing subjects (Simons et al. 2005). These observations suggest that the first step in glycitein metabolism is the reduction to dihydroglycitein. Then, dihydroglycitein can undergo to three different fates. It can be *O*-demethylated to dihydro-6, 7, 4'-trihydroxyisoflavone, the major metabolite occurring in fecal cultures, which may be reduced to 6-hydroxyequol or cleaved to 5'-hydroxy-*O*-desmethylangolensin. Diversely, it can be subject to C-ring cleavage or to reductive transformations, yielding 5'-O-methoxyl-*O*-desmethylangolensin and 6-methoxyequol, respectively, two metabolites occurring in minor amounts in fecal cultures (Simons et al. 2005) (Figure 5).

Minor transformation pathways of glycitein include demethylation and demethoxylation. Some intestinal bacteria, such as *Eubacterium limosum* of *Clostridium* cluster XII (Figure 1), can demethylate glycitein *in vitro* to 6, 7, 4'-trihydroxyisoflavone (Hur and Rafii 2000), but this metabolite has never been detected in human urine. It was proposed that it may be reduced to dihydro-6, 7, 4'-trihydroxyisoflavone, although this reaction has not been demonstrated to occur in fecal cultures (Simons et al. 2005). The direct demethoxylation of glycitein at the 6-position results in daidzein, but this reaction does not occur commonly *in vivo* and is rare in human fecal cultures (Simons et al. 2005, Setchell et al. 2001) (Figure 5).

Red Clover Isoflavones: Formononetin and Biochanin A

The main isoflavones of red clover (*Trifolium pratense*) are formononetin, biochanin A (Figure 2), and their glycosidic conjugates, ononin, and sissotrin, respectively (He et al. 1996, Kleydus et al. 2001) . Since estrogenic potency and antioxidant activity of isoflavones is significantly affected by the position and number of hydroxyl groups, biochanin A and formononetin are less active than genistein and daidzein (Arora et al. 1998, Kuiper et al. 1998).

Due to the limited occurrence in the human diet, the metabolism of red clover isoflavones has not been studied in great detail. It is known that formononetin and biochanin A are readily demethylated by the intestinal microbiota to daidzein and genistein, respectively (Batterham et al. 1971). Then, daidzein and genistein go through their respective metabolic pathways. The extent of demethylation of the methoxy substituents of isoflavones depends on its position in the isoflavone structure: the demethylation of the methoxy group occurs more easily in ring B than in ring A (Hur et al. 2000). Similar observations were made for flavonoids as well, where demethylation occured more easily at C6 than at C4' (Nielsen et al. 2000). However, minor amounts of reduced metabolites of formononetin and biochanin A, still bearing the methoxy group at the 4'-position, were identified in urine at low levels, suggesting that demethylation of the methoxy group at the 4'-position is the major transformation, which occurs almost quantitatively (Heinonen et al. 2004a, Heinonen et al. 2004b).

Puerarin

Puerarin is an isoflavone *C*-glucoside (daidzein-8-*C*-glucoside) occurring abundantly in the roots of the legume genus *Pueraria* (e.g. *P. thunbergiana* and *P. lobata* or Kudzu). Although puerarin does not occur commonly in food, especially in western diet, it is of great interest for herbal medicine, due to its potent biological activity against cardiovascular diseases (Tan et al. 2008). Certain human intestinal isolates can remove the *C*-glucosyl with high substrate specificity through reductive cleavage (Kim et al. 1998, Park et al. 2006). Hence, puerarin is transformed into daidzein, which may further converted into equol (Jin et al. 2008).

Conclusion

Likewise hundreds of non-nutritional phytochemicals occurring in diet, isoflavones undergo several transformations upon ingestion by the commensal microbiota harbored within the gut. Bacterial transformations may greatly increase or compromise the biological activity of dietary isoflavones, thus they have been subject of considerable interest, particularly those yielding valuable products with superior biological activity, such as equol. This chapter aimed to summarize what is known about the bacterial metabolism of dietary isoflavones. Although it is being increasingly unraveled for few bacterial isolates, an exhaustive knowledge of this topic is not available so far, due to the complexity of the intestinal bacterial ecosystem. In fact, this bacterial community is composed of hundreds of bacterial species which may encode approximately 100 times as many genes as the human genome and may carry out a number of biochemical metabolic reactions with complex reciprocal synergies. The task of disclosing all the interactions occurring *in vivo* among the diverse bacterial species and dietary isoflavones is daunting. To address this issue, it is necessary a coordinated work in the genetics, physiology, and metabolism of intestinal bacteria and in enzymology and chemistry as well. Recent advances in omic approaches, allowing the simultaneous analysis of huge numbers of genes and proteins, are now available to achieve a more comprehensive knowledge of the pathways occurring in isoflavones metabolism.

References

Aminov, RI; Walker, AW; Duncan, SH; Harmsen, HJM; Welling, GW; Flint, HJ. Molecular diversity, cultivation, and improved detection of fluorescent *in situ* hybridisation of a dominant group of human gut bacteria related to *Roseburia* spp. or *Eubacterium rectale*. *Appl Environ Microbiol*, 2006, 72, 6371-6376.

Arora, A; Nair, MG; Strasburg, GM. Antioxidant activities of isoflavones and their biological metabolites in a liposomal system. *Arch Biochem Biophys*, 1998, 356, 133-141.

Atkinson, C; Berman, S; Humbert, O; Lampe, JW. In vitro incubation of human feces with daidzein and antibiotics suggests interindividual differences in the bacteria responsible for equol production. *J Nutr*, 2004, 134, 596-599.

Atkinson, C; Frankenfeld, CL; Lampe, JW. Gut bacterial metabolism of the soy isoflavone daidzein: exploring the relevance to human health. *Exp Biol Med*, 2005, 230, 155-170.

Aura, AM. Microbial metabolism of dietary phenolic compounds in the colon. *Phytochem Rev*, 2008, 7, 407-429.

Bae, EA; Park, SY; Kim, DH. Constitutive beta-glucosidases hydrolyzing ginsenoside Rb1 and Rb2 from human intestinal bacteria. *Biol Pharm Bull*, 2000, 23, 1481-1485.

Bae, EA; Han, MJ; Choo, MK; Park, SY; Kim, DH. Metabolism of 20(S), and 20(R)-ginsenoside Rg3 by human intestinal bacteria and its relation to in vitro biological activities. *Biol Pharm Bull*, 2002, 25, 58-63.

Bae, EA; Han, MJ; Kim, EJ; Kim, DH. Transformation of ginseng saponins to ginsenoside Rh2 by acids and human intestinal bacteria and biological activities of their transformants. *Arch Pharm Res*, 2004, 27, 61-67.

Banerjee, S; Li, Y; Wang, Z; Sarkar, FH. Multi-targeted therapy of cancer by genistein. *Cancer Lett*, 2008, 269, 226-242.

Batterham, TJ; Hart, NK; Lamberton, JA; Braden AWH. Metabolism of Oestrogenic Isoflavones in Sheep. *Nature*, 1965, 206, 509.

Batterham, TJ; Shutt, DA; Hart, NK; Braden, AWH; Tweeddale, HJ. Metabolism of intraruminally administered [4-^{14}C]formononetin and [4-^{14}C]biochanin a in sheep. *Australian Journal of Agricultural Research*, 1971, 22(1), 131-138.

Blaut, M; Schoefer, L; Braune, A. Transformation of flavonoids by intestinal microorganisms. *Int J Vitam Nutr Res*, 2003, 73(2), 87-79.

Bokkenheuser, VD; Suzuki, JB; Polovsky, SB; Winter, J; Kelly, WG. Metabolism of deoxycorticosterone by human fecal flora. *Appl Microbiol*, 1975, 30, 82-90.

Bowey, E; Adlercreutz, H; Rowland, I. Metabolism of isoflavones and lignans by the gut microflora: a study in germ-free and human flora associated rats. *Food Chem Toxicol*, 2003, 41, 631-636.

Cederroth, CR; Nef, S. Soy, phytoestrogens and metabolism: A review. *Mol Cell Endocrinol*, 2009, 304, 30-42.

Chang, Y; Nair, GN. Metabolism of daidzein and genistein by intestinal bacteria. *J Nat Prod*, 1995, 58(12), 1892-1896.

Cheng, DL; Hashimoto, K; Uda, Y. *In vitro* digestion of sinigrin and glucotropaeolin by single strains of Bifidobacterium and identification of the digestive products. *Food Chem Toxicol,* 2004, 42, 351-357.

Chun, J; Kim, G; Lee, K; Choi, ID; Know, GH; Park, JY; Jeong, SJ; Kim, JS; Kim, JH. Conversion of Isoflavone Glucosides to Aglycones in Soymilk by Fermentation with Lactic Acid Bacteria. *Journal of Food Science*, 2007, 72, M39-44.

Clavel, T; Borrmann, D; Braune, A; Doré, J; Blaut, M. Occurrence and activity of human intestinal bacteria involved in the conversion of dietary lignans. *Anaerobe,* 2006a, 12, 140-147.

Clavel, T; Henderson, G; Engst, W; Doré, J; Blaut, M. Phylogeny of human intestinal bacteria that activate the dietary lignan secoisolariciresinol diglucoside. *FEMS Microbiol Ecol*, 2006b, 55, 471-478.

Coldham, NG; Howells, LC; Santi, A; Montesissa, C; Langlais, C; King, LJ; Macpherson, DD; Sauer, MJ. Biotransformation of genistein in the rat: elucidation of metabolite structure by product ion mass fragmentology. *Journal of Steroid Biochemistry and Molecular Biology*, 1999, 70, 169-184.

Coldham, NG; Darby, C; Hows, M; King, LJ; Zhang, AQ; Sauer, MJ. Comparative metabolism of genistin by human and rat gut microflora: detection and identification of the end-products of metabolism. *Xenobiotica*, 2002 32(1), 45-62.

Collins, MD; Lawson, PA; Willems, A; Cordoba, JJ; Fernandez-Garayzabal, J; Garcia, P; Cai, J; Hippe, H; Farrow, JAE. The phylogeny of the genus *Clostridium*: proposal of five new genera and eleven new species combinations. *Int J Syst Bacteriol*, 1994, 44, 812-826.

Crozier, A; Jaganath, IB; Clifford, MN. Dietary phenolics: chemistry, bioavailability and effects on health. *Nat Prod Rep*, 2009, 26(8), 1001-43.

Cummings, JH; Macfarlane, GT; Englyst, HN. Prebiotic digestion and fermentation. *Am J Clin Nutr*, 2001, 73, 415S-420S.

Dabek, M; Sheila I. McCrae, SI; Stevens, VJ; Duncan, SH; Louis, P. Distribution of β-glucosidaseand and β-glucuronidase activity and of β-glucuronidase gene *gus* in human colonic bacteria. *FEMS Microbiol Ecol*, 2008, 66, 487-495.

Decroos, K; Vanhemmens, S; Cattoir, S; Boon, N; Verstraete, W. Isolation and characterization of an equol-producing mixed microbial culture from a human fecal sample and its activity under gastrointestinal conditions. *Arch Microbiol*, 2005, 183, 45-55.

Dixon, RA; Ferreira, D. Molecules of Interest, Genistein. *Phytochemistry*, 2002, 60, 205-211.

Duncan, SH; Barcenilla, A; Stewart, CS; Pryde, SE; Flint, HJ. Acetate utilization and butyryl coenzyme A (CoA): acetate-CoA transferase in butyrate-producing bacteria from the human large intestine. *Appl Environ Microbiol*, 2002, 68, 5186-5190.

Eckburg, PB; Bik, EM; Bernstein, CN; Purdom, E; Dethlefsen, L; Sargent, M; Gill, SR; Nelson, KE; Relman, DA. Diversity of the human intestinal microbial flora. *Science*, , 2005, 308(5728), 1635-8.

Franke, AA; Custer, LJ; Cerna, CM; Narala, K. Rapid HPLC analysis of dietary phytestrogens from legumes and from human urine. *Proc Soc Exp Biol Med*, 1995, 208(1), 18-26.

French, CE; Bruce, NC. Bacterial morphinone reductase is related to old yellow enzyme. *Biochem J*, 1995, 312, 671-678.

Gibson, GR; Roberfroid, MB. Dietary modulation of the human colonic microbiota-introducing the concept of prebiotics. *J Nutr*, 1995, 125, 1401-1412.

Griffiths, LA; Smith, GE. Metabolism of Apigenin and Related Compounds in the Rat. *Biochem J*, 1972, 128, 901-911.

Harmsen, HJ; Raangs, GC; He, T; Degener, JE; Welling, GW. Extensive set of 16S rRNA-based probes for detection of bacteria in human feces. *Appl Environ Microbiol*, 2002, 68, 2982-2990.

He, X; Lin, L; Lian, L. Analysis of flavonoids from red clover by liquid chromatography-electrospray mass spectrometry. *J Chromatogr A*, 1996, 755, 127-132.

Heinonen, SM; Wähälä, K; Adlercreutz, H. Identification of Isoflavone Metabolites Dihydrodaidzein, Dihydrogenistein, 6'-OH-*O*-dma, and *cis*-4-OH-equol in Human Urine by Gas Chromatography-Mass Spectroscopy Using Authentic Reference Compounds. Analytical Biochemistry, 1999, 274, 211-219.

Heinonen, SM; Hoikkala, A; Wähälä, K; Adlercreutz, H. Metabolism of the soy isoflavones daidzein, genistein and glycitein in human subjects. Identification of new metabolites having an intact isoflavonoid skeleton. *J Steroid Biochem Mol Biol*, 2003, 87, 285-299.

Heinonen, SM; Wähälä, K; Adlercreutz, H. Identification of Urinary Metabolites of the Red Clover Isoflavones Formononetin and Biochanin A in Human Subjects. *J Agric Food Chem*, 2004a, 52, 6802-6809.

Heinonen, SM; Wähälä, K; Liukkonen, KH; Aura, AM; Poutanen, K; Adlercreutz, H. Studies of the In Vitro Intestinal Metabolism of Isoflavones Aid in the Identification of Their Urinary Metabolites. *J Agric Food Chem*, 2004b, 52, 2640-2646.

Hopkins, MJ; Macfarlane, GT.Nondigestible oligosaccharides enhance bacterial colonization resistance against *Clostridium difficile* in vitro. *Appl Environ Microbiol*, 2003, 69, 1920-1927.

Hur, HG; Lay, Jr. JO; Beger, RD; Freeman, JP; Rafii, F. Isolation of human intestinal bacteria metabolizing the natural isoflavone glycosides daidzin and genistin. *Arch Microbiol*, 2000, 174, 422-428.

Hur, HG; Rafii, F. Biotransformation of the isoflavonoids biochanin A, formononetin, and glycitein by *Eubacterium limosum*. *FEMS Microbiology Letters*, 2000, 192, 21-25.

Hur, HG; Beger, RD; Heinze, TM; Lay Jr, JO; Freeman, JP; Dore, J; Rafii, F. Isolation of an Anaerobic Intestinal Bacterium Capable of Cleaving the C-ring of the Isoflavonoid Daidzein. *Arch Microbiol*, 2002, 178, 8-12.

Iqbal, MF; Zhu, WY. Characterization of newly isolated *Lactobacillus delbrueckii*-like strain MF-07 isolated from chicken and its role in isoflavone biotransformation. *FEMS Microbiol Lett*, 2009, 291, 180-187.

Jin, JS; Nishihata, T; Kakiuchi, N; Hattori, M. Biotransformation of *C*-Glucosylisoflavone Puerarin to Estrogenic (3*S*)-Equol in Co-culture of Two Human Intestinal Bacteria. *Biol Pharm Bull,* 2008 31(8), 1621-1625.

Joannou, GE; Kelly, GE; Reeder, AY; Waring, M; Nelson, C. A urinary profile study of dietary phytestrogens. the identification and mode of metabolism of new isoflavonoids. *J Steroid Biochem Mol Biol*, 1995, 54, 167-184.

Kelly, GE; Nelson, C; Waring, MA; Joannou, GE; Reeder, AY. Metabolites of dietary (soya) isoflavones in human urine. *Clin Chim Acta*, 1993, 223, 9-22.

Kim, DH; Yu, KU; Bae, EA; Han, MJ. Metabolism of puerarin and daidzin by human intestinal bacteria and their relation to in vitro cytotoxicity. *Biol Pharm Bull*, 1998, 21(6), 628-30.

Kim, M; Kim, SI; Han, J; Wang, XL; Song, DG; Kim, SU. Stereospecific biotransformation of dihydrodaidzein into (3S)-equol by the human intestinal bacterium Eggerthella strain Julong 732. *Appl Environ Microbiol*, 2009, 75, 3062-3068.

Klejdus, B; Vitamvasova-Sterbova, D; Kuban, V. Identification of isoflavone conjugates in red clover (*Trifolium pratense*) by liquid chromatography-mass spectrometry after two-dimensional solid-phase extraction. *Anal Chim Acta*, 2001, 450, 81-97.

Kris-Etherton, PM; Hecker, KD; Bonanome, A; Coval, SM; Binkoski, AE; Hilpert, KF; Griel, AE; Etherton, TD. Bioactive compounds in foods: their role in the prevention of cardiovascular disease and cancer. *Am J Med*, 2002 113, 71S-88S.

Kuiper, GGJM; Lemmen, JG; Carlsson, B; Corton, JC; Safe, SH; Saag, PTVD; Burg, BVD; Gustafsson, JA. Interaction of estrogenic chemicals and phytestrogens with estrogen receptor. L. *Endocrinology*, 1998, 139, 4252-4263.

Lampe, JW; Karr, SC; Hutchins, AM; Slavin, JL. Urinary equol excretion with a soy challenge: influence of habitual diet. *Proc Soc Exp Biol Med*, 1998, 217(3), 335-339.

Lampe, JW; Skor, HE; Li, S; Wähälä, K; Howald, WN; Chen, C. Wheat bran and soy protein feeding do not alter urinary excretion of the isoflavon equol in premenopausal women. *J Nutr*, 2001, 131, 740-744.

Leser, TD; Mølbak L. Better living through microbial action: the benefits of the mammalian gastrointestinal microbiota on the host. *Environ Microbiol*, 2009, 11, 2194-2206.

Lu, LJ; Lin, SN; Grady, JJ; Nagamani, M; Anderson, KE. Altered kinetics and extent of urinary daidzein and genistein excretion in women during chronic soya exposure. *Nutr Cancer*, 1996, 26(3), 289-302.

Manach, C; Scalbert, A; Morand, C; Rémésy, C; Jiménez, L. Polyphenols: food sources and bioavailability. *Am J Clin Nutr*, 2004, 79, 727-747.

Mariat, D; Firmesse, O; Levenez, F; Guimarães, V; Sokol, H; Doré, J; Corthier, G; Furet, JP. The Firmicutes/Bacteroidetes ratio of the human microbiota changes with age. *BMC Microbiol*, 2009, 9, 123.

Maruo, T; Sakamoto, M; Ito, C; Toda, T; Benno, Y. *Adlercreutzia equolifaciens* gen. nov., sp. nov., an equol-producing bacterium isolated from human faeces, and emended description of the genus *Eggerthella*. *Int J Syst Evol Microbiol*, 2008, 58, 1221-1227.

Matthies, A; Clavel, T; Gütschow, M; Engst, W; Haller, D; Blaut, M; Braune, A. Conversion of daidzein and genistein by a newly isolated anaerobic bacterium from mouse intestine. *Appl Environ Microbiol*, 2008, 74, 4847-4852.

Matthies, A; Blaut, M; Braune, A. Isolation of a human intestinal bacterium capable of daidzein and genistein conversion. *Appl Environ Microbiol*, 2009, 75, 1740-1744.

Marotti, I; Bonetti, A; Biavati, B; Catizone, P; Dinelli, G. Biotransformation of common bean (*Phaseolus vulgaris* L.) flavonoid glycosides by *Bifidobacterium* species from human intestinal origin. *J Agric Food Chem*, 2007, 55, 3913-3919.

McBain, AJ; Macfarlane, GT. Ecological and physiological studies on large intestinal bacteria in relation to production of hydrolytic and reductive enzymes involved in formation of genotoxic metabolites. *J Med Microbiol*, 1998, 47, 407-416.

McMahon, LG; Nakano, H; Levy, MD; Gregory, JF 3rd. Cytosolic pyridoxine β-D-glucoside hydrolase from porcine jejunal mucosa. Purification, properties and comparison with broad specificity b-glucosidase. *J Biol Chem*, 1997, 272, 32025-32033.

McMullen, MH; Hamilton-Reeves, JM; Bonorden, MJ; Wangen, KE; Phipps, WR; Feirtag, JM; Kurzer, MS. Consumption of *Lactobacillus acidophilus* and *Bifidobacterium longum* does not alter phytestrogen metabolism and plasma hormones in men: a pilot study. *J Altern Complement Med*, 2006, 12, 887-94.

Minamida, K; Tanaka, M; Abe, A; Sone, T; Tomita, F; Hara, H; Asano, K. Production of equol from daidzein by gram-positive rod-shaped bacterium isolated from rat intestine. *J Biosci Bioeng*, 2006, 102(3), 247-250.

Minamida, K; Ota, K; Nishimukai, M; Tanaka, M; Abe, A; Sone, T; Tomita, F; Hara, H; Asano, K. *Asaccharobacter celatus* gen. nov., sp. nov., isolated from rat caecum. Int *J Syst Evol Microbiol*, 2008, 58, 1238-1240.

Mueller, S; Saunier, K; Hanisch, C; Norin, E; Alm, L; Midtvedt, T; Cresci, A; Silvi, S; Orpianesi, C; Verdenelli, MC; Clavel, T; Koebnick, C; Zunft, HJ; Doré, J; Blaut, M. Differences in fecal microbiota in different European study populations in relation to age, gender, and country: a cross-sectional study. *Appl Environ Microbiol*, 2006, 72, 1027-1033.

Nakamura, J; Kubota, Y; Miyaoka, M; Saitoh, T; Mizuno, F; Benno, Y. Comparison of four microbial enzymes in *Clostridia* and *Bacteroides* isolated from human feces. Microbiol Immunol, 2002, 46, 487-490.

Nielsen, SE; Breinholt, V; Cornett, C; Dragsted, LO. Biotransformation of the citrus flavone tangeretin in rats. Identification of metabolites with intact flavane nucleus. *Food Chem Toxicol*, 2000, 38(9), 739-46.

O'Keefe, SJ. Nutrition and colonic health: the critical role of the microbiota. *Curr Opin Gastroenterol*, 2008, 24(1), 51-8.

O'Keefe, SJ; Ou, J; Aufreiter, S; O'Connor, D; Sharma, S; Sepulveda, J; Fukuwatari, T; Shibata, K; Mawhinney, T. Products of the colonic microbiota mediate the effects of diet on colon cancer risk. *J Nutr*, 2009, 139(11), 2044-8.

Park, EK; Shin, J; Bae, EA; Lee, YC; Kim, DH. Intestinal bacteria activate estrogenic effect of main constituents puerarin and daidzin of *Pueraria thunbergiana*. Biol. *Pharm Bull*, 2006, 29, 2432-2435.

Rafii, F; Jackson, LD; Ross, I; Heinze, TM; Lewis, SM.; Aidoo, A; Lyn-Cook, L; Manjanatha, M. Metabolism of Daidzein by Fecal Bacteria in rats. *Comparative Medicine*, 2007, 57(3), 282-286.

Raimondi, S; Roncaglia, L; De Lucia, M; Amaretti, A; Leonardi, A; Pagnoni, UM; Rossi, M. Bioconversion of soy isoflavones daidzin and daidzein by *Bifidobacterium* strains. *Appl Microbiol Biotechnol*, 2009, 81, 943-950.

Rechner, AR; Smith, MA; Kuhnle, G; Gibson, GR; Debnam, ES; Srai, SK; Moore, KP; Rice-Evans, CA. Colonic metabolism of dietary polyphenols: influence of structure on microbial fermentation products. *Free Radic Biol Med*, 2004, 36, 212-225.

Rowland, IR; Wiseman, H; Sanders, TA; Adlercreutz, H; Bowey, EA. Interindividual variation in metabolism of soy isoflavones and lignans: influence of habitual diet on equol production by the gut microflora. *Nutr Cancer*, 2000, 36(1), 27-32.

Rüfer, CE; Glatt, H; Kulling, SE. Structural elucidation of hydroxylated metabolites of the isoflavan equol by gas chromatography-mass spectrometry and high-performance liquid chromatography-mass spectrometry. *Drug Metab Dispos*, 2006, 34(1), 51-60.

Selma, MV; Espín, JC; Tomás-Barberán, FA. Intertaction between Phenolics and Gut Microbiota: Role in Human Health. *J Agric Food Chem*, 2009, 57, 6485-6501.

Scalbert, A; Williamson, G. Dietary intake and bioavailability of polyphenols. *J Nutr*, 2000 130, 2073S-85S.

Scheline, RR. Metabolism of foreign compounds by gastrointestinal microorganisms. *Pharmacol Rev*, 1973, 25, 451-523.

Schell, MA; Karmirantzou, M; Snel, B; Vilanova, D; Berger, B; Pessi, G; Zwahlen, MC; Desiere, F; Bork, P; Delley, M; Pridmore, RD; Arigoni, F. The genome sequence of *Bifidobacterium longum* reflects its adaptation to the human gastrointestinal tract. *Proc Natl Acad Sci*, USA, 2002, 99, 14422-14427.

Schoefer, L; Mohan, R; Braune, A; Birringer, M; Blaut, M. Anaerobic C-ring cleavage of genistein and daidzein by *Eubacterium ramulus*. FEMS Microbiology Letters, 2002, 208, 197-202.

Setchell, KD; Borriello, SP; Hulme, P; Kirk, DN; Axelson, M. Nonsteroidal estrogens of dietary origin: possible roles in hormone-dependent disease. *Am J Clin Nutr*, 1984, 40, 569-78.

Setchell, KDR; Cassidy, A. Dietary isoflavones: biological effects and relevance to human health. *J Nutr*, 1999, 129, 758S-767S.

Setchell, KDR; Brown, NM; Desai, P; Zimmer-Nechemias, L; Wolfe, BE; Brashear, WT; Kirschner, AS; Cassidy, A; Heubi, JE. Bioavailability of Pure Isoflavones in Healthy Humans and Analysis of Commercial Soy Isoflavone Supplements. *J. Nutr*, 2001, 131, 1362S-1375S.

Setchell, KD; Cole, SJ. Variations in isoflavone levels in soy foods and soy protein isolates and issues related to isoflavone databases and food labeling. *J Agric Food Chem*, 2003, 51, 4146-4155.

Setchell, KD; Clerici, C; Lephart, ED; Cole, SJ; Heenan, C; Castellani, D; Wolfe, BE; Nechemias-Zimmer, L; Brown, NM; Lund, TD; Handa, RJ; Heubi, JE. S-equol, a potent ligand for estrogen receptor beta, is the exclusive enantiomeric form of the soy isoflavone metabolite produced by human intestinal bacterial flora. *Am J Clin Nutr*., 2005, 81(5), 1072-1079.

Setchell, KD; Zhao, X; Shoaf, SE; Ragland, K. The pharmacokinetics of S-(-)equol administered as SE5-OH tablets to healthy postmenopausal women. *J Nutr*, 2009, 139(11), 2037-2043.

Simons, AL; Renouf, M; Hendrich, S; Murphy, PA. Metabolism of Glycitein (7, 4'-Dihydroxy-6-methoxy-isoflavone) by Human Gut Microflora. *J Agric Food Chem*, 2005, 53, 8519-8525.

Song, TT; Barua, K; Buseman, G; Murphy, PA. Isoflavone analysis: New internal standard and quality control. *Am J Clin Nutr*, 1998, 68, 1474S-1479S.

Song, TT; Hendrich, S; Murphy, PA. Estrogenic activity of glycitein, a soy isoflavone. *J Agric Food Chem*, 1999, 47, 1607-1610. *Erratum J Agric Food Chem*, 2002, 50, 2470.

Steer, TE; Johnson, IT; Gee, JM; Gibson, GR. Metabolism of the soyabean isoflavone glycoside genistin in vitro by human gut bacteria and the effect of prebiotics. *British Journal of Nutrition*, 2003 90, 635-642.

Tamura, M; Tsushida, T; Shinohara, K. Isolation of an isoflavone-metabolizing, *Clostridium*-like bacterium, strain TM-40, from human faeces. *Anaerobe*, 2007, 13, 32-35.

Tan, Y; Liu, M; Wu, B. Puerarin for acute ischaemic stroke. *Cochrane Database Syst Rev*, 2008, 23(1), CD004955.

Tsangalis, D; Ashton, JF; McGill, AEJ; Shah, NP. Enzymic transformation of isoflavone phytestrogens in soymilk by β-glucosidase-producing bifidobacteria. *J Food Sci*, 2002, 67, 3104-3113.

Tsangalis, D; Wilcox, G; Shah, NP; McGill, AE; Stojanovska, L. Urinary excretion of equol by postmenopausal women consuming soymilk fermented by probiotic bifidobacteria. *Eur J Clin Nutr*, 2007, 61, 438-441.

Uehara, M; Ohta, A; Sakai, K; Suzuki, K; Watanabe, S; Adlercreutz, H. Dietary Fructooligosaccharides Modify Intestinal Bioavailability of a Single Dose of Genistein and Daidzein and Affect Their Urinary Excretion and Kinetics in Blood of Rats. *J. Nutr.*, 2001, 131, 787-795.

Uchiyama, S; Ueno, T; Imaizumi, K; Kumemura, M; Masaki, K; Shimizu, S. Isoflavone-containing compositions. *Patent: Publication No*. WO9907392, 1999.

Uchiyama, S; Ueno, T; Suzuki, T. Composition containing lactic acid bacterium producing equol. Patent: Publication No. WO2005000042, 2005.

Uchiyama, S; Ueno, T; Suzuki, T. Identification of a Newly Isolated Equol-Producing Lactic Acid Bacterium from the Human Feces. *J Intestinal Microbiol*, 2007, 21, 217-220.

Wiseman, H; Casey, K; Bowey, EA; Duffy, R; Davies, M; Rowland, IR; Lloyd, AS; Murray, A; Thompson, R; Clarke, DB. Influence of 10 wk of soy consumption on plasma concentrations and excretion of isoflavonoids and on gut microflora metabolism in healthy adults. *Am J Clin Nutr*, 2004, 80, 692-699.

Wang, XL; Hur, HG; Lee, JH; Kim, KT; Kim, SI. Enantioselective Synthesis of S-Equol from Dihydrodaidzein by a Newly Isolated Anaerobic Human Intestinal Bacterium. *AEM*, 2005a, 71(1), 214-219.

Wang, XL; Shin, KH; Hur, HG; Kim, SI. Enhanced biosynthesis of dihydrodaidzein and dihydrogenistein by a newly isolated bovine rumen anaerobic bacterium. *J Biotechnol*, 2005b, 115, 261-269.

Wang, XL; Kim, HJ; Kang, SI; Kim, SI, Hur, HG. Production of phytestrogen S-equol from daidzein in mixed colture of two anaerobic bacteria. *Arch Microbiol*, 2007, 187, 155-160.

Xu, X; Harris, KS; Wang, HJ; Murphy, PA; Hendrich, S. Bioavalability of soybean isoflavones depends upon gut microflora in women. *J Nutr*, 1995, 125, 2307-2315.

Yasuda, T; Ueda, J; Ohsawa, K. Urinary Metabolites of Genistein Administered Orally to *Rats. Chem Pharm Bull*, 2001, 49(11), 1495-1497.

Yee, S; Berdock, GA; Kurata, Y; Enomoto, Y; Narumi, K; Hamada, S; Itoh, T; Shimomura, Y; Ueno, T. Acute and subchronic toxicity and genotoxicity of SE5-OH, an equol-rich product produced by *Lactococcus garvieae. Food and Chemical Toxicology*, 2008, 46, 2713-2720.

Yokoyama, SI; Suzuki, T. Isolation and characterization of novel equol-producing bacterium from human feces. *Biosci Biotechnol Biochem*, 2008, 72, 2660-2666.

Yokoyama, S; Niwa, T; Osawa, T; Suzuki, T. Characterization of an *O*-desmethylangolensin-producing bacterium isolated from human feces. *Arch Microbiol*, 2009, 192, 15-22.

Yu, ZT; Yao, W; Zhu, WY. Isolation and identification of equol-producing bacterial strains from cultures of pig faeces. *FEMS Microbiol Lett*, 2008, 282, 73-80.

Yuan, JP; Wang, JH; Liu, X. Metabolism of dietary soy isoflavones to equol by human intestinal microflora-implications for health. *Mol Nutr Food Res*, 2007, 51, 765-781.

Zhang, Y; Wang, G; Song, TT; Murphy, PA; Hendrich, S. Urinary Disposition of the Soybean Isoflavones Daidzein, Genistein and Glycitein Differs among Humans with Moderate Fecal Isoflavone. *J Nutr*, 1999, 129, 957-962.

Zhang, L; Zuo, Z; and Lin, G. Intestinal and hepatic glucuronidation of flavonoids. *Mol Pharm*, 2007 4, 833-845.

In: Isoflavones Biosynthesis, Occurence and Health Effects ISBN: 978-1-61728-113-6
Editor: Melanie J. Thompson, pp. 163-180 © 2010 Nova Science Publishers, Inc.

Chapter V

Bioactive Isoflavones: Implications in Cancer Treatment and Prevention

Alok Bhushan*[*], Vikas Bhardwaj, Cassandra Frandsen and James C. K. Lai

Department of Biomedical and Pharmaceutical Sciences, College of Pharmacy and ISU Biomedical Research Institute, Idaho State University, Pocatello, ID 83209

Abstract

Flavonoids are bioactive agents present in several plants that are used for food. These foods include legumes, soy, fruits, vegetables and different parts of plants. Vast amount of literature exits implicating the role of flavonoids in influencing changes in biological processes in cancer cells. The ability of flavonoids to induce changes in receptor activation, anti-estrogenic properties, tyrosine phosphorylation and cell signaling makes these agents important in cancer prevention and treatment and development of anticancer drugs. The antioxidant properties of flavonoids have also been extensively investigated. The above-mentioned cellular mechanisms are known to play critical roles in carcinogenesis, including metastasis and angiogenesis. Several isoflavones are in clinical trials and there has been epidemiological evidence for their role in cancer prevention and in decreasing incidence of several cancer types. Development of derivatives of isoflavones and associating these natural compounds with nanomaterials to give better efficacy and pharmacokinetic properties is some of the future areas of research which will have impact on incidence and treatment of an array of cancer types. This review will summarize some of the critical findings in the field and discuss the future directions in use of isoflavones in treatment and prevention of several selected cancer types.

[*] Corresponding author: Department of Bimedical and Pharmaceutical Sciences, College of Pharmacy, 970 S 5th Avenue, Idaho State University, Pocatello, ID 83209-8334, abhushan@pharmacy.isu.edu, Phone: 208-282-4408

Over half a million people in US die of cancer every year. Lung cancer is the major contributor of deaths due to smoking. The progression of cancer is a multi-step process and can be interrupted at several steps. The impact of consumption of isoflavones has been a topic of research and discussion for decades. To understand how isoflavones may block progression of cancer, it is necessary to discuss the process of carcinogenesis.

Keywords: Isoflavones, signaling and cancer

Carcinogenesis and Cancer Intervention

The process of carcinogenesis is conversion of a normal cell to cancer cell. It is a multi-step process sometimes taking years to decades before a mutated cell progresses to metastatic disease. The progression of cancer development involves an array of biochemical events from receptor activation, phosphorylation of several cell signaling events, activation of oncogenes and decreased activity of tumor suppressor genes. Of critical importance is the phosphorylation of proteins that drive the intracellular signaling processes that result in tumor progression.

The multi-step process of carcinogenesis includes initiation, promotion, malignant conversion and tumor progression. Bioactive compounds can have impact on any of these steps. Clonal expansion of initiated cells is the hallmark of promotion. Promotion results in accumulation of cells with mutation. The cellular proliferation during promotion is associated with phosphorylation of proteins at their tyrosine, serine and/or threonine moieties. The phosphorylation of these amino acids is critical in propagation of signals to the nucleus and in initiating a response.

Treatment of cancer cells with genistein, an isoflavone, results in inhibition of tyrosine kinase activity of several receptors. Populations that consume soybean, which contains genistein, have lower rates of cancer. Epidemiological studies from Asian countries whose populations consume beans and lentils also indicate lower cancer incidences in these countries. The exact molecular mechanisms underlying the actions of isoflavones are not well understood. This chapter reviews the literature focusing in the actions of isoflavones on various cancers.

Isoflavones and other Diseases

It was in 1967 when the cholesterol-lowering property of isoflavones was first observed (1). Later, the hypolipidemic effect of soy protein was noted by various researchers (2, 3).

In 1999, FDA claimed that daily consumption of soy protein prevents heart disease (4). A study published in 1999 showed that a mean consumption of 50g/day of soy proteins containing isoflavones lowered cholesterol LDL level by 3% compared to milk and other proteins. Although, the effect of soy proteins on high density lipoprotein (HDL), triglycerides and blood pressure was not significant, the diet reduced LDL and other lipid risk factors (5).

Epidemiological studies have shown that soy intake increases bone mineral density. The study reported that isoflavones increase bone-specific alkaline phosphatase (bone formation marker) in menopausal women (6, 7). Similar effects were observed in postmenopausal women (8, 9). These observations suggest that isoflavones play a role in reducing coronary artery disease as well as osteoporosis.

Effect of Isoflavones on Signal Transduction

Isoflavones such as genistein have been known to influence intracellular signaling. Some of the targets for their mechanisms of action include:

- Tyrosine kinase receptors
- Topoisomerase II
- Estrogen receptor
- ABC transporter
- Enzymes involved in phosphatidylinositol turnover.

Inhibitor of protein tyrosine kinase (PTK): This class of kinases play an important role in signal transduction from the extracellular to the intracellular region. Tyrosine kinases are of two types — transmembrane receptor PTK and cytoplasmic PTK. Members of transmembrane PTK family include epidermal growth factor receptor (EGFR), insulin receptor, fibroblast growth factor receptor (FGFR), vascular endothelial growth factor receptor (VEGFR), platelet-derived growth factor receptor (PDGFR) and others. Upon activation by growth factors, these receptors phosphorylate target proteins, which carry out the signal transduction pathway(s) leading to enzyme activation, second messenger generation and transcription of genes. Activation of these receptors leads to proliferation and differentiation of cells. Mutation of these receptors resulting in their constitutive activation can cause malignant transformation of cells. The activity of these receptors is known to be upregulated in many forms of cancer. The upregulated activity could be either due to upregulation in the amount of growth factors in the serum or mutation of the receptor leading to its constitutive activation. Up-regulated activity of these receptors has been implicated in cancer cell proliferation, inhibition of apoptosis, invasion and metastasis. A few studies also implicated the role of receptor tyrosine kinases (RTKs) in initiation of cancer. Hence these receptors serve as a good target for anti-cancer therapies.

In 1987, genistein was shown to have anti-protein tyrosine kinase property. Genistein was shown to inhibit activity of EGFR and oncogenic src *in vitro*. EGF-induced activation of EGFR was also inhibited by genistein in a cellular system. Although genistein inhibited EGFR-induced serine threonine kinase activity, there was no direct effect on the activity of serine threonine kinase (10). Later in 1990, Linassier *et al.* observed that the genistein is cytostatic at lower concentration and cytotoxic at higher concentrations when tested on NIH-3T3 cells. Genistein was also observed to inhibit EGF- and insulin-induced mitogenicity (11). However, the efficiency of genistein as a potent PTK inhibitor is controversial because of the high concentrations required for its anti-PTK activity in an intact cellular system. Various studies have shown that high concentration of genistein (100 μM or above) is required to observe the anti-PTK activity (10, 12-14). Peterson and Barnes, in 1993, observed that genistein failed to inhibit EGF-induced activation of EGFR; however, tyrphostin (EGFR inhibitor) was able to inhibit EGFR activation in prostate cancer cells at IC_{50} concentrations (15). On the other hand, genistein did not affect the activities of insulin receptor (16), platelet derived growth factor receptor (17) and nerve growth factor receptor (18) at concentration above 100 μM. A study by Matsukawa *et al.* compared the effects of genistein and daidzein (both isoflavones) on the cell cycle. They found that although genistein mainly arrest cells at G2/M phase, daidzein causes G1 cell cycle arrest (19).

Some of the studies discussed above suggest that genistein is a potent inhibitor of purified PTK *in vitro*; however, the inhibitory effect in an intact cellular system is observed at a high concentration or not observed at all. The latter observation implies limited penetration of genistein into the cell and its intracellular concentration did not attain those that are inhibitory to the PTK.

Topoisomerase II inhibitor: The physiological functions of DNA depend on its tertiary configuration. DNA replication and progression of the replication fork cause supercoiling of the DNA due to multiple twisting. However, for gene expression, the DNA molecule is required to unwind. This untangling and relaxation of the supercoiled DNA is mediated by DNA topoisomerase enzymes. These enzymes can modify the tertiary conformation of the DNA without changing its primary structure. In eukaryotic cells, topoisomerases are of two types, topoisomerase I and topoisomerase II (topo I and II). Type I catalyzes the breaking and joining of a single strand of DNA. Topo II helps in cleaving of both strands of the DNA (20, 21). A Topo II enzyme is present in two isoforms- topo IIα (encoding a 170 kDa protein) and topo IIβ (encoding a 180 kDa protein) (22). A study in the early nineties suggested that the expression of topo IIα is regulated with the cell cycle, while that for the β form is not (23). The expression of topo IIα increases in the proliferating cells peaking at G2/M phase. In contrast, the level of this enzyme is downregulated in non-dividing cells or cells in quiescent stage (24). Upregulation of topo IIα levels in proliferating and downregulation in non-dividing cells suggest a role of the enzyme in cellular proliferation.

Different studies have shown high levels of topoisomerase IIα in cancer cells. A study in 1992 found that monoclonal antibody KiS1 recognizes proliferating cells. Increased staining with this antibody was found to correlate with poor prognosis of breast cancer patients (25). This antibody recognizes a 170 kDa protein later identified as DNA topoisomerase IIα (26). Increased expression of topo IIα in breast and non-small cell lung carcinoma also correlates with elevated levels of erbB2 protein (27- 29). Recent studies have correlated high topo IIα levels with decreased survival (30- 32), malignancy (33), and poor differentiation (34) in several tumor types. Turley *et al.* analyzed the levels of the two isoforms of topoisomerase II in normal and neoplastic tissues and noted that topo IIβ is present in all normal and neoplastic cells. However, the expression of topo IIα was higher in proliferating compared to quiescent cells. Topo IIα was also found in tumors unresponsive to anti-topoisomerase II treatment suggesting that low level of topo II is not responsible for resistance to anti-topo II treatment (35). Due to its higher expression and importance in proliferating cells, topo II has been used as a target for anti-cancer drugs. Some of the potent anti-topo II agents such as etoposide, adriamycin etc are used as anti-tumor agents (36- 38).

Markovits *et al.* found that Chinese hamster lung cell line DC-3F/9-OH-E was resistant to genistein induced cytotoxicity compared to their parent cell line DC-3F. The DC-3F/9-OH-E cell line has altered topoisomerase II activity. Those researchers also observed that genistein stimulates the topo II-induced DNA strand breakage at a site different from that induced by etoposide without intercalating with DNA and inhibits decatenation activity of topo II at concentrations of 20 μM and above (39). Similarly, other studies have shown the anti-topoisomerase activity of genistein in leukemia and thrombocytes (40, 41).

Inhibitor of phosphatidyl-inositol turnover: Inositol-1,4,5-triphosphate (IP3) is involved in signal transduction. The levels of IP3 increase with the activation of phospholipase C (PLC). PLC hydrolyses phosphatidyl-inositol 4,5- bisphosphate (PIP2) to IP3 and di-acyl glycerol (DAG). PIP2 is restored by the action of the enzymes PI kinase and PIP kinase. The levels of IP3 and PLC are found to be upregulated in cancer (42). Genistein induces lowering of the concentration of IP_3 concentration by inhibiting PI kinase and PIP kinase (43). A supporting evidence for this effect came from Ozaki *et al.* who showed genistein attenuated intracellular Ca^{2+} levels. On the other hand, genistein induces reduction in the PIP2 levels without affecting the levels of PIP suggesting that genistein acted at the level of PIP kinase which converts PIP into PIP2 (44). Previously, a study by Imoto *et al.* also showed that Psi-tectorigenin, an isoflavone, reduces phosphatidyl-inositol turnover with IC_{50} of 1 μg/ml in A431 cells without affecting EGFR protein kinase (45).

Inhibitor of proteins involved in multidrug resistance: Overexpression of P-glycoprotein (Pgp) and multidrug resistant protein (MRP) has been associated with multidrug resistance. Drug resistance has been recognized as an important reason for failure of chemotherapy (46). These proteins transport the drugs from intracellular to extracellular regions utilizing ATP and are also known as ABC transporters (ATP binding cassettes) (47). Treatment with genistein has resulted in increased accumulation of drugs in the resistant cells. Earlier, genistein was thought to specifically block actions of MRP because treatment with genistein resulted in increased concentration of intracellular daunorubicin in non-Pgp MRP cell lines, but, did not affect the drug concentration in Pgp MRP cells. Three other isoflavones biochanin A, apigenin and quercetin were also effective in increasing the concentration of the drug in the cell (48). However, Castro *et al.* showed that genistein increases the levels of daunorubicin and rhodamine 123 in Pgp MRP cells (49). In 1998, Conceil *et al.* reported that the isoflavone- binding region overlaps with the ATP-binding region on Pgp thereby inhibiting Pgp activity (50). Later, in 2008, Chung *et al.* showed that the isoflavone biochanin A decreased the efflux of dounomycin from Pgp-expressing breast cancer cells (51). A recent study by Li *et al.* noted that isoflavone reverses the epithelial mesenchymal transition (EMT) characteristics in pancreatic cancer cells which were resistant to gemcitabine compared to their parent cell line (52).

Isoflavones as Anti-Cancer Agents

Effect on apoptosis and cell cycle: The effect of isoflavones, especially that of genistein, on signal transduction pathways has prompted their role as anti-cancer agents. Genistein treatment induces apoptosis in rapidly growing cells. The induction of apoptosis can be due to:

1) Inhibition of topo II causing DNA strand breakage;

2) Inhibition of PTK which protects cell against apoptosis; and/or
3) Interfering with the IP3 induction

A study by Tilly *et al.* reported that genistein suppresses the anti-apoptotic property of EGF on rat ovarian granulose cells kept in serum free medium (53). Later in 1992, Traganos *et al.* reported the effect of genistein on apoptosis and cell cycle progression in two leukemia cell lines. Genistein decreased the cell survival by about 50% at 8.5 and 13 μg/ml in HL-60 (acute myeloid leukemia) and MOLT-4 (acute lymphoblastic leukemia) cells, respectively (54). Short-term (4-8 hr) genistein exposure led to suppression of cell progression from S phase to G2 phase. The progression through the early G1 phase was also suppressed. Long-term genistein exposure (24 hr) led to S phase arrest (54). Similarly, studies with other cancer types also provided evidence regarding the proapoptotic effect of genistein (55- 57).

Induction of cellular differentiation: Genistien treatment has shown to induce diffentiation in various cancer cell types. Genistein-treated neuroblastoma cells exhibit morphological features of differentiation (e.g., development of dendritic extention) and caused cell growth inhibition and increase in the acetylcholinesterase activity (58). Other studies have also shown that genistein induce cellular differentiation in leukemia and melanoma cells (59, 60).

Antioxidant effects: Cancer cells are known to have high levels of reactive oxygen species (ROS), which play an important role in carcinogenesis and tumor promotion [61]. Isoflavones such as genistein have been shown to lower the cellular pool of ROS. In HL-60 leukemia cells, hydrogen peroxide formation was inhibited by genistein in a dose-dependent manner (61). Furthermore, genistein blocked superoxide ion formation and also scavenged exogenously-added hydrogen peroxide (61). In one animal study, oral administration of genistein (250 ppm) for 30 days was shown to significantly increase the activity of antioxidant enzymes such as superoxide dismutase, glutathione peroxidase and catalase (62).

Anti-angiogenesis effect: Tumors require glucose and oxygen to survive and proliferate. Upon reaching a size of a few millimeters, a tumor requires the formation of blood vessels for transport of the nutrients to its core. The process of angiogenesis (growth of a new blood vessel from an existing one) is also important in metastasis of tumor cells from one location to another. Treatment with genistein decreases angiogenesis in tumor cells. For example, Fotsis *et al* (1993) showed that genistein reduces the proliferation of vascular endothelial cells and *in vitro* angiogenesis at 5 and 150 μM, respectively (63). The microvascular density of the tumors in mice injected with bladder carcinoma cells was found to be reduced upon treatment with genistein (64).

Prostate Cancer and Isoflavones

Prostate cancer is the third most prevalent form of cancer in the US and is estimated to account for more than 27,000 deaths in the US in 2009 (65). The incidence rate of prostate cancer is higher in the US and Europe compared to Asia. Different genetic, epigenetic and environmental factors have been suggested to contribute to the development of prostate cancer. Among the environmental factors, diet is presumed to play an important role in prostate cancer development.

Prostate cancer and diet: Diet high in vegetables and vitamins is associated with a lower risk of prostate cancer while diet high in red meat is associated with a higher risk of prostate cancer. High intake of antioxidants sources such as tomatoes, soy isoflavones and Vitamin E decreases such a risk. For example, Giovannucci *et al.* (66) reported that high intake of tomato sauce reduces the incidence of advanced prostate cancer by 35%. Tomatoes contain large amounts of lycopene, a carotenoid, which can efficiently quench singlet oxygen, thus acting as an antioxidant. Diet supplemented with Vitamin E is associated with a decreased risk of prostate cancer by 30-40% when compared to that of placebo. Other epidemiological studies suggest that γ-tocopherol may also have a protective effect against prostate cancer. Vegetables such as cauliflower, cabbage and broccoli are high in sulforaphane and indole-3 carbinol which are anti-carcinogenic. These compounds can induce antioxidant enzymes and are also suggested to have anti-apoptotic, anti-proliferative and anti-metastatic activity Although increased intake of cruciferous vegetables is known to reduce the risk of prostate cancer, other studies reported the absence of such beneficial effects (67). A diet high in red meat and milk is associated with an increased risk of prostate cancer. However, other studies have shown that there is no correlation between meat intake and prostate cancer risk.

Isoflavones and Prostate Cancer

Molecular effects of isoflavones: Isoflavones are known to act on estrogen receptors. Upon binding to its receptor estradiol produces a feedback response to the hypothalamus, thereby inhibiting the release of leutinizing hormone (LH) and subsequently decreasing the production and secretion of testosterone. Shen *et al.* in 2000 noted that, at physiological concentrations, genistein reduces the proliferation of LNCaP prostate cancer cells by halting cell cycle progression at the G1 phase. The levels of p21, p27 and cyclin dependent kinase (CDK) inhibitors were increased (68). Other studies have also shown the inhibition of cell cycle progression at G2/M phase by genistein in PC3 and LNCaP prostate cancer cells. In a separate report, Handayani *et al.* (2006) observed that isoflavones alter the expression of various genes involved in metabolism, metastasis, adhesion, angiogensis and cellular growth in PC3 prostate cancer cells (69). Other studies have also shown a reduction in the metastatic potential and invasive markers upon treatment of prostate cancer cells with genistein.

Effect of isoflavones on animal models: Zhou *et al.* (1999) showed that increasing concentration of genistien in the diet reduces the tumor volume in immune-deficient mice injected with LnCaP prostate cancer cells (70). Fritz *et al.* (2002) reported that presence of genistein in a diet reduced the prostate androgen receptor mRNA level by 85% in a 70-day study. Furthermore, the levels of estrogen receptor α and β were also significantly reduced in the prostate after treatment with genistein (71).

Clinical effects of isoflavones in humans: Several studies have documented the effect of soy isoflavones on prostate specific antigen (PSA) levels in the humans but few have shown a positive effect. Adams *et al.* (2004) gave subjects a drink containing either high or low isoflavone content for a 12 month period and found the level of PSA in the two groups of subjects differed by 0.5%, which was not, however, statistically significant (72). Similarly, in 52 men with prostate cancer given an isoflavone-rich diet for 6 months, (DeVere *et al.* [2004]) only one patient had his PSA level reduced by more than 50%, with seven others with reduction in PSA level at less than 50% (73). In contrast with the above two studies Dalais *et al.* (2004) observed a reduction on PSA levels of prostate cancer patients fed with soy grits whereas the PSA levels increased in the placebo group. Shroder *et al.* (2005) showed that feeding with large amounts of isoflavones reduces the doubling time for PSA in prostate cancer patients by 2.6 fold compared to that in the control group (75).

Isoflavones and Breast Cancer

Estrogen and Breast Cancer

Estrogens are steroid hormones that play an important role in the development and progression of breast cancer. Estrogen has been shown to encourage proliferation of breast cancer cells *in vitro* (76). The importance of estrogens was first established over 100 years ago when researchers found that removing the ovaries of women with breast cancer increased their chances of survival (77). Although the exact mechanism behind estrogens' ability to promote tumor growth is not yet fully understood, there are at least two potential explanations. Environmental factors produce free radicals that damage DNA and estrogens then act as a promoter (76). Another mechanism may be that estrogens are themselves mutagens, implicating them as the initiators and promoters of cancer (Cordera and Jordan, 2006). Estrogens interact with the breast epithelial cells via the estrogen receptor (76). The estrogen receptor (ER) is a ligand-activated transcription factor that belongs to a superfamily of nuclear receptors. The two main forms of ER are ERα and ERβ. ERα is expressed in bone, hypothalamus and major female reproductive organs, including the breast ERβ may be expressed in these areas, but it is highly expressed in the ovaries. Both forms of the receptor bind to the same DNA response elements and exhibit similar ligand-binding characteristics (78). Both receptor types bind estrogens to the same degree; however, they differ in their affinity for binding other ligands, and in the responses that are produced (79). ER levels are normally low in breast tissue and increased levels have been found to correlate with an increased risk of breast cancer (76). As early as 1953, the isoflavones found in soy act were found to be weak estrogens, with the ability to increase the weight of the uterus of young rodents (80); thus, this mechanism led to the classification of isoflavones as phytoestrogens. During the 1960's, Folman and Pope elaborated upon this finding in determining that genistein, an isoflavone found in soy, inhibited the action of more potent estrogens, possibly by competing for receptor binding sites (81). The relative affinity of genistein for the estrogen receptor is 100-1000 fold less than that of estogen. However, it can compete for binding sites. During the 1990's, multiple studies further examined the interaction of isoflavones with estrogen receptors and found a preference for isoflavones to bind ERβ relative to ERα. This classification as both an agonist and an antagonist may, in part, explain some of the seeming contradictions that are encountered when examining the relationship between isoflavones and breast cancer.

Epidemiological Studies

Historically, breast cancer rates are higher in Western countries when compared with those in Asian countries; however, after migration to the United States, the rate of breast cancer occurrence among Asians approaches that of Caucasian American women within a few generations (82). Early studies indicated that this increase was not obvious until the second generation of Asians born within the United States (83). Several factors could account for this observation. First, the protective effect against increased breast cancer may be due to modifiable lifestyle and environmental factors and not genetics. Secondly, the exposure to the Western lifestyle at an early age may be important to breast carcinogenesis.

The focus on lifestyle factors led to the notion that diet may, in part, be responsible for the decrease in breast cancer rates in Asian women when compared with those in American women. In 1991, Lee *et al.* first reported an inverse association between soy intake and breast cancer risk among Chinese women in Singapore (84). More recently, epidemiological studies have focused on isoflavones, found in soybeans and other food sources derived from plants, as being a contributor to decreased breast cancer incidence (85, 86). The isoflavones found in soy are the most extensively studied in relation to breast cancer due to the epidemiological findings of decreased breast cancer incidence among the Asian populations where a distinguishable and notable intake of soy foods is exhibited. For example, among older adults in Japan and China soy intake provides approximately 25-50 mg of isoflavones/day. However, in the United States and Europe, the average daily isoflavone intake is < 3 mg/day (87). The main isoflavones found in the soybean are genestin and diadzin which are then metabolized to genistein and daidzein (82).

One meta-analysis by Trock et al. (2006) based on epidemiological data from 18 studies published noted that among all women, high soy intake was modestly associated with decreased breast cancer risk; however, this association did not hold when the focus was narrowed to include only women in Asian countries. The protective effect was more pronounced in pre- than in postmenopausal women (88). In 2008, a review by Wu *et al.* also examined the epidemiological body of evidence relating isoflavone exposure to breast cancer risk. Given that Asians differ from Westerners in the types and amounts of soy consumed, this review looked at the evidence from two different meta-analyses. One examined the risk association studies among high soy consuming Asians and the other examined risk association studies among low soy consuming Western populations. For the Asian population a statistically significant 29% reduction of breast cancer was observed among women with a high amount (> 20 mg/day) of isoflavone intake when compared with breast cancer incidence in population with the lowest levels of intake (<5 mg/day). A 12% reduction in breast cancer risk was observed among women with an intake of approximately 10 mg/day (85). When the relationship between soy intake and breast cancer risk was examined among 11 studies involving Westerners, there was no association with decreased breast cancer risk when the high soy intake group (median approximately 0.8 mg/day) when compared with the low intake group (median intake of 0.15 mg or less per day). However, results from those who consumed relatively low levels in the West were found to be similar to those with corresponding consumption levels in Asians. Furthermore, little evidence was found to support a modifying effect by either menopausal status or body weight (89). The conflicting results between the two studies discussed above may be attributed to differences between the cases used and in the methods for determining isoflavone intake.

Isoflavones and Breast Cancer *in Vivo*

A number of *in vivo* studies have suggested that isoflavones exert a protective effect against chemically-induced mammary carcinogenesis. However, other studies have suggested that genistein acts as a tumor promoter. In one study published in 1998 by Hsieh *et al.*, ovariectomized athymic mice were fed a diet which contained either 2 mg of estradiol or 750 ppm of genistein. These mice were then compared with a third, control group which was fed neither. MCF-7 breast cancer cells were implanted into 35-day-old mice and after a growth period of 5-6 weeks, the resulting tumors were measured weekly. Mammary gland growth was significantly ($p < 0.05$) increased in mice fed either genistein or estradiol when compared with the control. The genistein group and the estradiol group did not differ significantly. The uterine weight of the genistein-treated mice was also increased when compared with the controls. These findings suggest that genistein does have the ability to stimulate estrogenic responses, and therefore stimulate the growth of estrogen dependent breast cancer (90). However, an earlier study published in 1995 by Lamartiniere *et al.* examined the effects of genistein on chemically-induced mammary cancer in Sprague-Dawley rats. In this study, rats were injected with genistein at 2, 4, and 6 days postpartum. At day 50, they were exposed to Dimethylbenz[a]anthracene [DMBA]. The rats that had been treated with genistein, exhibited an increased latency and decreased incidence and multiplicity of mammary tumors. After examination of the terminal end buds, these researchers postulated that genistein acted in a chemo-protective manner to enhance maturation of the terminal end buds and by altering the endocrine system to reduce cellular proliferation within the mammary glands (91). This observation is consistent with findings of Barnes *et al.* (1997) who also reported a longer latency period and decreased number of induced mammary tumors in rats exposed to genistein neonatally. However, in rats treated after 35 days of age, the reduction in breast cancer risk was diminished (82). These combined findings further the idea that timing to isoflavone exposure may be critical to the ability to prevent breast cancer and may help partially explain the rapid increase in breast cancer rates of Asian immigrants within a few generations.

Breast Cancer Cell Proliferation

Since isoflavones may act both as an agonist and an antagonist of the estrogen receptor, much research has been undertaken to determine under what circumstances isoflavones act as an agonist or antagonist of the estrogen receptor. Genistein has been shown to elicit a concentration-dependent dual threshold effect on the growth of estrogen-dependent breast cancer (MCF-7) cells *in vitro*. At concentrations from 0.01 to 1 μM, a dose-dependent increase in cell proliferation was observed. At 1 μM, maximal growth stimulation occurred and was sustained up to 10 μM. At concentrations from 25-100 μM, a dose-dependent decrease in proliferation was observed (90).

While phytoestrogens, such as genistein, are able to stimulate the growth of ER+ breast cancer cells at lower concentrations, they have not been shown to stimulate the growth of ER- breast cancer cells. However, at higher concentrations the growth inhibitory effect observed with the ER+ cells is also observed with ER- cell types. These results suggest that multiple mechanisms may exist by which genistein can alter the growth of cells. At low doses the mechanism of action is probably ER-mediated while at the higher doses, some other mechanism of action is exerted (92).

Apoptosis and the Cell Cycle

Many studies have examined the role of isoflavones in apoptosis. Treatment with genistein induces G_2/M cell cycle arrest: this effect is more pronounced in malignant breast cells versus normal breast cells (93). Furthermore, when PARP (Poly (ADP-ribose) polymerase) cleavage (an indicator of apoptosis) was examined, a similar differential effect was noted. These researchers noted apoptotic cleavage fragments within the malignant MDA-MB-231 breast cell line but not in the normal MCF-10A breast epithelial cell line, suggesting that cancerous cells are more sensitive to genistein-mediated cell cycle arrest and apoptosis compared with normal breast cells (93). Furthermore, the number of apoptotic cells increases with increased dose. Upregulation of Bax (a pro-apoptotic protein) and $p21^{WAF1}$ expression but downregulation of p53 and Bcl-2 were observed in genistein-treated MDA-MB-231 cells (94). A more recent study published in 2007 delved further into the genistein-Bcl-2 interaction. These results seemed to indicate that when exposed to genistein, Bcl-2 switches from an anti-apoptotic to a pro-apoptotic protein. The mechanism of action involves an altered Bcl-2/Bax ratio and increased cytochrome c release resulting in modified cell regulation (95).

ANGIOGENESIS and INVASION

Genistein is known to inhibit invasion in both MCF-7 and MDA-MB-231 breast cancer cell lines by downregulating MMP-9 and upregulating tissue inhibitor of MMP-1 (92). Using nude mouse xenografts, these researchers also showed that genistein induced inhibition of angiogenesis by decreasing vessel density and decreasing the levels of vascular endothelial growth factor and transforming growth factor-β1 (96).

Inhibition of Tyrosine Kinase

Genistein is known to inhibit tyrosine phosphorylation as well as total HER-2 protein expression in the BT-474 cell line. However, estradiol did not exhibit this effect. These observations strengthen the notion that the inhibitory effect of genistein is an estrogen-independent mechanism more attributable to inhibition of tyrosine kinase activity (97).

Future Directions

Because the biological effects of isoflavones have not been fully elucidated, it is important to emphasize that their mechanisms of action and therapeutic implications in cancer treatment and prevention have yet to be comprehensively realized. Current literature is however mainly focused on genistein. Thus, the other isoflavones in plants clearly need to be much more systematically and comprehensively investigated. Another worthwhile and productive line of inquiry is the study of synthetic analogs of isoflavones, which may have therapeutic effects on several cancer types. Furthermore, the bioactive nature of isoflavones deserves more research emphasis as their low toxicity and effects on processes involved in cancer progression may play important role(s) in prevention as well as treatment of multiple cancer types.

References

[1] Hodges, R. E., Krehl, W. A., Stone, D. B. & Lopez, A. (1967). Dietary carbohydrates and low cholesterol diets: effects on serum lipids on man. *Am J Clin Nutr. 20*, 198-208

[2] Carroll, K. K. (1991) Review of clinical studies on cholesterollowering response to soy protein. *J. Am. Diet. Assoc., 91,* 820-827.

[3] Erdman, J. W., Jr. & Fordyce, E. (1989). Soy products and the human diet. Am. *J. Clin. Nutr., 49,* 725-737.

[4] U.S. Food and Drug Administration. Food labeling health claims: soy protein and coronary heart disease. Food and Drug Administration, HHS. Final rule. *Fed Regist.*, 1999, 64, 57700-33.

[5] Ma, D. F., Qin, L. Q., Wang, P. Y. & Katoh, R. (2007). Soy isoflavone intake increases bone mineral density in the spine of menopausal women: Meta-analysis of randomized controlled trials. *Clin Nutr. 27,* 57-64.

[6] Ma, D. F., Qin, L. Q., Wang, P. Y. & Katoh, R. (2008). Soy isoflavone intake inhibits bone resorption and stimulates bone formation in menopausal women: meta-analysis of randomized controlled trials. *Eur J Clin Nutr., 62,* 155-61.

[7] Morabito, N., Crisafulli, A., Vergara, C., Gaudio, A., Lasco, A., Frisina, N., D'Anna, R., Corrado, F., Pizzoleo, M. A., Cincotta, M., Altavilla, D., Ientile, R. & Squadrito, F. (2002). Effects of genistein and hormone-replacement therapy on bone loss in early postmenopausal women: a randomized double-blind placebo-controlled study. *J Bone Miner Res., 17*, 1904-1912.

[8] Marini, H., Minutoli, L., Polito, F., Bitto, A., Altavilla, D., Atteritano, M., Gaudio, A., Mazzaferro, S., Frisina, A., Frisina, N., Lubrano, C., Bonaiuto, M., D'Anna, R., Cannata, M. L., Corrado, F., Adamo, E. B., Wilson, S. & Squadrito, F. (2007). Effects of the phytoestrogen genistein on bone metabolism in osteopenic postmenopausal women: a randomized trial. *Ann Intern Med., 146,* 839-47.

[9] Sacks, F. M., Lichtenstein, A., Van Horn, L., Harris, W., Kris-Etherton, P. & Winston, M. (2006). Soy protein, isoflavones, and cardiovascular health. An American Heart Association Science Advisory for Professionals from the *Nutrition Committee. Circulation., 113*, 1034-44.

[10] Akiyama, T., Ishida, J., Nakagawa, S., Ogawara, H., Watanabe, S., Itoh, N., Shibuya, M. & Fukami, Y. (1987). Genistein, a specific inhibitor of tyrosine-specific protein kinases. *J Biol Chem., 262(12)*, 5592-5.

[11] Linassier, C., Pierre, M., Le Pecq, J. B. & Pierre, J. (1990). Mechanisms of action in NIH-3T3 cells of genistein, an inhibitor *Biochem Pharmacol., 39(1)*, 187-93.

[12] Barnes, S. & Peterson, T. G. (1995). Biochemical targets of the isoflavone genistein in tumor cell lines. *Proc Soc Exp Biol Med., 208(1)*, 103-8.

[13] Spinozzi, F., Pagliacci, M. C., Migliorati, G., Moraca, R., Grignani, F., Riccardi, C. & Nicoletti, I. (1994). The natural tyrosine kinase inhibitor genistein produces cell cycle arrest and apoptosis in Jurkat T-leukemia cells. *Leuk Res. 18(6),* 431-9.

[14] Pagliacci, M. C., Smacchia, M., Migliorati, G., Grignani, F., Riccardi, C. & Nicoletti, I. (1994). Growth-inhibitory effects of the natural phyto-oestrogen genistein in MCF-7 human breast cancer cells. *Eur J Cancer., 30A(11),* 1675-82.

[15] Peterson, T. G. & Barnes, S. (1993). Isofalvones inhibit the growth of human prostate cancer cell lines without inhibiting epidermal growth factor receptor autophosphorylation. *Prostate, 22,* 335-345.

[16] Alber, A., Smith, J., Randazzo, P. A., Rothenburg, P. L. & Jarett, L. (1992). Genistein differentially inhibits postreceptor effects of insulin in rat adipocytes without inhibiting the insulin receptor kinase. *J Biol Chem, 267,* 3946-3951.

[17] Davidai, G., Lee, A., Schvart, Z. I. & Hazum, E. (1992). PDGF induces tyrosine phosphorylation in osteoblast-like cells: relevance to mitogenesis. *Am J Physiol, 263,* E205-E209.

[18] Nakafuku, M., Satoh, T. & Kaziro, Y. (1992). Differentiation factors, including nerve growth factor, fibroblast growth factor and interleukin-6 induce an accumulation of an active Ras-GTP complex in rat pheochromocytome PC12 cells. *J Biol Chem, 267*, 19448-19454.

[19] Matsukawa, Y., Marui, N., Sakai, T., Satomi, Y., Yoshida, M., Matsumoto, K., Nishino, H. & Aoike, A. (1993). Genistein arrests cell *Cancer Res., 53(6),* 1328-31.

[20] Rose, K. M. (1988). DNA topoisomerases as targets for chemotherapy. *FASEB J. 2(9)*, 2474-8.

[21] Kellner, U., Sehested, M., Jensen, P. B., Gieseler, F. & Rudolph, P. (2002). Culprit and victim - DNA topoisomerase II. *Lancet Oncol., 3(4),* 235-43.

[22] Jenkins, J. R., Ayton, P., Jones, T., Davies, S. L., Simmons, D. L., Harris, A. L., Sheer, D. & Hickson, I. D. (1992). Isolation of cDNA clones encoding the beta isozyme of human DNA topoisomerase II and localisation of the gene to chromosome 3p24. *Nucleic Acids Res, 20,* 5587-92.

[23] Woessner, R. D., Mattern, M. R., Mirabelli, C. K., Johnson, R. K. & Drake, F. H. (1991). Proliferation- and cell cycle-dependent differences in expression of the 170 kilodalton and 180 kilodalton forms of topoisomerase II in NIH-3T3 cells. *Cell Growth Differ, 2*, 209-14.

[24] Adachi, N., Nomoto, M., Kohno, K. & Koyama, H. (2000). Cell-cycle regulation of the DNA topoisomerase IIα promoter is mediated by proximal CCAAT boxes: possible involvement of acetylation. *Gene, 245,* 49-57.

[25] Sampson, S. A., Kreipe .H., Gillett, C. E., Smith, P., Chaudary, M. A., Khan, A., Wicks, K., Parwaresch, R. & Barnes, D. M. (1992). KiS1–a novel monoclonal antibody which recognizes proliferating cells: evaluation of its relationship to prognosis in mammary carcinoma. *J Pathol, 168,* 179-85.

[26] Kellner, U., Heidebrecht, H. J., Rudolph, P., Biersack, H., Buck, F., Dakowski, T., Wacker, H. H., Domanowski, M., Seidel, A., Westergaard, O. & Parwaresch, R. (1997). Detection of human topoisomerase II alpha in cell lines and tissues: characterization of five novel monoclonal antibodies. *J Histochem Cytochem, 45*, 251-63.

[27] Järvinen, T. A., Kononen, J., Pelto-Huikko, M. & Isola, J. (1996). Expression of topoisomerase IIα is associated with rapid cell proliferation., aneuploidy, and c-erbB2 overexpression in breast cancer. *Am J Pathol, 148,* 2073-82.

[28] Keith, W. N., Tan, K. B. & Brown, R. (1992). Amplification of the topoisomerase IIα gene in a non-small cell lung cancer cell line and characterisation of polymorphisms at the human topoisomerase IIα and β loci in normal tissue. *Genes Chromosomes Cancer 4,* 169-75.

[29] Smith, K., Houlbrook, S., Greenall, M., Carmichael, J. & Harris, A. L. (1993). Topoisomerase IIα co-amplification with erbB2 in human primary breast cancer and breast cancer cell lines: relationship to m-AMSA and mitoxantrone sensitivity. *Oncogene, 8,* 933-8.

[30] Albadine, R., Wang, W., Brownlee, N. A., Toubaji, A., Billis, A., Argani, P., Epstein, J. I., Garvin, A. J., Cousi, R., Schaeffer, E. M., Pavlovich, C. & Netto, G. J. (2009). Topoisomerase II alpha status in renal medullary carcinoma: immuno-expression and gene copy alterations of a potential target of therapy. *J Urol., 182(2),* 735-4.

[31] Faggad, A., Darb-Esfahani, S., Wirtz, R., Sinn, B., Sehouli, J., Könsgen, D., Lage, H., Weichert, W., Noske, A., Budczies, J., Müller, B. M., Buckendahl, A. C., Röske, A., Eldin Elwali, N., Dietel, M. & Denkert, C. (2009). Topoisomerase IIalpha mRNA and protein expression in ovarian carcinoma: correlation with clinicopathological factors and prognosis. *Mod Pathol., 22(4),* 579-88.

[32] Chen, C. C., Gau, J. P., You, J. Y., Lee, K. D., Yu, Y. B., Lu, C. H., Lin, J. T., Lan, C., Lo, W. H., Liu, J. M. & Yang, C. F. (2009). Prognostic significance of beta-catenin anfood d topoisomerase IIalpha in de novo acute myeloid leukemia. *Am J Hematol., 84(2),* 87-92

[33] Diaz-Rubio, J. L., Duarte-Rojo, A., Saqui-Salces, M., Gamboa-Dominguez, A. & Robles-Diaz, G. (2004). Cellular proliferative fraction measured with topoisomerase IIalpha predicts malignancy in endocrine pancreatic tumors. *Arch Pathol Lab Med., 128(4),* 426-9.

[34] Willman, J. H. & Holden, J. A. (2000). Immunohistochemical staining for DNA topoisomerase II-alpha in benign, premalignant, and malignant lesions of the prostate. *Prostate., 42(4)*, 280-6

[35] Turley, H., Comley, M., Houlbrook, S., Nozaki, N., Kikuchi, A., Hickson, I. D., Gatter, K. & Harris, A. L. (1997). The distribution and expression of the two isoforms of DNA topoisomerase II in normal and neoplastic human tissues. *Br J Cancer. 75(9)*,1340-6.

[36] Björkholm M. (1990). Etoposide and teniposide in the treatment of acute leukemia *Med Oncol Tumor Pharmacother., 7(1),* 3-10.

[37] Gliński, B., Pawlicki, M., Reinfuss, M., Skołyszewski, J., Brandys, A., Krzemieniecki, K., Zuchowska, B., Mitus, J., Stelmach, A. & Walasek, T. (1997). Multimodality treatment of noninflammatory stage IIIb breast cancer *J Surg Oncol. 66(3)*, 179-85.

[38] Lucci, A., Han, T. Y., Liu, Y. Y., Giuliano, A. E. & Cabot, M. C. (1999). Multidrug resistance *Cancer., 86(2),* 300-11.

[39] Markovits, J., Linassier, C., Fossé, P., Couprie, J., Pierre, J., Jacquemin-Sablon, A., Saucier, J. M., Le Pecq, J. B. & Larsen, A. K. (1989). Inhibitory effects of the tyrosine *Cancer Res., 49(18),* 5111-7.

[40] McCabe, M. J. Jr & Orrenius, S. (1993). Genistein induces apoptosis in immature human thymocytes by inhibiting topoisomerase-II. *Biochem Biophys Res Commun., 194(2),* 944-50.

[41] Finlay, G. J., Holdaway, K. M. & Baguley, B. C. (1994). Comparison of the effects of genistein and amsacrine *Oncol Res., 6(1),* 33-7.

[42] Weber, G., Shen, F., Prajda, N., Yang, H., Li, W., Yeh, A., Csokay, B., Olah, E. & Look, K. Y. (1997). Regulation of the signal transduction *Adv Enzyme Regul., 37,* 35-55.

[43] Makishima, M., Honma, Y., Hozumi, M., Sampi, K., Hattori, M., Umezawa, K. & Motoyoshi, K. (1991). Effects of inhibitors of protein tyrosine *Leuk Res., 15(8),* 701-8.

[44] Ozaki, Y., Yatomi, Y., Jinnai, Y. & Kume, S. (1993). Effects of genistein, a tyrosine *Biochem Pharmacol., 46(3),* 395-403.

[45] Imoto, M., Yamashita, T., Sawa, T., Kurasawa, S., Naganawa, H., Takeuchi, T., Bao-quan, Z. & Umezawa, K. (1988). Inhibition of cellular phosphatidylinositol turnover *FEBS Lett., 230(1-2),* 43-6.

[46] Gottesman, M. M. & Pastan, I. (1993). Biochemistry of multidrug resistance mediated by the multidrug transporter. *Annu RevBiochem, 62,* 385-427.

[47] Nooter, K. & Stoter, G. (1996). Molecular mechanisms of multidrug resistance *Pathol Res Pract., 192(7),* 768-80.

[48] Versantvoort, C. H., Schuurhuis, G. J., Pinedo, H. M., Eekman, C. A., Kuiper, C. M., Lankelma, J. & Broxterman, H. J. (1993). Genistein modulates the decreased drug accumulation in non-P-glycoprotein mediated multidrug resistant tumour cells. *Br J Cancer., 68(5),* 939-46.

[49] Castro, A. F. & Altenberg, G. A. (1997). Inhibition of drug transport *Biochem Pharmacol. 53(1),* 89-93.

[50] Conseil, G., Baubichon-Cortay, H., Dayan, G., Jault, J. M., Barron, D. & Di Pietro, A. (1998). Flavonoids: a class of modulators with bifunctional interactions at vicinal ATP- and steroid-binding sites on mouse P-glycoprotein. *Proc Natl Acad Sci* U S A*. 95(17),* 9831-6.

[51] Chung, S. Y., Sung, M. K., Kim, N. H., Jang, J. O., Go, E. J. & Lee, H. J. (2005). Inhibition of Pglycoproteinby natural products in human breast cancer cells. *Arch Pharm Res., 28(7),* 823-8.

[52] Li, Y., VandenBoom, T. G ., Kong, D., Wang, Z., Ali, S., Philip, P. A. & Sarkar, F. H. (2009). Up-regulation of miR-200 and let-7 by natural agents leads to the reversal of epithelial-to-mesenchymal transition in gemcitabine-resistant pancreatic cancer cells. *Cancer Res., 69(16),* 6704-12.

[53] Tilly, J. L., Billig, H., Kowalski, K. I. & Hsueh, A. J. (1992). Epidermal growth factor and basic fibroblast growth factor suppress the spontaneous onset of apoptosis in cultured rat ovarian granulosa cells and follicles by a tyrosine kinase-dependent mechanism. *Mol Endocrinol., 6(11),* 1942-50.

[54] Traganos, F., Ardelt, B., Halko, N., Bruno, S. & Darzynkiewicz, Z. (1992). Effects of genistein on the growth and cell cycle progression of normal human lymphocytes and human leukemic MOLT-4 and HL-60 cells. *Cancer Res., 52(22),* 6200-8.

[55] Onozawa, M., Fukuda, K., Ohtani, M., Akaza, H., Sugimura, T. & Wakabayashi, K. (1998). Effects of soybean *Jpn J Clin Oncol., 28(6),* 360-3.

[56] Alhasan, S. A., Pietrasczkiwicz, H., Alonso, M. D., Ensley, J. & Sarkar, F. H. (1999). Genistein-induced cell *Nutr Cancer., 34(1)*, 12-9.
[57] Iwashita, K., Kobori, M., Yamaki, K. & Tsushida, T. (2000). Flavonoids inhibit cell *Biosci Biotechnol Biochem., 64(9)*, 1813-20.
[58] Rocchi, P., Ferreri, A. M., Simone, G., Magrini, E., Cavallazzi, L. & Paolucci, G. (1995). Inhibitors of protein kinases *Anticancer Res. 15(4)*, 1381-5.
[59] Watanabe, T., Shiraishi, T., Sasaki, H. & Oishi, M. (1989). Inhibitors for protein-tyrosine *Exp Cell Res., 183(2)*, 335-42.
[60] Kiguchi, K., Constantinou, A. I. & Huberman, E. (1990). Genistein-induced cell *Cancer Commun., 2(8)*, 271-7.
[61] Wei, H., Wei, L., Frenkel, K., Bowen, R. & Barnes, S. (1993). Inhibition of tumor promoter-induced hydrogen peroxide formation in vitro and in vivo by genistein. *Nutr Cancer., 20(1),* 1-12.
[62] Cai, Q. & Wei, H. (1996). Effect of dietary genistein on antioxidant enzyme activities in SENCAR mice. *Nutr Cancer., 25(1),* 1-7.
[63] Fotsis, T., Pepper, M., Adlercreutz, H., Fleischmann, G., Hase, T., Montesano, R. & Schweigerer, L. (1993). Genistein, a dietary-derived inhibitor *Proc Natl Acad Sci,* U S A. *90(7)*, 2690-4.
[64] Zhou, J. R., Mukherjee, P., Gugger, E. T., Tanaka, T., Blackburn, G. L. & Clinton, S. K. (1998). Inhibition of murine bladder *Cancer Res., 58(22)*, 5231-8.
[65] American, Cancer Society. Cancer Facts & Figures (2008). Atlanta: American Cancer Society; 2008.
[66] Giovannucci, E. (2002). A review of epidemiologic studies *Exp Biol Med (Maywood)., 227(10)*, 852-9.
[67] Chan, J. M., Gann, P. H. & Giovannucci, E. L. (2005). Role of diet *J Clin Oncol., 23(32)*, 8152-60.
[68] Shen, J. C., Klein, R. D., Wei, Q., Guan, Y., Contois, J. H., Wang, T. T., Chang, S. & Hursting, S. D. (2000). Low-dose genistein induces cyclin-dependent kinase inhibitors and G(1) cell-cycle arrest in human prostate cancer cells. *Mol Carcinog. 29(2),* 92-102.
[69] Handayani, R., Rice, L., Cui, Y., Medrano, T. A., Samedi, V. G., Baker, H. V., Szabo, N. J. & Shiverick, K. T. (2006). Soy isoflavones alter expression of genes *J Nutr., 136(1),* 75-82.
[70] Zhou, J. R., Gugger, E. T., Tanaka, T., Guo, Y., Blackburn, G. L. & Clinton, S. K. (1999). Soybean phytochemicals inhibit the growth *J Nutr., 129(9)*, 1628-35.
[71] Fritz, W. A., Wang, J., Eltoum, I. E. & Lamartiniere, C. A. (2002). Dietary genistein down-regulates androgen *Mol Cell Endocrinol., 186(1)*, 89-99.
[72] Adams, K. F., Chen, C., Newton, K. M., Potter, J. D. & Lampe, J. W. (2004). Soy isoflavones do not modulate prostate-specific antigen concentrations in older men in a randomized controlled trial. *Cancer Epidemiol Biomarkers Prev., 13(4),* 644-8.
[73] deVere White, R. W., Hackman, R. M., Soares, S. E., Beckett, L. A., Li, Y. & Sun. (2004). Effects of a genistein-rich extract on PSA levels in men with a history of prostate cancer. *Urology., 63(2),* 259-63.
[74] Dalais, F. S., Meliala, A., Wattanapenpaiboon, N., Frydenberg, M., Suter, D. A., Thomson, W. K. & Wahlqvist, M. L. (2004). Effects of a diet rich in phytoestrogens on prostate-specific antigen and sex hormones in men diagnosed with prostate cancer. *Urology., 64(3),* 510-5.

[75] Schröder, F. H., Roobol, M. J., Boevé, E. R., de Mutsert, R., Zuijdgeest-van Leeuwen, S. D., Kersten, I., Wildhagen, M. F. & van Helvoort, A. (2005). Randomized, double-blind, placebo-controlled crossover study in men with prostate cancer and rising PSA: effectiveness of a dietary supplement. *Eur Urol., 48(6)*:922-30.

[76] Cordera, F. & Jordan, C. V. (2006). Steroid receptors and their role in the biology and control of breast cancer growth. *Seminars in Oncology, 33*, 631-641.

[77] Love, R. R. & Philips, J. (2002). Oophorectomy for breast cancer: history revisited. J *Natl Cancer Inst, 94*, 1433-1434.

[78] Duffy, M. J. (2005). Predictive markers in breast and other cancers: a review. *Clinical Chemistry, 51*, 494-503.

[79] Payne, S. J., Bowen, R. L., Jones, J. L. & Wells, C. A. (2008). Predictive markers in breast cancer-the present. *Histopathology, 52*, 82-90.

[80] Cartee, M. W., Smart, W. W. & Matrone, A. (1953). Estimation of estrogenic activity of genistein obtained from soybean meal. *Proc.Soc. Exp. Biol. Med, 84*, 506-507.

[81] Folman, Y. & Pope, G. S. (1966). The interaction in the immature mouse of potent oestrogens with coumestrol, genestein and other utero-vaginotrophic compounds of low potency. *J Endocrinol, 34*, 215-25.

[82] Barnes, S. (1997). The chemopreventative properties of soy isoflavonoids in animal models of breast cancer. *Breast Cancer Res Treat, 46*, 169-179.

[83] Lamartiniere, C. A. (2000). Protection against breast cancer with genestein: a component of soy. *Am J Clin Nutr, 71(suppl)*, 1705S-1707S.

[84] Lee, H. P., Gourley, L., Duffy, S. W., Esteve, J., Lee, J. & Day, N. E. (1991). Dietary effects on breast cancer risk in Singapore. *Lancet, 337*, 1197-2000.

[85] Lee, S. A., Shu, X. O., Li, H., Yang, G., Hui, C., Wen, W., Ji, B. T., Gao, J., Gao, Y. T. & Zheng, W. (2009). Adolescent and adult soy food intake and breast cancer risk: results from the Shanghai women's health study. *Am J Clin Nutr, 89*, 1920-1926.

[86] Wu, A. H., Wan, P., Hankin, J., Tseng, C. C., Yu, M. C. & Pike, M. C. (2002). Adolescent and adult soy intake and risk of breast cancer in Asian-Americans. *Carcinogenesis, 23,* 1491-1496.

[87] Messina, M. & Wu, A. H. (2009). Perspectives on the soy-breast cancer relation. Am *J Clin Nutr, 89(suppl)*, 1673S-1679S.

[88] Trock, B. J., Hilakivi-Clarke, L. & Clarke, R. (2006). Meta-Analysis of Soy Intake and Breast Cancer Risk. *J Natl Cancer Inst, 98*, 459-71.

[89] Wu, A. H., Yu, M. C., Tseng, C. C. & Pike, M. C. (2008). Epidemiology of soy exposures and breast cancer risk. *Br J Cancer, 98*, 9-14.

[90] Hsieh, C. Y., Santell, R. C., Haslam, S. Z. & Helferich, W. G. (1998). Estrogenic effects of genestein on growth of estrogen receptor-positive human breast cancer (MCF-7) cells in vitro and in vivo. *Cancer Res, 58*, 3833-3838.

[91] Lamartiniere, C. A., Moore, J. B., Brown, N. M., Thompson, R., Hardin, M. J. & Barnes, S. (1995). Genestein suppresses mammary cancer in rats. *Carcinogenesis, 16,* 2833-2840.

[92] Magee, P. J. & Rowland, I. R. (2004). Phyto-oestrogens, their mechanism of action: current evidence for a role in breast and prostate cancer. *Br J Nutr, 91*, 513-531.

[93] Upadhyay, S., Neburi, M., Chinni, S. R., Alhasan, S., Miller, F. & Sarkar, F. H. (2001). Differential sensitivity of normal and malignant breast epithelial cells to genistein is partly mediated by p21WAF1. *Clin Cancer Res, 7,* 1782-1789.

[94] Li, Y., Upadhyay, S., Bhuiyan, M. & Sarkar, F. H. (1999). Induction of apoptosis in breast cancer cells MDA-MB-231 by genistein. *Oncogene, 18*, 3166-3172.

[95] Tophkhane C, Yang S, Bales W, Archer L, Osunkoya A, Thor, A. D. & Yang, X. (2007). Bcl-2 overexpression sensitizes MCF-7 cells to genistein by multiple mechanisms. *Int J Oncol, 31*, 867-874.

[96] Shao, Z. M., Wu, J., Shen, Z. Z. & Barsky, S. H. (1998) Genistein exerts multiple suppressive effects on human breast carcinoma cells. *Cancer Res, 58,* 4851-4857.

[97] Sakla, M. S., Shenouda, N. S., Ansell, P. J., Macdonald, R. S. & Lubahn, D. B. (2007). Genestein affects HER-2 protein concentration, activation, and promoter regulation in BT-474 human breast cancer cells. *Endocrine, 32*, 69-78.

In: Isoflavones Biosynthesis, Occurence and Health Effects ISBN: 978-1-61728-113-6
Editor: Melanie J. Thompson, pp. 181-193 © 2010 Nova Science Publishers, Inc.

Chapter VI

Comparative in Vitro Effects of Soybean Isoflavones and Resveratrol on Human Osteoblastic Cells

Arancha R. Gortázar, Verónica Alonso and Pedro Esbrit*
Bone and Mineral Metabolism Laboratory, Fundación Jiménez Díaz (Capio Group), Madrid, Spain.

Abstract

Estrogen, the main factor responsible for the maintenance of bone mass in women, inhibits bone resorption and increases bone density at least in part by interacting with osteoblasts. Estrogen replacement therapy effectively prevents bone loss after menopause, but also has adverse effects. Natural plant phytoestrogens, which have structural similarity to 17β-estradiol, are of growing interest by their potential role in osteoporosis prevention and treatment. In the present study, we compared several osteoblastic effects of the soy isoflavones genistein and daidzein, two well known phytoestrogens, and resveratrol, a phenolyc phytoestrogen abundant in grapes and wine. We used human osteosarcoma MG-63 cells, a characterized model for their nontransformed counterparts in a variety of studies on osteoblast function. We found that all these phytoestrogens, similarly to 17β-estradiol, at 100 nM, maximally increased alkaline phosphatase activity, as well as osteocalcin mRNA and osteoprotegerin (both mRNA and protein) expression. The protein levels of receptor activator of NF-κB ligand (RANKL) in MG-63 cell membrane extracts were also increased by either genistein or 17ß-estradiol but decreased by daidzein and resveratrol. However, the latter phytoestrogens, similarly to genistein and 17β-estradiol, decreased NF-κB-dependent interleukin-6 mRNA expression in MG-63 cells. Our findings support previous findings in other nonhuman osteoblastic cell preparations, and indicate that relatively low

* Corresponding author: Laboratorio de Metabolismo Mineral y Óseo, Fundación Jiménez Díaz (Capio Group), Avda. Reyes Católicos, 2; 28040 Madrid, Spain, Phone: 34 91 5504894; Fax: 34 91 5498075, e-mail: pesbrit@fjd.es

concentrations of the main soy isoflavones and resveratrol display anabolic features apparently through estrogen receptors in human osteoblastic cells.

Keywords: Human osteoblasts, phytoestrogens, resveratrol, soy isoflavones.

Introduction

Estrogen deficiency induces rapid bone loss during the first decade after menopause. In this setting, an imbalance of bone resorption and bone formation towards the former results from an increased rate of initiation of new bone remodeling cycles (1). Moreover, in postmenopausal osteoporosis, a poorly mineralized bone replaces old bone during bone remodeling, which contributes to fracture risk (2).

The beneficial effects of estrogen on bone metabolism appear to be complex. Estrogen acts directly on osteoclastic cells to inhibit bone resorption, but it can also exert inhibitory effects on some osteolytic cytokines and their receptors in osteoblasts (3). In these cells, 17β-estradiol modulates the expression of receptor activator of NF-κB ligand (RANKL) –a potent osteoclast activator- and its decoy receptor osteoprotegerin (OPG), and thus inhibits osteoclastogenesis (4-6). However, a putative anabolic action of estrogen based on its direct effect(s) on bone formation remains controversial (7). Estrogen replacement therapy has been the traditional method to prevent bone loss after menopause, but has concomitant adverse side-effects such as high breast and endometrial cancer risk (8, 9).

Phytoestrogens are a heterogeneous group of plant-derived compounds with structural and functional similarities to estrogen but lacking the oncogenic effects of the latter. Among them, the isoflavones abundant in soy products genistein and daidzein are considered as a potential alternative to hormone replacement therapy for preventing and/or treating postmenopausal osteoporosis (10, 11). Various reports have shown that these isoflavones can prevent trabecular bone loss in ovariectomized rats (12-14). In addition, the majority of observational studies as well as randomized controlled trials in menopausal women fed isoflavone-rich diets for ≤3 years have shown a favorable effect of such diets on the skeleton (15-18). However, the existence of some confounding data in this regard (19), has raised some concerns on a generalized use of isoflavones as therapeutic agents to prevent bone loss.

The mechanisms whereby dietary isoflavones might induce their beneficial effects in bone are still poorly understood. *In vitro* studies performed in either rodent osteoblastic cells or human osteoblast-like cells mainly expressing estrogen receptor (ER) α indicate that these agents might stimulate bone formation at least in part by directly affecting osteoblastic function (20-23). Genistein and daidzein have lower affinity than 17β-estradiol for both ER subtypes α and β, but their relative binding affinity for the latter receptor is higher than for the former receptor. Thus, these isoflavones appear to have similar transcriptional activities than 17β-estradiol at least in some cell systems, but they might induce their effects by differentially interacting with ERβ (24, 25). In this regard, a recent study has shown that both 17β-estradiol and daidzein, at low concentration (1 nM), can induce osteoblast differentiation apparently by selective interaction with ERα and ERβ, respectively, in porcine osteoblasts (26). On the other hand, these phytoestrogens might act through alternative nongenomic pathways in bone cells (27, 28).

The present study was undertaken in an attempt to coordinate and compare several effects of individual phytoestrogens in a human osteoblastic cell model. Considering the ligand specificity of ERβ toward phytoestrogens, we used human osteoblastic osteosarcoma MG-63 cells displaying a high constitutive expression of this ER subtype (29, 30), a well characterized cell model for their nontransformed counterparts (31-36). We compared the effects of genistein, daidzein and resveratrol -a phenolyc phytoestrogen abundant in red wine (37), which is similar to the synthetic estrogen diethylstilbestrol and display osteogenic features in several murine osteoblastic cell preparations (38,39)- on various osteoblast differentiation markers in MG-63 cells. In addition, we assessed the actions of these phytoestrogens on the expression of osteoclastogenesis-modulatory cytokines –namely interleukin (IL)-6, OPG and RANKL- by these cells.

Materials and Methods

Materials

Genistein, daidzein, trans-resveratrol and 17β-estradiol were supplied by Sigma-Aldrich (St. Louis, MO). ICI 182,780 was obtained from Tocris Cookson (Bristol, UK), and raloxifene from Lilly-Spain (Madrid). Rabbit polyclonal antibodies against human OPG (H-249), RANKL (FL-317), p50 (sc-1190X), p65 (sc-372X), and p52 (sc-298X) were from Santa Cruz Biotechnology (Santa Cruz, CA). T4 polymerase was from Promega (Madison, WI), and [γ^{32}P]ATP (3000Ci/mmol) from Amersham (Buckinghamshire, UK).

Cell Culture

The human osteosarcoma cell line MG-63 (ATCC CRL 1427) was grown in Dulbecco's Modified Eagle's Medium (DMEM) (without fenol red) with 10% charcoal-treated fetal bovine serum (FBS), 1% nonessential aminoacids, and antibiotics (100 IU/ml of penicillin and 100 μg/ml of streptomycin), in 5 % CO_2 at 37°C. Confluent cells were incubated for 3 days in serum-depleted medium with or without (control) the different agonists. Genistein and daidzein were diluted in dimethyl sulfoxide, and resveratrol and 17β-estradiol in ethanol. Final concentration of these diluents in the cell incubation medium was < 0.01 %.

Cell total RNA and protein were isolated using Tri-Reagent© (Molecular Research Center, Cincinnati, OH), according to the manufacturer´s instructions. Nuclear extracts were prepared according to a commercially available procedure (NE-PER®, Pierce, Rockford, IL), as previously described (34). To obtain membrane extracts, cells were lysed with phosphate-buffered saline (PBS)/0.1% Triton X-100, centrifuged at 2,000xg for 10 min, and the resulting supernatant was spun at 100,000x g for 1 h. The resulting pellet was resuspended in 50 mM Tris-HCl, pH 7.4, containing 150 mM NaCl, 1 mM EDTA, 0.2 mM phenylmethylsulphonyl fluoride, 1% Triton X-100, 1% sodium deoxycolate, and 0.1% SDS (membrane protein extract) (35). Protein was determined by a Coomasie reagent-based micromethod (Pierce), using bovine serum albumin (BSA) as standard.

Semiquantitative Reverse Transcription-Polymerase Chain Reaction (Rt-Pcr)

Cell total RNA (10-100 ng) was reverse transcribed, and the resulting cDNA was amplified using a standard method (Titanium™ One-step RT-PCR kit; Clontech, Palo Alto, CA), with the following specific primers: 5'-CATGAGAGCCCTCACACTCC-3' (sense) and 5'-CAGCAGAGCGACACCCTAGACC-3' (antisense), corresponding to nucleotides 18-37 and 315-336, respectively, in the human osteocalcin (OC) gene (Genebank accession number X51699); 5'-GGAATAGATGTTACCCTGTG-3' (sense) and 5'-TCAATGTCTTCTG CTCCC-3' (antisense), corresponding to nucleotides 680-699 and 1015-1032, respectively, in the human OPG gene (Genebank accession number U94332); and 5′-TTCGGTCCAGTTGCCTTCT-3′(sense), and 5′-GTACTCATCTGGACAGCTC-3′(antisense), corresponding to nucleotides 88-106 and 485-503, respectively, in the human interleukin (IL)-6 gene (Genebank accession number M14584). These primers yield PCR amplification products of 319-bp (OC), 353-bp (OPG), and 416-bp (IL-6) products, respectively. The reaction mixture (10 μl) was incubated for 45 min at 48°C, and 2 min at 95°C, followed by 32-35 cycles of 1 min at 95 °C, 1 min at 58-62°C, and 2 min at 68°C, with a final extension of 7 min at 68°C (34, 35). As constitutive controls, either the housekeeping gene glyceraldehyde-3-phosphate dehydrogenase (GAPDH) was amplified using specific primers, or modified 18S primers (QuantumRNA™ 18S Internal Standards; Ambion) were added to the reaction mixture for 18S coamplification (34-36). Preliminary experiments proved that the PCR conditions used provide submaximal amplification, corresponding to the linear segment in titration curves for PCR amplification of each gene. PCR products were separated on 2% agarose gels, and bands visualized by ethidium bromide staining. Values obtained after densitometric scanning of PCR products were normalized against those of the corresponding GAPDH and 18S PCR products.

Western Blot Analysis

Cell membrane or total cell extracts (50 or 20 μg protein, respectively) were transferred onto nitrocellulose membranes (Amersham), blocked with 5% defatted milk in PBS with 0.05% Tween-20, and then incubated overnight at 4°C with either 0.4 μg/ml anti-RANKL or 80 ng/ml anti-OPG antibody, or with a monoclonal antibody against α-tubulin (Sigma) as constitutive control. After extensive washing, the membranes were incubated with peroxidase-conjugated goat anti-rabbit or anti-mouse IgG, and developed by enhanced chemiluminescence (Amersham). The corresponding fluorogram bands were quantitated by densitometric scanning.

Electrophoretic Mobility Shift Assay (EMSA)

A double-stranded oligonucleotide containing a consensus sequence specific for NF-κB, 5'-AGTTGAGGGGACTTTCCCAGGC-3' (34) was 5'-end-labeled with 10 μCi [γ^{32}P]ATP and T4 polymerase. Nuclear extracts (3 μg protein) were incubated with 200,000 dpm of ^{32}P-labeled oligonucleotide probe in 10 μl of a reaction mixture containing: 10 mM Tris-HCl (pH 7.5), 100 mM NaCl, 1 mM $MgCl_2$, 0.5 mM EDTA, 4% glycerol, and 50 μg/μl poly(dI-dC) for 20 min at 4°C. Protein-DNA complexes were resolved on native 5% polyacrilamide/0.25xTBE gels. Dried gels were exposed to radiosensitive film. As control for specificity of the binding reaction, nuclear extracts were pre-incubated with 100-fold excess of unlabeled NF-κB oligonucleotide for 10 min at 4°C before addition of the labeled probe. In some experiments, cell nuclear extracts were preincubated for 15 min at room temperature with 1 μl (2 μg) of anti-p65, -p50 or -p52 antibodies.

Alkaline Phosphatase (ALP) Activity

ALP activity was measured in cell extracts obtained with 0.1% Triton X-100. The enzymatic activity was assayed for 30 min at 37°C, with 5 mM sodium p-nitrophenylphosphate as substrate in 1 M diethanolamine, 0.5 mM $MgCl_2$, pH 9.8. Absorbance was read at 405 nm (35). One unit of enzymatic activity was expressed as nmol/min/mg protein.

Statistical analysis

The data throughout the text are expressed as mean ± SEM. Statistical significance ($p<0.05$) was assessed by nonparametric ANOVA (Kruskal-Wallis) followed by Dunn's *post-hoc* test.

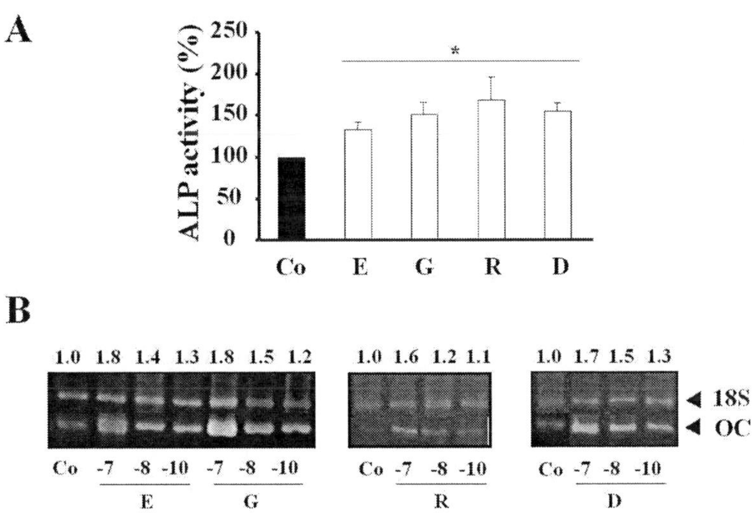

Figure 1. Effect of different phytoestrogens and 17β-estradiol on ALP activity (A) and OC mRNA expression (B) in confluent MG-63 cells. (A) ALP activity was measured in cell extracts obtained with 0.1% Triton X-100. Values are means ± SEM of four determinations in duplicate. *$p<0.05$ versus nonstimulated control (Co) value (100% or 21.0 ± 3.1 nmol/min/mg protein). (B) The autoradiograms show changes in OC mRNA and 18S mRNA (a constitutive control) which were coamplified by RT-PCR. Relative intensities of OC/18S mRNA ratio from 3 independent measurements are indicated at the top. The phytoestrogens and 17β-estradiol were added at 100 nM, except when stated otherwise (B; the numbers at the bottom of each autoradiogram are M). E = 17β-estradiol; G = genistein; D = daidzein; R = resveratrol.

Results

We first analyzed the effect of the tested phytoestrogens on ALP activity, an early marker of osteoblast differentiation (40,41). We found that genistein, daidzein, and resveratrol, in a similar manner to 17β-estradiol, at 100 nM, stimulated this enzymatic activity at 72 h in MG-63 cells (Figure 1A). Moreover, both isoflavones, similarly to 17ß-estradiol, dose-dependently (between 0.1 and 100 nM) increased the gene expression of OC, a late osteoblast differentiation marker (40,42), in these cells (Figure 1B). A similar but less efficient effect was observed for resveratrol in MG-63 cells (Figure 1B).

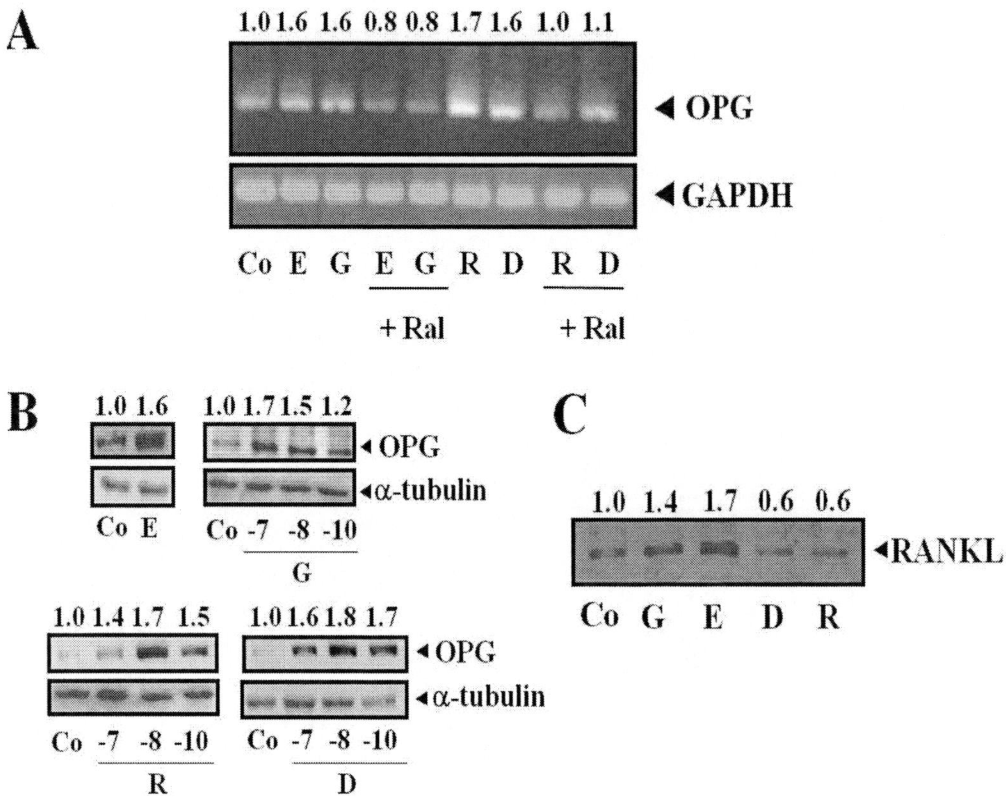

Figure 2. Effect of different phytoestrogens and 17β-estradiol on the expression of OPG (A,B) and RANKL protein (C) in MG-63 cells. (A) The autoradiograms show changes in OPG mRNA and GAPDH mRNA (as constitutive control) evaluated by RT-PCR. Total cell protein extracts (B) or cell membrane extracts (C) were analyzed by Western blot using a polyclonal anti-OPG antibody detecting a single band corresponding to an apparent molecular weight of 55 kDa (B), or a polyclonal anti-RANKL antibody which detects a single band corresponding to an apparent molecular weight of 42 kD (C). Protein loading was similar in each well, as assessed by using α-tubulin as internal control (B) or Ponceau S staining (not shown) (C). Relative intensities of corresponding OPG and RANKL signals from 3 independent measurements are indicated at the top. The phytoestrogens and 17β-estradiol were added at 100 nM, except when stated otherwise (B; the numbers at the bottom of each autoradiogram are M). When present, raloxifene (Ral) was added, at 10 nM, 1 h before the other agonists. Other abbreviations are as in the legend to figure 1.

OPG, a member of the tumor necrosis factor superfamily which acts as a decoy receptor for RANKL, is predominantly expressed by the more differentiated osteoblasts (43, 44). In the present study, we found that the three tested phytoestrogens as well as 17ß-estradiol, at 100 nM, similarly induced OPG expression (at both mRNA and protein levels) in MG-63 cells (Figure 2A and B). Moreover, the selective estrogen receptor modulator raloxifene, which can interact with both ERα and ERβ (24, 45), abolished the effect of each of these agonists on OPG mRNA in these cells (Figure 2A). Induction of OPG protein was even observed after stimulation with each phytoestrogen at 0.1 nM (Figure 2B). On the other hand, the expression of RANKL protein in MG-63 cell membrane extracts was increased by either genistein or 17ß-estradiol, but decreased by daidzein and resveratrol, each at 100 nM (Figure 2C).

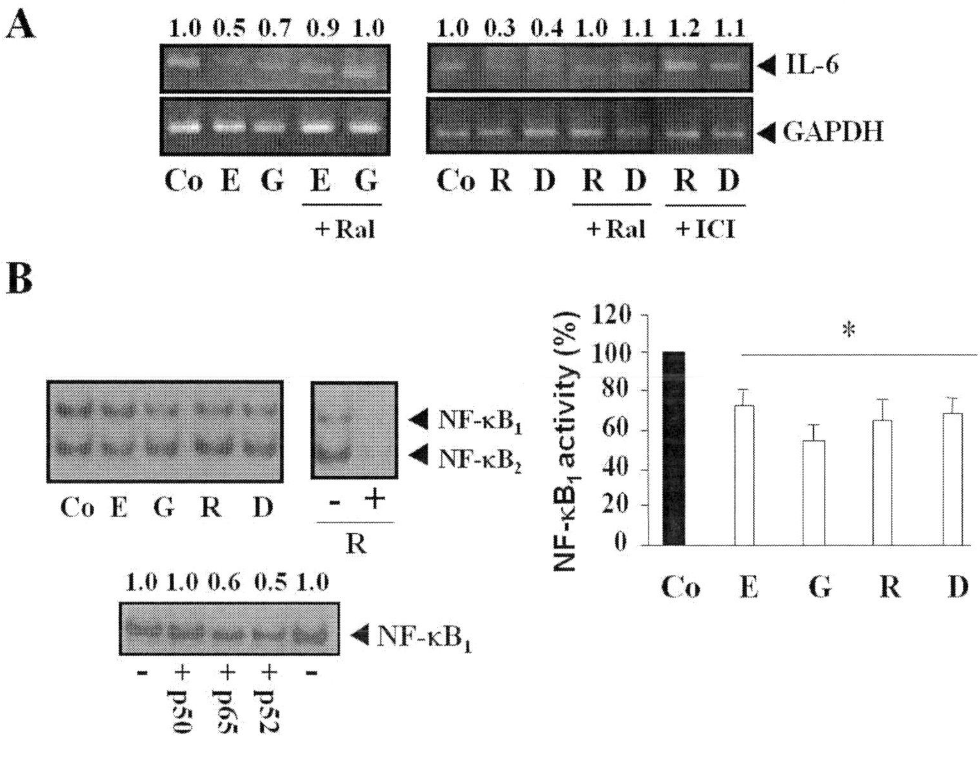

Figure 3. Effect of different phytoestrogens and 17β-estradiol on IL-6 mRNA expression (A) and NF-κB activation (B) in MG-63 cells. (A) The autoradiograms show changes in IL-6 mRNA and GAPDH mRNA (as constitutive control) evaluated by RT-PCR. Relative intensities of IL-6 signal from 3 independent measurements are indicated at the top. (B) NF-κB activity was determined in cell nuclear extracts by EMSA. Two specific NF-κB-DNA bands were evident, since they disappeared with an excess of unlabeled NF-κB consensus oligonucleotide (left, +). Densitometric values corresponding to the relative intensities of the upper NF-κB-DNA band (NF-κB_1), as mean ± SEM (n=4), are shown (right). *$p<0.05$ versus nonstimulated corresponding control (Co) value (100%). Some nuclear extracts from nonstimulated cells were pretreated with specific anti-p50, -p52 or -p65 antibodies. The phytoestrogens and 17β-estradiol were added at 100 nM. When present, raloxifene (Ral) and ICI 182,780 were added at 10 nM and 100 nM, respectively, 1 h before the other agonists. Other abbreviations are as in the legend to figure 1.

We previously demonstrated that MG-63 cells constitutively express IL-6, a cytokine which seems to have an important role on the increased bone remodeling associated with estrogen deficiency (34, 46). In the present study, we found that the three tested phytoestrogens and 17ß-estradiol inhibited the constitutive gene expression of this cytokine in these cells (Figure 3); and this inhibition was abolished by either raloxifene or the ER antagonist ICI 182,780 (24) (Figure 3A). Furthermore, we assessed the effect of these phytoestrogens on the activity of the transcription factor NF-κB, which is a major modulator of IL-6 gene transcription in MG-63 cells (34). By using EMSA, we observed two specific NF-κB-DNA bands in MG-63 nuclear extracts, but only the upper band was consistently affected (decreased) by the different phytoestrogens (Figure 3B). A significant decrease in this band was evident after preincubating MG-63 cell nuclear extracts with the anti-p65 or –p52 antibodies, although not with the anti-p50 antibody (Figure 3B).

Discussion

Phytoestrogens are a heterogeneous group of plant-derived ER ligands (11). Dietary soybean is a rich source of these compounds, namely genistein and daidzein, which likely account for the bone-sparing effect of soy protein extracts (47,48). However, this beneficial action of these soy isoflavones might depend on factors such as different dietary schemes and estrogen status (49, 50).

Previous *in vitro* studies in the murine osteoblastic cell line MC3T3-E1 showed that soybean isoflavones and resveratrol, at high concentrations (≥ 1 μM), increased DNA synthesis and the early stages of osteoblast differentiation (20, 38, 51, 52). Moreover, in another study, daidzein (within 10-100 μM) was found to stimulate osteoblast differentiation at various stages in primary osteoblastic cells from newborn rat calvaria (23). However, fewer studies have examined the action of phytoestrogens on osteoblast function in human osteoblastic cell preparations. Using a human fetal osteoblastic cell line or primary osteoblastic cells from healthy young subjects showing predominant expression of ERα as compared to ERβ, lower concentrations (1-100 nM) of genistein and daidzein were shown to stimulate OPG expression and secretion (21, 22). More recently, genistein and resveratrol have been shown to stimulate several osteoblastic markers during osteogenic differentiation of human bone marrow stromal cells (53,54).

In the present study, addition of a single dose (in the range 0.1-100 nM) of either genistein or daidzein stimulated ALP activity and OC mRNA levels after 3 days. These effects were similar to those observed with either 17β-estradiol or the structurally unrelated phytoestrogen resveratrol. In contrast, a previous report has shown that continuous exposure to the latter phytoestrogen, added at μM doses every other day for 4 days, failed to increase these osteoblastic markers in growing MG-63 cells (33). This discrepancy might be explained by differences in the temporal pattern of its exposure at different stages of MG-63 cell differentiation. Of note in this regard, continuous exposure of another human osteoblastic osteosarcoma cell line to 17β-estradiol had no effect on either ALP activity or mineralized bone nodule formation (55).

The effects of estrogen on osteoblasts and osteoclasts appear to involve its interaction with the OPG/RANKL system to modulate osteoclast activity (1,4,5). Interestingly, in the context of the present study, it has recently been reported a decrease in the RANKL/OPG ratio in serum related to changes in trabecular bone density in postmenopausal women after 1-year treatment with genistein (56). We found herein that the three tested phytoestrogens, in a similar manner to 17β-estradiol, stimulated OPG (both mRNA and protein), within a similar dose range as that observed for their effects on the other osteoblast differentiation markers studied in MG-63 cells. Previous studies failed to detect RANKL gene expression in some nontransformed human osteoblastic cell lines (21, 22). In the present study, genistein and 17ß-estradiol induced a similar increase in the levels of RANKL protein which was detectable in MG-63 cell membranes. On the other hand, both daidzein and resveratrol inhibited this expression in the membrane of these cells. This suggests that the latter phytoestrogens can induce a higher decrease in the RANKL/OPG ratio than genistein (or 17β-estradiol) in MG-63 cells. However, it can be speculated that daidzein and resveratrol might have mainly enhanced RANKL secretion into the MG-63 cell-conditioned medium. In fact, such an effect of daidzein has been reported to occur in primary piglet osteoblastic cells (26). However, membrane-bound RANKL seems to work more efficiently than its soluble form in activating osteoclastogenesis (56); our findings would thus provide a rationale to explain the different effects of daidzein and genistein on cancellous bone in ovariectomized rats (14).

Genistein and daidzein as well as resveratrol, in a similar manner to 17β-estradiol, inhibited the constitutive IL-6 mRNA expression in MG-63 cells. Considering the key role of this cytokine on the increased osteoclastogenesis in estrogen-deficient states, this finding further supports the anti-resorptive action of these phytoestrogens by interaction with osteoblastic cells (1, 3, 46). Our present results also support the notion that coupling of these phytoestrogens to ER leads to interaction with an NF-κB complex -likely consisting of p52/p65 heterodimer-, and prevents IL-6 and OPG transcription in MG-63 cells (58).

In summary, the present in vitro findings are consistent with and integrate previous data showing the osteogenic effects of genistein and daidzein as well as resveratrol in nontransformed human osteoblastic and bone marrow stromal cell preparations. Our results confirm the validity of the MG-63 cell model to study the effects of phytoestrogens on osteoblast function.

Acknowledgments

This work was supported by a grant from Fundación MAPFRE Medicina and Instituto de Salud Carlos III (RETICEF, RD06/0013/1002). A.R. de Gortázar, Ph.D., is presently the recipient of a contract from Universidad San Pablo-CEU (Departamento de Ciencias Médicas, Sección de Histología), Boadilla del Monte (Madrid, Spain). V. Alonso, Ph.D., is currently the recipient of a post-doctoral contract at the Pharmacology and Chemical Biology Department (Pittsburgh University, Pittsburgh, PA).

References

[1] Riggs, B. L., Khosla, S. & Melton, L. J. III (2002). Sex steroids and the construction and conservation of the adult skeleton. *Endocrine Rev, 23*, 279-302.

[2] Cummings, S. R., Browner, W. S., Bauer, D., Stone, K., Ensrud, K., Jamal, S. & Ettinger, B. (1998). Endogenous hormones and the risk of hip and vertebral fractures among older women. Study of Osteoporotic Fractures Research Group. *N Engl J Med, 339*, 733-738.
[3] Spelsberg, T. C., Subramaniam, M., Riggs, B. L. & Khosla, S. (1999). The actions and interactions of sex steroids and growth factors/cytokines on the skeleton. *Mol Endocrinol 13*, 819-828.
[4] Hofbauer, L. C., Khosla, S., Dunstan, C. R., Lacey, D. L., Spelsberg, T. C. & Riggs, B. L. (1999). Estrogen stimulates gene expression and protein production of osteoprotegerin in human osteoblastic cells. *Endocrinology, 140*, 4367-4370.
[5] Theolyre, S., Wittrant, Y., Tat, S.K., Fortun, Y., Redini, F. & Heyman, D. (2004). The molecular triad OPG/RANK/RANKL: involvement in the orchestration of pathophysiological bone remodeling. *Cytokine Growth F Res, 15*, 457-475.
[6] Zallone A. (2006). Direct and indirect estrogen actions on osteoblasts and osteoclasts. *Ann N Y Acad Sc, 1068*, 173-179.
[7] Lorenzo, J. (2003). A new hypothesis for how sex steroid hormones regulate bone mass. *J Clin Invest, 111*, 1641-1643.
[8] Mahavni, V. & Sood, A. K. (2001). Hormone replacement therapy and cancer risk. *Current Opinion in Oncology, 13*, 384-389.
[9] Emons, G., Huschmand-Nia, A., Krauss, T. & Hinney, B. (2004). Hormone replacement therapy and endometrial cancer. *Onkologie, 27,* 207-210.
[10] Anderson, J. J. & Garner, S. C. (1998). Phytoestrogens and bone. *Baillière's Clin Endocrinol Metab, 12*, 543-557.
[11] Anderson, J. J. B., Messina, M. & Garner, S. C. (1999). Effects of phyto-estrogens on tissues. *Nutr Res Rev, 12*, 75-116.
[12] Arjmandi, B. H., Alekel, L., Hollis, B. W., Amin, D., Stacewicz-Sapuntzakis, M., Guo, P. & Kukreja, S. C. (1996). Dietary soybean protein prevents bone loss in an ovarectomized rat model of osteoporosis. *J Nutr, 126*, 161-167.
[13] Fanti, P., Monier-Faugere, M. C., Geng, Z., Schmidt, J., Morris, P. E., Cohen, D. & Malluche, H. H. (1998). The phytoestrogen genistein reduces bone loss in short term ovariectomized rats. *Osteopor Int, 8,* 274-281.
[14] Picherit, C., Coxam, V., Bennetau-Pelissero, C., Kati-Coulibaly, S., Davicco, M. J., Lebecque, P. & Barlet, J. P. (2000). Daidzein is more efficient than genistein in preventing ovariectomy-induced bone loss in rats. *J Nutr, 130*, 1675-1681.
[15] Ma, D. F, Qin, L. Q., Wang, P. Y. & Katoh, R. (2008). Soy isoflavone intake inhibits bone resorption and stimulates bone formation in menopausal women: meta-analysis of randomized controlled trials. *Eur J Clin Nutr, 62*, 155-161.
[16] Alekel, D. L., Germain, A. S., Peterson, C. T., Hanson, K. B., Stewart, J. W. & Toda, T. (2000). Isoflavone-rich soy protein isolate attenuates bone loss in the lumbar spine of perimenopausal women. *Am J Clin Nutr, 72*, 844-852.
[17] Ho, S. C., Chan, S. G., Yi, Q., Wong, E. & Leung, P. C. (2001). Soy intake and the maintenance of peak bone mass in Hong Kong Chinese women. *J Bone Min Res, 16*, 1363-1369.
[18] Marini, H., Minutoli, L., Polito, F., Bitto, A., Altavilla, D., Atteritano, M., Gaudio, A., Mazzaferro, S., Frisina, A., Frisina, N., Lubrano, C., Bonaiuto, M., D'Anna, R., Cannata, M. L., Corrado, F., Adamo, E. B., Wilson, S. & Squadrito, F. (2007). Effects

of the phytoestrogen genistein on bone metabolism in osteopenic postmenopausal women: a randomized trial. *Ann Intern Med, 146*, 839-847.

[19] Brink, E., Coxan, V., Robins, S., Wahala, K., Cassidy, A. & Branka, F. (2008). Long-term consumption of isoflavone-enriched foods does not affect bone mineral density, bone metabolism, or hormonal status in early postmenopausal women: a randomized, double-blind, placebo controlled study. *Am J Clin Nutr, 87*, 761-70.

[20] Sugimoto, E. & Yamaguchi, M. (2000). Anabolic effect of genistein in osteoblastic MC3T3-E1 cells. *Int J Mol Med, 5*, 515-520.

[21] Chen, X. W., Garner, S. C. & Anderson, J. J. B. (2002). Isoflavones regulate interleukin-6 and osteoprotegerin synthesis during osteoblast cell differentiation via an estrogen-receptor-dependent pathway. *Biochem Biophys Res Commun, 295*, 417-422.

[22] Viereck, V., Gründker, C., Blaschke, S., Siggelkow, H., Emons, G. & Hofbauer, L.C. (2002). Phytoestrogen genistein stimulates the production of osteoprotegerin by human trabecular osteoblasts. *J Cel Biochem, 84*, 725-735.

[23] Jia, T. L., Wang, H. Z., Xie, L. P., Wang, X. Y. & Zhang, R. Q. (2003). Daidzein enhances osteoblast growth that may be mediated by increased bone morphogenetic protein (BMP) production. *Biochem Pharmacol, 65*, 709-715.

[24] Kuiper, G. G., Lemmen, J. G., Carlsson, B., Corton, J. C., Safe, S. H., van der Saag, P. T., van der Burg, B. & Gustafsson, J.-Å. (1998). Interaction of estrogenic chemicals and phytoestrogens with estrogen receptor beta. *Endocrinology, 139*, 4252-4263.

[25] An, J., Tzagarakis-Foster, C., Scharschmidt, T. C., Lomri, N. & Leitman, D. C. (2001). Estrogen receptor β-selective transcriptional activity and recruitment of coregulators by phytoestrogens. *J Biol Chem, 276*, 17808-17814.

[26] De Wilde, A, Lieberherr, M., Colin, C. & Pointillart, A. (2004). A low dose of daidzein acts as an ERβ-selective agonist in trabecular osteoblasts of young female piglets. *J Cel Physiol, 200*, 253-262.

[27] Akiyama, T., Ishida, J., Nagakawa, S., Ogawara, H., Watanabe, S., Itoh, N., Shibuya, M. & Fukami, Y. (1987). Genistein, a specific inhibitor of tyrosine-specific protein kinases. *J Biol Chem, 262*, 5592-5595.

[28] Yamagishi, T., Otsuka, E. & Hagiwara, H. (2001). Reciprocal control of expression of mRNAs for osteoclast differentiation factor and OPG in osteogenic stromal cells by genistein: Evidence for the involvement of topoisomerase II in osteoclastogenesis. *Endocrinology, 142*, 3632-3637.

[29] Vidal, O., Kindblom, L. G. & Ohlsson, C. (1999). Expression and localization of estrogen receptor-β in murine and human bone. *J Bone Miner Res, 14*, 923-929.

[30] Chen, F. P., Hsu, T., Hu, C. H., Wang, W. D., Wang, K. C. & Teng, L. F. (2004). Expression of estrogen receptors alpha and beta in human osteoblasts: identification of exon-2 deletion variant of estrogen receptor beta in postmenopausal women. *Chang Gung Med J, 27*, 107-115.

[31] Bilbe, G., Roberts, E., Birch, M. & Evans, D. B. (1996). PCR phenotyping of cytokines, growth factors and their receptors and bone matrix proteins in human osteoblast-like cell lines. *Bone, 19*, 437-445.

[32] Carpio, L., Gladu, J., Goltzman, D. & Rabbani, S. A. (2001). Induction of osteoblast differentiation indexes by PTHrP in MG-63 cells involves multiple signaling pathways. *Am J Physiol Endocrinol Metab, 281*, E489-E499.

[33] Boyan, B. D., Lohmann, C. H., Sisk, M., Liu, Y., Sylvia, V. L., Cochran, D. L., Dean, D. D. & Schwartz, Z. (2001). Both cyclooxygenase-1 and cyclooxygenase-2 mediate osteoblast response to titanium surface roughness. *J Biomed Mater Res, 55*, 350-359.

[34] Guillén, C., Martínez, P., de Gortázar, A. R., Martínez, M. E. & Esbrit, P. (2002). Both N- and C-terminal domains of parathyroid hormone-related protein increase interleukin-6 by nuclear factor-κB activation in osteoblastic cells. *J Biol Chem, 277*, 28109-28117.

[35] De Gortázar, A. R., Alonso, V., Alvarez-Arroyo, M. V. & Esbrit, P. (2006). Transient exposure to PTHrP (107-139) exerts anabolic effects through vascular endothelial growth factor receptor 2 in human osteoblastic cells in vitro. *Calcif Tissue Int, 79*, 360-9.

[36] Alonso, V., de Gortázar, A. R., Ardura, J. A. &rade-Zapata, I., Alvarez-Arroyo, M. V. & Esbrit, P. (2008). Parathyroid hormone-related protein (107-139) increases human osteoblastic cell survival by activation of vascular endothelial growth factor receptor-2. *J Cell Physiol, 217*, 717-727.

[37] Soleas, G. J., Diamandis, E. P. & Goldberg, D. M. (1997). Wine as a biological fluid: history, production, and role in disease prevention. *J Clin Lab Anal, 11*, 287-313.

[38] Mizutani, K., Ikeda, K., Kawai, Y. & Yamori, Y. (1998). Resveratrol stimulates the proliferation and differentiation of osteoblastic MC3T3-1 cells. *Biochem Biophys Res Commun, 253,* 859-863.

[39] Rucinski, M., Ziolkowska, A., Hochol, A., Pucher, A., Macchi, C., Belloni, A. S., Nussdorfer, G. G. & Malendowicz, L. K. (2006). Estradiol and resveratrol stimulating effect on osteocalcin, but not osteonectin and collagen-1alpha gene expression in primary culture of rat calvarial osteoblast-like cells. *Int J Mol Med, 18*, 565-570.

[40] Stein, G. S. & Lian, J. B. (1993). Molecular mechanisms mediating proliferation/differentiation: Interrelationships during progressive development of the osteoblast phenotype. *Endocr Rev, 14*, 424-442.

[41] Martínez, M. E., del Campo, T., Medina, S., Sánchez, M., Sánchez-Cabezudo, M. J., Esbrit, P., Martínez, P., Moreno, I., Rodrigo, A., Garcés, M. V. & Munuera, L. (1999). Influence of skeletal site of origin and donor age on osteoblastic cell growth and differentiation. *Calcif Tissue Int, 64*, 280-286.

[42] Martínez, P., Moreno, I., de Miguel, F., Vila, V., Esbrit, P. & Martínez, M. E. (2001). Changes in osteocalcin response to 1, $25(OH)_2$ D_3 stimulation and basal VDR expression in human osteoblastic cells according to donor age and skeletal origin. *Bone, 29*, 35-41.

[43] Gori, F., Hofbauer, L. C., Dunstan, C. R., Spelsberg, T. C., Khosla, S. & Riggs, B. L. (2000). The expression of osteoprotegerin and RANK ligand and the support of osteoclast formation by stromal-osteoblast lineage cells is developmentally regulated. *Endocrinology, 141*, 4768-4776.

[44] Thomas, G. P., Baker, S. U. K., Eisman, J. A. & Gardiner, E. M. (2001). Changing RANKL/OPG mRNA expression in differentiating murine primary osteoblasts. *J Endocrinol, 170*, 451-460.

[45] Wijayaratne, A. L., Nagel, S. C., Paige, L. A., Christensen, D. J., Norris, J. D., Fowlkes, D. M. & McDonnell, D. P. (1999). Comparative analyses of mechanistic differences among antiestrogens. *Endocrinology, 140*, 5828-5840.

[46] Jilka, R. L., Hangoc, G., Girasole, G., Passeri, G., Williams, D. C., Abrams, J. S., Boyce, B., Broxmeyer, H. & Manolagas, S. C. (1992). Increased osteoclast development after estrogen loss: Mediation by interleukin-6. *Science, 257*, 88-91.

[47] Arjmandi, B. H., Birnbaum, R., Goyal, N. V., Getlinger, M. J., Juma, S., Alekel, L., Hasler, C. M., Drum, M. L., Hollis, B. W. & Kukreja, S. C. (1998). Bone-sparing effect of soy protein in ovarian hormone-deficient rats is related to its isoflavone content. *Am J Clin Nutr, 68*, 1364S-1368S.

[48] Nieves, J. W. (2009). Nutritional therapies (including fosteum). *Curr Osteoporos Rep, 7*, 5-11.

[49] Chen, Y. M., Ho, S. C., Lam, S. S., Ho, S. S. & Woo, J. L. (2004). Beneficial effect of soy isoflavones on bone mineral content was modified by years since menopause, body weight, and calcium intake: A double-blind, randomized, controlled trial. *Menopause, 11*, 246-254.

[50] Fitzpatrick, L. A. (2003). Soy isoflavones: hope or hype?. *Maturitas, 44 (Suppl.1)*, S21-S29.ç

[51] Choi, E. M., Suh, K. S., Kim, Y. S., Choue, R. W. & Koo, S. J. (2001). Soybean ethanol extract increases the function of osteoblastic MC3T3-E1 cells. *Phytochemistry, 56*, 733-739.

[52] Suh, K. S., Koh, G., Park, C. Y., Woo, J. T., Kim, S. W., Kim, J. W., Park, I. K. & Kim, Y. S. (2003). Soybean isoflavones inhibit tumor necrosis factor-α-induced apoptosis and the production of interleukin-6 and prostaglandin E2 in osteoblastic cells. *Phytochemistry, 63*, 209-215.

[53] Hein, M., Frank, O., Kampmann, G., Sochocky, N., Pennimpede, T., Fuchs, P., Hunziber, W., Weber, P., Martin, I. & Bendik, I. (2004). The phytoestrogen genistein enhances osteogenesis and represses adipogenic differentiation of human primary bone marrow stromal cells. *Endocrinology, 145*, 848-859.

[54] Dai, Z., Li, Y., Quarles, L. D., Song, T., Pan, W., Zhou, H. & Xiao, Z. (2007), Resveratrol enhances proliferation and osteoblastic differentiation in human mesenchymal stem cells via ER-dependent ERK1/2 activation. *Phytomedicine, 14*, 806-814.

[55] Gómez-Rao, L., Liu, L. J. F., Murray, T. McC., McDermott, E. & Zhang, X. (2003). Estrogen added intermittently, but not continuously, stimulates differentiation and bone formation in SaOS-2 cells. *Biol Pharmacol Bul, 26*, 936-945.

[56] Crisafulli, A., Altavilla, D., Squadrito, G., Romeo, A., Adamo, E. B., Marini, R., Inferrera, M. A., Marini, H., Bitto, A., D'Anna, R., Corrado, F., Bartolone, S., Frisina, N. & Squadrito, F. (2004). Effects of the phytoestrogen genistein on the circulating soluble receptor activator of nuclear factor kappaB ligand-osteoprotegerin system in early postmenopausal women. *J Clin Endocrinol Metab, 89*,188-192.

[57] Nakashima, T., Kobayashi, Y., Yamasaki, S., Kawakami, A., Eguchi, K., Sasaki, H. & Sakai, H. (2000). Protein expression and functional difference of membrane-bound and soluble receptor activator of NF-κB ligand: Modulation of the expression by osteotropic factors and cytokines. *Biochem Biophys Res Commun, 275*, 768-775.

[58] Galien, R. & García, T. (1997). Estrogen receptor impairs interleukin-6 expression by preventing protein binding on the NF-κB site. *Nucl Acids Res, 25*, 2424-2429.

In: Isoflavones Biosynthesis, Occurence and Health Effects ISBN: 978-1-61728-113-6
Editor: Melanie J. Thompson, pp. 195-224 © 2010 Nova Science Publishers, Inc.

Chapter VII

Health Effects of *Dioscorea Bulbifera* of Aerial-Potato (Yam)

Noboru Motohashi*
Meiji Pharmaceutical University, 2-522-1, Noshio, Kiyose-shi, 204-8588 Tokyo, Japan

Abstract

Dioscorea spp. (yam) is a rich source of carbohydrate crop in the diet and also resource of different phytohormones. Among *Dioscorea* spp, especially, *Dioscorea bulbifera* of aerial-potato has widely been used as their traditional folklore medicines for their treatments such as thyroid disease, tumor, diretic disturbance and inflamation in the tropical and subtropical areas. The diverse compounds in *Dioscorea bulbifera* have been found and their functionality has been examined in various researches such as their antibacterial and antitumor activity. The purpose of this review paper is to focus and describe their compositions found and functional activities.

Keywords: *Dioscorea bulbifera*, diosgenin, xanthophylls, steroid saponins, clerodane diterpenoids, bafoudiosbulbins, diosbulbins

1. Source and Scientific Name

It is thought that the aeial-potato (aerial potato, potato yam, aerial yam, air yam) might be a native of tropical Africa, Asia and Australia. Aerial-potatoes also have been spread into subtropical and temperate zones such as Japan, the United States and Europa from these native regions.

Scientific name of aerial-potato is *Dioscorea bulbifera* L. and a plant belonging to *Dioscoreales* (order), *Dioscoreacea*e (family), and *Dioscorea* (genus) (photo1) [1].

* Corresponding author: < noborumotohashi@jcom.home.ne.jp>

Photo 1. Aerial tubers of air-potao of *Dioscorea bulbifera* (at home garden of author. Photographed by Noboru Motohashi) <406. IMGP5642(aerial tuber of aerial-potato, 11/29/2009)>

Around 600 varieties of *Dioscorea* species might be widely eaten in tropical, subtropical and temperate zones in the world [2, 3].

The cultivated yams such aerial-potatoes with their edible starchy tubers and aerial tubers are very important nutrient and economical root crops in especially both tropical and subtropical areas [3].

Because aerial-potato contains steroid diosgenin (**1**) has been used as starting material of their manufactures for diverse artificial steroidal hormones such as hormonal contraception and other sex hormones. Moreover, aerial-potato also has been their improvement or treatment for the conjunctivitis, diarrhea and dysentery as a folklore medicine. For example, aerial tubers of aerial-potato have abundantly diosgenin (**1**) of steroid saponins and therefore these tubers both aerial-potato tubers and underground tubers have been actually used in the partial synthesis of hormones having a steroid structure such as pregnenolone (**2**) for treatment of rheumatoid arthritis and progesterone (**3**) for their treatment by orally and intramusculary administrations on functional uterine bleeding, abnormalities of the menstrual cycle, and threatened abortion, administered (Figure 1).

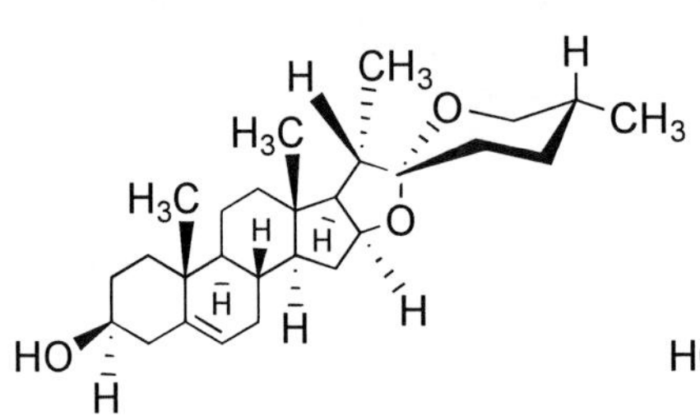

diosgenin (**1**)

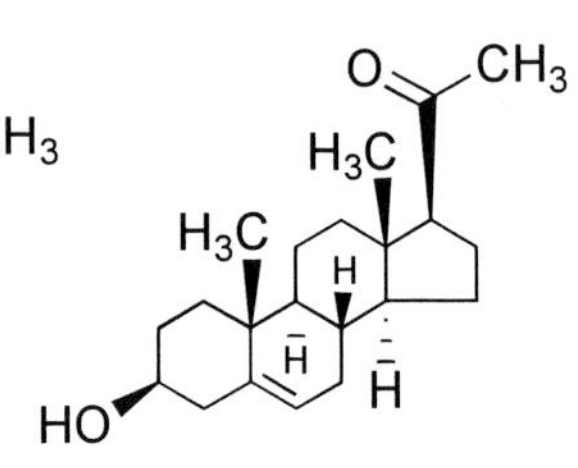

pregnenolone (**2**)

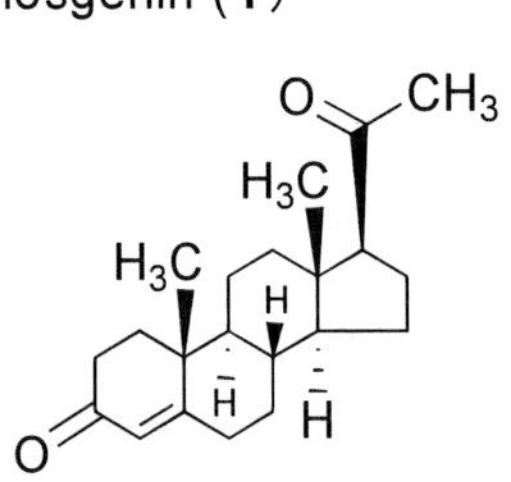

progesterone (**3**)

Fig.1 Diosgenin (**1**), pregnenolone (**2**), progesterone (**3**)

2. Components, Safety, Nutritive and Anti-nutritive Values

2.1. Components of Aerial-Potato (Yam)

1991, the composition per 100 edible portions of aerial-potato was analysed by a group of Dep. of Foods and Nutrition, College of Home Science, G. B. Plant University of Agriculture & Technology, Pantnagar, India [4].

Energy value	: 100.03kcal
Water	: 74.8g
Crude protein	: 3.55g
Crude fat	: 0.38g
Carbohydrate	: 20.17g
Total ash	: 0.89g
Bio-minerals	: Ca 152.2mg; P 65.7mg; Fe 6.23mg
Antioxidative β-caroten	: 56.3μg
Vitamin B complex	: vitamin B_1 (Thiamine) 0.031mg; vitamin B_z (riboflavin) 0.084mg; Niacin (Nicotinic acid) 0.169mg
Antioxidative vitamin C	: 5.27mg
Crude fiber	: 1.08mg

In 1994, an Indian reseach group (Dep. of Foods and Nutrition, Postgraduate and Research Centre, Andhra Pradesh Agricultural University, Rajendranagar, Hyderabad) has studied the components per 100 g edible parts of aerial-potato and the componets were almost similar to their composition values measured in 1991 [5].

2.2. Safety of Aerial-Potato (Yam)

In 2005, the bitterness and toxicity of wild yam *Dioscorea spp.* tubers in Nepal were studied. Two main bitter components were identified as their furanoid norditerpenes of diosbulbin A (**4**) and diosbulbin B (**5**) and were found in the range of 0.023-0.046 and 0.151-0.442g/kg, respectively. Diosbulbin B (**5**) with an average value of 0.314g/kg will be the principal bitter component when compared to diosbulbin A (**4**) with 0.037g/kg. The cooking showed the considerable reduction of two bitter principles of diosbulbin A (**4**) and diosbulbin B (**5**), when compared to the uncooked condition. The boiling could reduce their bitterness in the range 75-100%, whereas pressure cooking and baking reduced the bitterness in the range 50-75% only, when compared to the uncooked condition. Then, the boiling might be the most excellent cooking method for these wild yams for the reduction of the bitterness on acceptable and satisfactory levels and supplying their final food palatable and tastes.

A toxic dioscorine (**6**) of piperidine alkaloids and histamine (**7**) of allergen could not be detected in their tubers of aerial-potatoes. Their cyanogens (**8**) as HCN equivalent in the tubers could be found ranging from 3.2 to 6.0 ppm, however their cyanogens (**8**) contents were satisfactorily below the safety limits. Because the lethal dose range of human for HCN taken by mouth is estimated to be only 0.5-3.5 mg/kg body weight [3, 6].

However, the inflammation and occasional toxicity observed by aerial-potato tubers might be due to the presence of oxalate (**9**) with their high content levels. Using the domestic cooking methods, it is very essential to remove their bitterness of aerial-potatoes (Figure 2) [3].

2.3. Nutritive and Anti-nutritive Components of Wild Inedible Yam

It is generally thought that some alkaloids with a nitrogen atom in the chemical structures having part of a heterocycle are very poisonous and toxic such as aconitine in acinite of *Conium maculatum* and many other *Aconitum* spp., strychnine in nux-vomica of *Strychnos nux-vomica*, (+)-coniine in hemlock of *Conium maculatum* towards man and animals. At the same time, these alkaloids could have been clinically used as many medicinal drugs.

The tubers of aerial-potatoes with these poisonous alkaloids are also found in the wild varieties of yam tubers [1, 7].

Additionally, aerial-potato tubers contains oxalic acid and oxalates (**9**) as raphides of the insoluble calcium salt in their cell vacuoles similar to other plants [1]. It is known that oxalates (**9**) at their high levels in our diets could cause paralysis when man orally took oxalates (**9**) in soluble form.

The nutritive compositions per 100g tubers of wild aerial-potatoes were examined. The uncooked and cooked aerial-potatoes were compared on their contents of saponins, alkaloids and oxalate per 100g aerial-potatoes tubers, respectively as follows:

saponins: before cooking 79.48mg/100g; after cooking 24.44mg/100g
alkaloids: before cooking 0.98%; after cooking 0.26%
oxalate : before cooking 2.46mg/100g; after cooking 0.69mg/100g

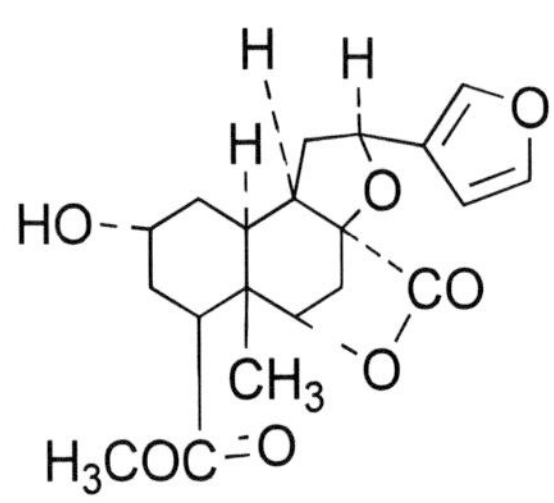

diosbulbin A (4)

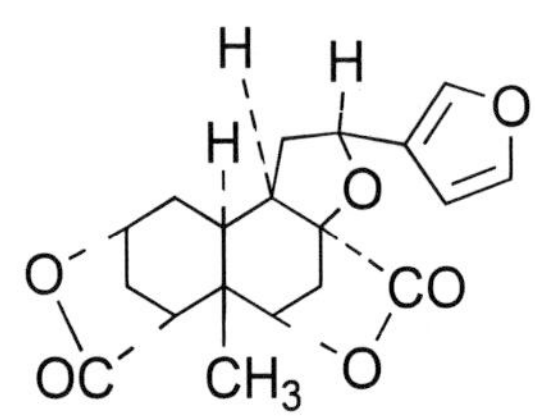

diosbulbin B (5)

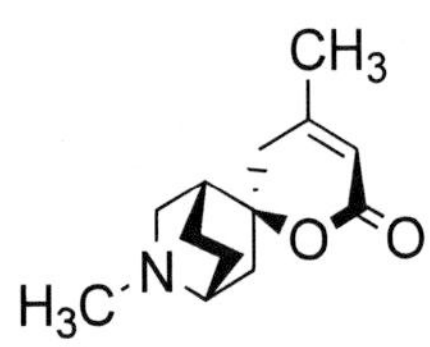

dioscorine (6)

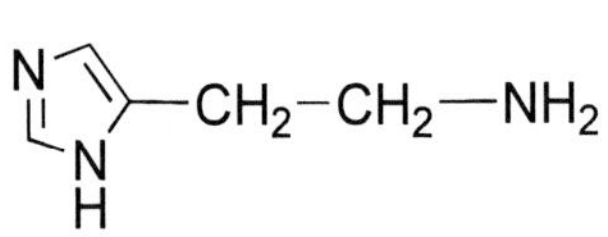

histamine (7)

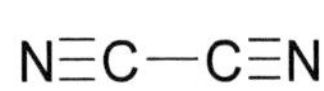

cyanogens (8)

oxalate (9)

Fig,2 Diosbulbin A (4), diosbulbin B (5), dioscorine (6), histamine (7), cyanogens (8), oxalate (9)

Their contents of three anti-nutritive compositions of saponins, alkaloids and oxalates of the aerial-potatoes showed significantly a reduction after cooking, respectively. In Nigeria, aerial-potato tubers have been traditionally processed several times before their consumption as food crops. The processes could reduce the contents of saponins, alkaloids and oxalate contents for a further possible consumption without their toxicity and poisons [1].

Aerial-potato tubers contains saponins of flavonoid polymers, are beneficial for many healthful properties to human [1].

Some saponins could show to be their powerful antioxidants and free radical scavengers. Saponins might inhibit the growth of cancer cells, help to low the blood cholesterol, and protect cardiovascular diseases [8, 9].

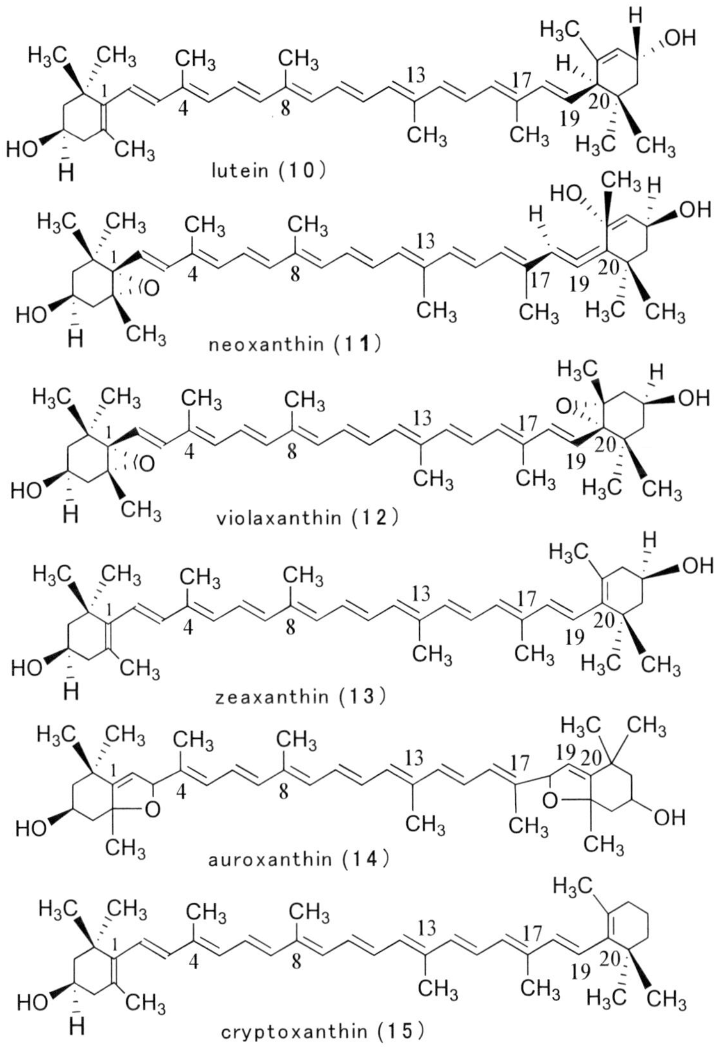

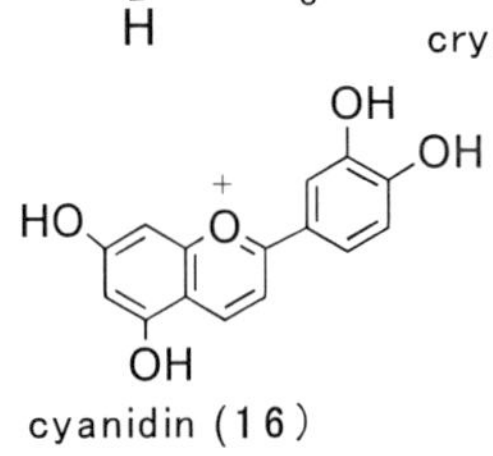

Fig.3 Lutein (10), neoxanthin (11), violaxanthin (12), zeaxanthin (13), auroxanthin (14), cryptoxanthin (15), cyanidin (16)

3. Components and Their Functionality

3.1. Aerial Tubers (Bulbils)

3.1.1. Xanthophylls: Antitumor Activity

The aerial tubers of aerial-potato are generally yellow and moreover. Their colorful components of the aerial tubers might relate a mixture of diverse plant pigments of phytochemicals such as carotenoids and flavonoids. The yellow color phytochemical of sweet potato of *Ipomoea batatas* (L.) Lam. principally come from β-carotene. However, other carotenoids of mainly carotenes might be found in much smaller quantities. Then, the phytochemical yellowish colors of the plant roots is generally thought to be a content indication of the β-carotene [10].

The yellow pigment of the tuberous roots of a potato, cassava of *Manihot esculenta* Crantz, was found as β-carotene.

Their yellow pigments on the extracts from the aerial tubers of ten varieties of aerial-potatoes were identified by their properties of the chromatographic mobilities, colors, stability, and adsorption spectra, and in comparison with two easily identified pigments of spinach leaves and carrot roots. Six xanthophylls such as lutein (**10**), neoxanthin (**11**), violaxanthin (**12**), zeaxanthin (**13**), auroxanthin (**14**), cryptoxanthin (**15**) were identified (Figure 3).

Including these xanthophylls, a cyanidin (**16**) of anthocyanins with its absorption maxima of 535 nm was found in the aerial tubers of aerial-potatoes. However, any carotenes were not found in the aerial tubers of aerial-potatoes (Figure 3) [10].

3.1.2. Diosgenin, Steroid Saponins of Sexogen and Hormone Materials

It is well-known that diosgenin (**1**) has been used as the starting material for their partial synthesis of sex hormones such as progesterone (**3**) for contraceptive pills.

The diosgenin (**1**) content of sapogenin was analysed on their dry samples from forty-seven aerial tubers belonging to ten different wild *Dioscorea* species from the forest and forest-ege regions of Ghana in West Africa. Among ten species, the dry aerial tubers of *Dioscorea* bulbifera had the content of 0.08% diosgenin (**1**).

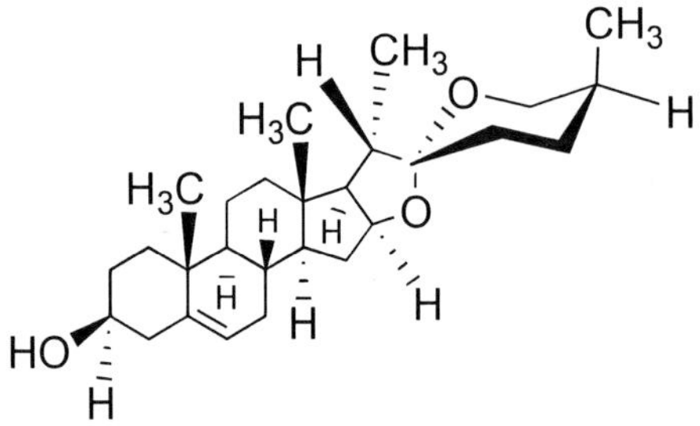

yamogenin (17)

Fig.4 Yamogenin (17)

Following, the fresh aerial tubers of *Dioscorea* bulbifera had a bound-diosgenin (**1**) 0.05% with steroidal sapogenin, and free diosgenin (**1**) 0.02%, respectively. Then, the total content of two bound-diosgenin (**1**) and free diosgenin (**1**) was 0.07%. Moreover, yamogenin (**17**), C-25 epimer of diosgenin (**1**), was also identified in *Dioscorea praehensilis* of *Dioscorea* spp. (Figure 4) [11].

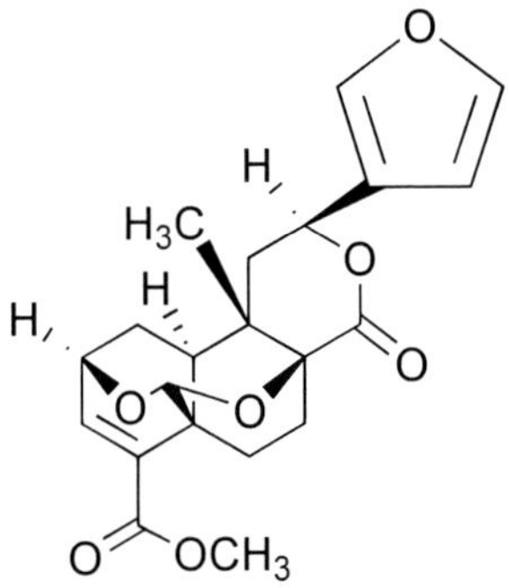

bafoudiosbulbin C (**18**)

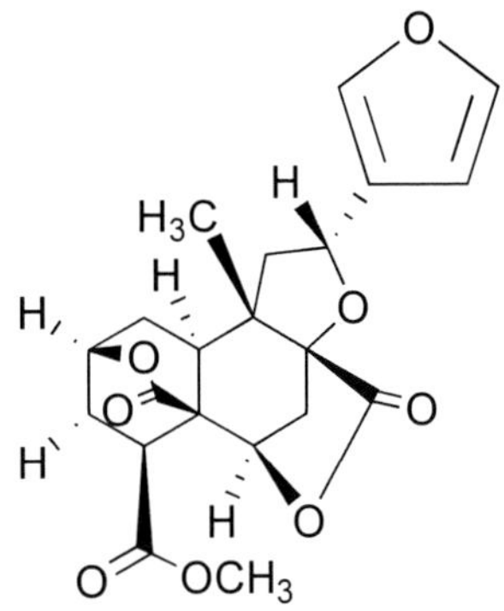

bafoudiosbulbin D (**19**)

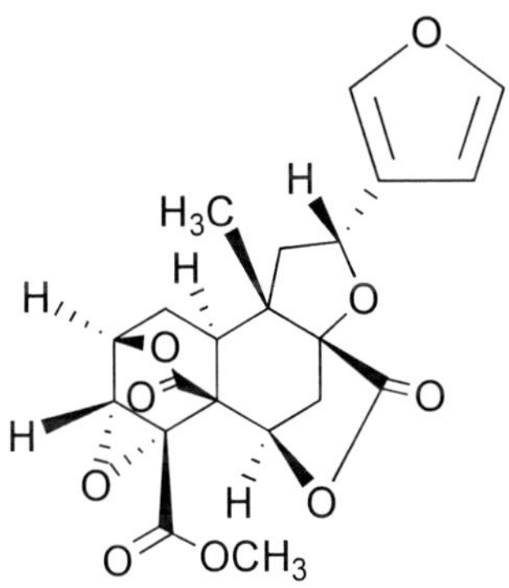

bafoudiosbulbin E (**20**)

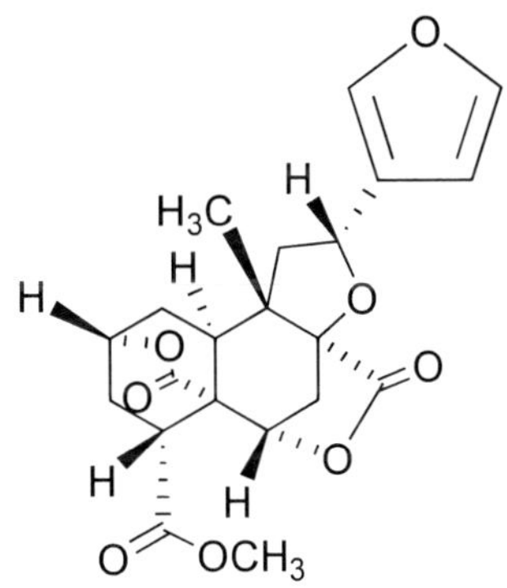

bafoudiosbulbin A (**21**)

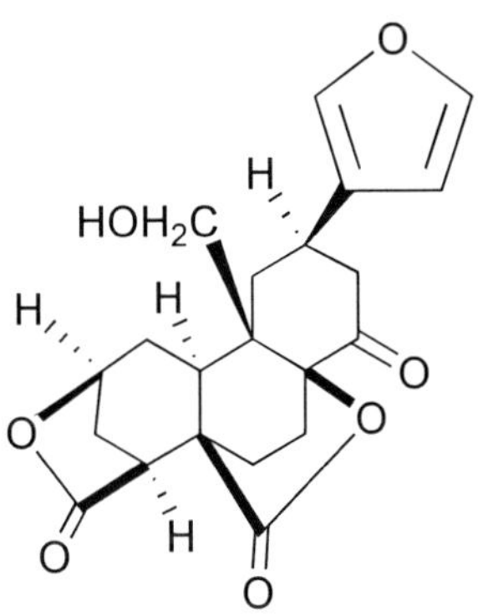

bafoudiosbulbin B (**22**)

Fig.5 Bafoudiosbulbin C (**18**), bafoudiosbulbin D (**19**), bafoudiosbulbin E (**20**), bafoudiosbulbin A (**21**), bafoudiosbulbin B (**22**)

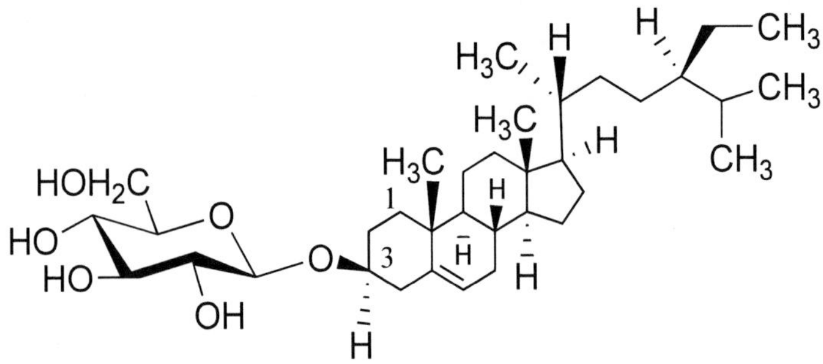

3-*O*-β-D-glucopyranosyl-β-sitosterol (23)

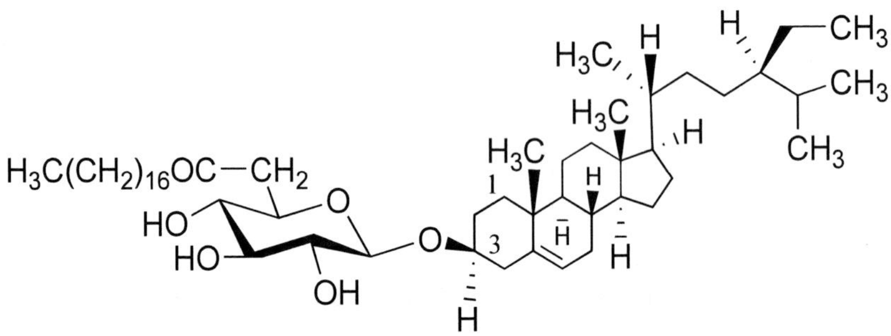

6'-stearoyl-3-*O*-β-D-glucopyranosyl-β-sitosterol (24)

Fig.5 (cont.1) 3-*O*-β-D-glucopyranosyl-β-sitosterol (23), 6'-stearoyl-3-*O*-β-D-glucopyranosyl-β-sitosterol (24)

3.1.3. Clerodane Diterpenoids

3.1.3.1. Three Clerodane Diterpenoids: Bafoudiosbulbin C, Bafoudiosbulbin D, Bafoudiosbulbin E: Antibacterial Activity

In 2007, three novel clerodane diterpenoids such as bafoudiosbulbin C (**18**), bafoudiosbulbin D (**19**) and bafoudiosbulbin E (**20**) were isolated together with four known compouds such as bafoudiosbulbin A, (**21**), bafoudiosbulbin B, (**22**), 3-*O*-β-D-glucopyranosyl-β-sitosterol, (**23**), 6'-stearoyl-3-*O*-β-D-glucopyranosyl-β-sitosterol (**24**) from the bulbils of *Dioscorea bulbifera* (Figures 5, 5(cont.1)) [12].

Diterpenoids with the clerodane skeleton have been stimulated by their biological activities as their insect antifeedant and antitumor, antimicrobial and antifungal agents [12-15].

3.1.3.2. Two Clerodane Diterpenoids: Bafoudiosbulbin F, Bafoudiosbulbin G: Antibacterial Activity

In 2008, two clerodane diterpenoids such as bafoudiosbulbin F (**25**) and bafoudiosbulbin G (**26**) together with three known bafoudiosbulbins such as bafoudiosbulbin A (**21**), bafoudiosbulbin B (**22**), bafoudiosbulbin C (**18**), a benzyls of 3,5,4'-trihydroxy-3'-methoxy-bibenzyl (tristin, **27**), and a flavonol of kaempferol (**28**) were isolated from bulbils of *Dioscorea bulbifera* L. var. sativa (Figure 6) [16].

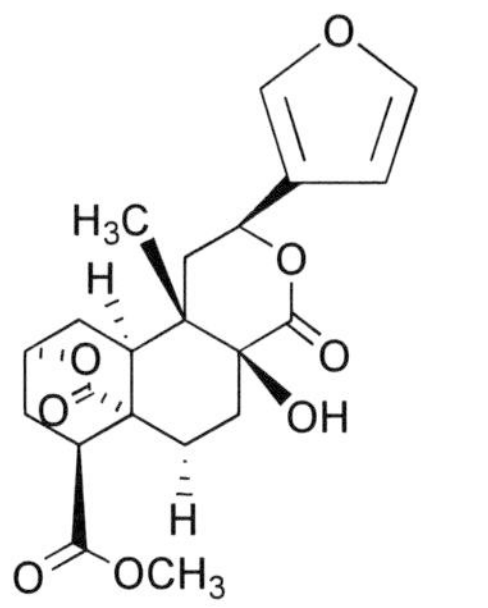

bafoudiosbulbin F (25)

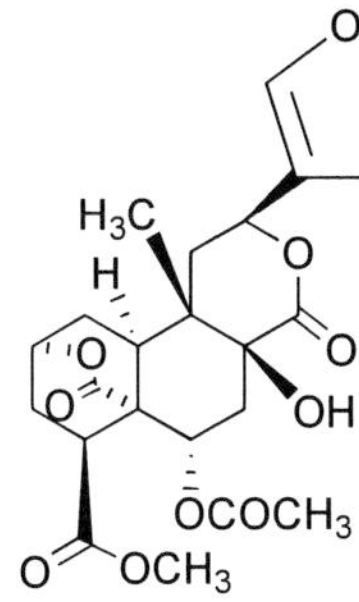

bafoudiosbulbin G (**26**)

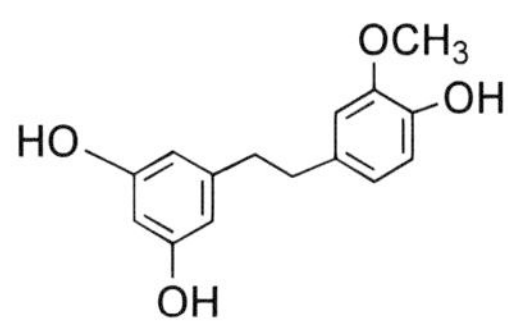

3,5,4'-trihydroxy-3'-methoxy-bibenzyl, tristin (27)

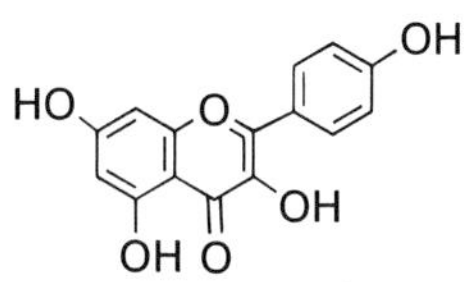

kaempferol (28)

Fig.6 Bafoudiosbulbin F (25), bafoudiosbulbin G (**26**), 3,5,4'-trihydroxy-3'-methoxy-bibenzyl, tristin (27), kaempferol (28)

3.2. Tubers

3.2.1. A Clerodane Diterpenoid of 8-Epidiosbulbin E Acetate: A Potential Plasmid-curing Agent and Antibacterial Activity Against Multidrug-resistant Bacteria

Today, the multidrug-resiatant (MDR) bacteria against the common antibiotics clinically used will be a big problem in their treatment of various bacterial diseases. Then, the novel drugs against MDR bacteria have been required for their treatment and therapy. Oxazolidinone linezolid of a synthetic antibiotics has been clinically used for the treatment of the severe infections which are caused by Gram-positive bacteria having their resistances to other antibiotics. Actually, oxazolidinone linezolid has been used as an antibiotic for curing the vancomycin-resistant bacterial infections. Their resistance to oxazolidinone linezolid has already been known because the resistance clinical isolated bacteria of *Staphylococcus aureus* and *Enterococcus* have been identified [17, 18].

It has been shown that the genetic determinations that confer the resistance to many antibiotics might be located on plasmids. Their extrachromosomal (plasmid) DNA sequences might be often transferable to other bacteria in the environment. Then, the extrachromosomal DNA sequences could begin to express the MDR-responsibility by transfering the resistance against other multiple antibiotics to normal bacteria [19]. The plasmid-mediated multi-drug resistance (MDR) might give the difficulty for the treatment of infectious diseases. A combination of phenothiazines of plasmid-curing agents with antibiotics might be effective against antibiotic resistance encoded by antibiotic resistance plasmids (R-plasmids) [20, 21].

However, it is known that their plasmid-curing agents such as acridine orange, ethidium bromide and sodium dodecyl sulfate are toxic or mutagenic. Therefore, the combination therapy might not be suitable for the *in vivo* testing or clinical remedy [22]. Thus, the entirely novel curing agents with the higher efficacy and safety might be now required the success with the ultra-speed [23].

There are many folklore herbal medicines such as Chinese traditional medicine (Kanpoh remedy medicine), Japanese traditional medicine (Honzoh medicine), African traditional medicines and many other traditional remedies in all countries and regions in the world. These informations could contribute to prepare the novel anti-MDR drugs [24]. In fact, the plant-derived phytochemicals for MDR plasmid bacteria and MDR resistance bacteria have been found for their treatments of many MDR bacteria [25-31].

Dioscorea bulbifera has also been traditionally used as a plant in folklore remedy of Ayurvedic medicine in India. 8-Epidiosbulbin E acetate (**29**) was isolated from bulbs of *Dioscorea bulbifera.* 8-Epidiosbulbin E acetate (**29**) of clerodane derivatives of diterpenes is very important compound because 8-Epidiosbulbin E acetate (**29**) represents a stereochemical structure with the diverse biological activities such as antitumor, antibacterial, antifeedant, antifungal and hallucinogenic activities [14, 32, 33].

8-Epidiosbulbin E acetate (**29**) exhibited their broad-spectrum plasmid-curing activity against MDR bacteria including vancomycin-resiatant enterococcus. 8-Epidiosbulbin E acetate (**29**) cured the antibiotic resistance plasmids (R-plasmid) from four clinical isolates of *Enterococcus faecalis, Escherichia coli, Shigella sonnei* and *Psedomonas aeruginosa* with 12-48% curing efficieny. Their reference plasmids without MDR plasmids of *Bacillus subtilis* (pUB110), *Escheria coli* (RP4), *Psedomonas aeruginosa* (RIP64) and *Salmonella typhi* (R136) showed their curing efficiencies ranging from 16% to 64%. The 8-epidiosbulbin E acetate (**29**)-mediated R-plasmid curing decreased apparently the minimal inhiobotory concentration (MIC) of antibiotics against each MDR bacterium. Then, the antibiotic resistance pattern represents that 8-epidiosbulbin E acetate (**29**) might be effective in their reversal of bacterial resistance towards the different antibiotics. Therefore, 8-Epidiosbulbin E acetate (**29**) could be a potential lead compound for the research of the novel anti-MDR drugs.

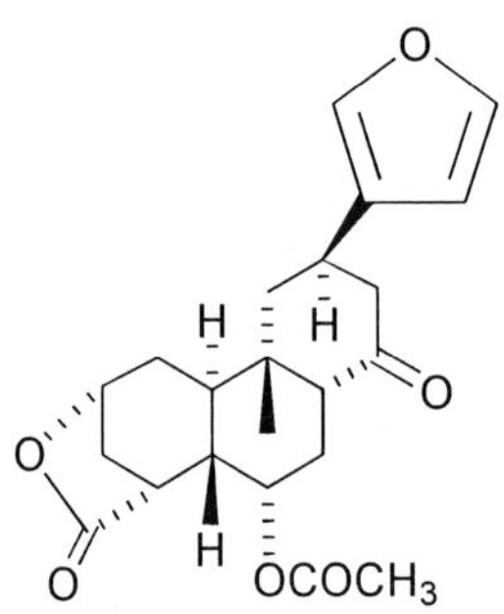

8-epidiosbulbin E acetate (29)

Fig.7 8-Epidiosbulbin E acetate (29)

Further, 8-epidiosbulbin E acetate (**29**) did not show any cytotoxic activity against three human cancer cell lines such as MCF-7 (breast cancer), SiHa (cervical cancer) and A431 (epidermal carcinoma) (Figure 7) [23].

3.2.2. Extracts of Dioscorea bulbifera Tubers: Anorexient Activity

In Ayurveda of Hindu Medicine in ancient India, the chips of aerial-potato rhizomes, kalio-kund in Indian name, have been used as a food supplement for leading the loss of appetite [34, 35].

Some anorexient compounds have been known as follows. Cocaine could happen to some anorexias [36, 37]. Methylphenidate of an antidepressant showed the anorexient activity as the side effects [38]. However, these two cocaine and methylphenidate showed only a very low anorexient activity. Additionally, there are no significant stimulant effect as a representive hyperactivity in the test of the spontaneous locomotor activity.

Based on these facts, the anorexient activity of tuber extracts of aerial-potatoes was examined. These six extracts was intraperitoneally injected to Norwegian fully grown male rats by using 5 to 15 rats per dose level of the extracts. First, among seven extracts, five petroleum ether, light ether, ether, ethanol and aqueous extracts at administration of all up to 40 mg/kg had no significant anorexient effect on the behavior of their food intake. Second, chloroform extract showed the reduction of food intakes in 300 min tests with a moderate extent. However, the anorexient effect was not reproducible. Third, acetone extract represented the increase of their food intakes at 0.5 mg/kg administration and the reduction of food intake at their administration from 5 to 45 mg/kg. However, the tolerance might preclude the loss of weight on chronic administration.

Furthermore, their results of ablation and stimulation on the administrated rats of the tuber extracts could express the lateral hypothalamus as the site of hunger center in a part of diencephalon. The ventromedial part on hunger center might seemingly participate in their actual reactions of the satiety. Because, the anorexient effect of amphetamine (**30**), a sympathomimetic amine that has a stimulating effect on both the central and peripheral nervous systems, might be due to their action on the central mechanisms concerning to the food intake [39, 40]. The action of hunger center might relate to the satiety center in the cat [41].

On the other hand, cocaine, methylphenidate and corticomedullary stimulants have not been understood well in this respect. However, the weak anorexient effects of cocaine and methylphenidate might indirectly be their central-mediated activity through the prefrontal areas.

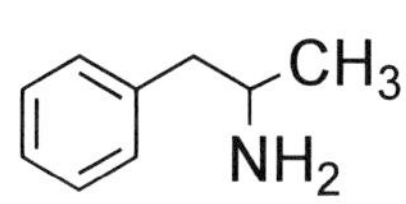

amphetamine (30)

Fig.8 Amphetamine (30)

There are further evidences on adequqte feeding stimulus. The heat production [42], the amount of the store fat in the body [43] and the "glucoreceptors" for the regulation of glucose metabolism in their hypothalamic areas [44] might participate in regulating their feeding stimulus (Figure 8) [35].

3.2.3. Steroid Saponins: Antiarthritic

In India and its surrounding countries, *Dioscorea* tubers (yams) have been used as materials in a traditional remedy. The dried tubers of *Dioscorea bulbifera* have been used as their treatments and remedy of ulcers, pilles, dysentery and syphilis. The tubers have been also used in the bitter, pungent, flattening, tonic, alterative, aphrodisiac, stomachic, anthelmintic, appetiser, useful in leukoderma, urinary discharges, dyspepsia, bronchitis and tumors. The powered tubers are used in the treaments of asthma and abdominal pain.

Other uses of the tubers are as famine food, for the preparation of starch, for washing the wool and silk and for attracting the fishes for the catching.

The occurrence of diosgenin (**1**) of a sapogenin in *Dioscorea* is now focused worldwide. Diosgenin (**1**) has been widely used their steroidal drugs such as oral contraceptives, antiarthritic, medications and many other climatic applications. Actually, diosgenin (**1**) has been applied as hormonal substitutes such as estrone (31), progesterone (**3**), androsterone (**32**) and testosterone (**33**). Moreover, *Dioscorea* tubers (yams) are the important materials of very cheaper diosgenin (**1**) and the starting materials of many other hormones (Figure 9).

Steroidal sapogenin is the material of diosgenin (**1**). Then, the content of steroidal sapogenin in the dry root tubers of *Dioscorea* (yams) was measured to be 0.5% [45].

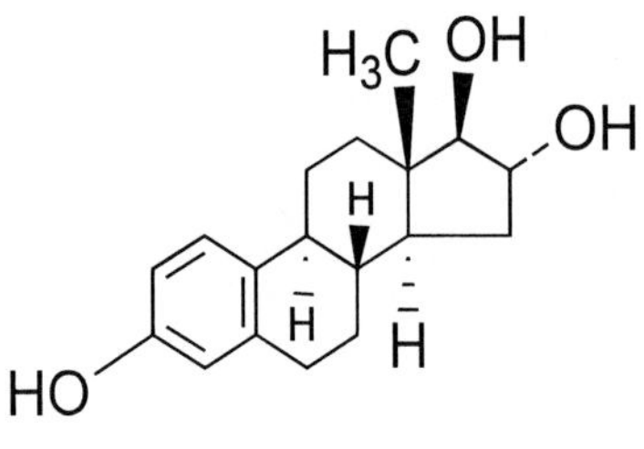

estrone (**31**)

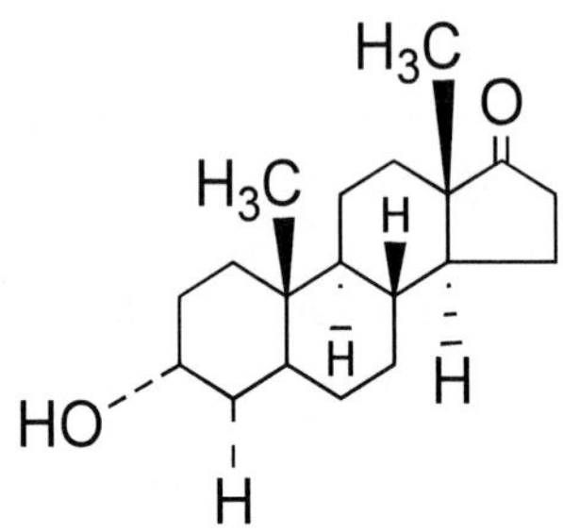

androsterone (**32**)

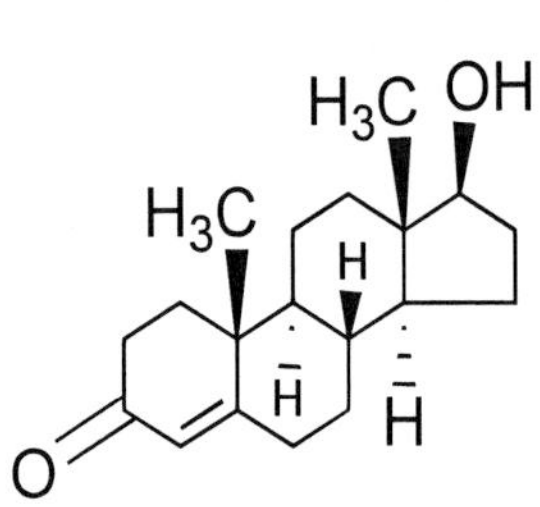

testosterone (**33**)

Fig.9 Estrone (**31**), androsterone (**32**), testosterone (**33**)

3.2.4. Cyanidin: Antioxidative Action

The yams have almost been eaten as a fresh product in West Africa. In some of areas in Western Africa such as Nigeria, yams have been traditionally processed by parboiling, sundrying and grinding a coarse flour. Today, yams are commercially supplied as their forms such as instant reconstitutable flakes or powder products.

Cyanidin (**16**) and polyhydric phenols were found in the root tubers of *Dioscorea bulbifera.* The content of cyanidin (**16**) in dry root tubers was 240 mg/kg. Total phenolic contents in dry root tubers was 4,233 mg/kg at harvest of Nov./1979 and 4,272 mg/kg at harvest of Oct./1980, respectively. The total phenolic content of *Dioscorea bulbifera* had higher values when compared to other four *Dioscorea spp.*

The fresh flesh of yams change white to brown by *o*-dihydroxyphenoloxidase (*o*-DPOase) of a oxidase which is contained in yams when exposed to air. Therefore, (+)-catechin levels in fresh *Dioscorea bulbifera* was 240mg/kg edible parts. The activity of *o*-DPOase was assayed by recording the rate of oxygen consumption. At the same time, cyanidin (**16**) released by the hydrolysis with HCl was also measured. Their levels of cyanidin (**16**) in *D. alata* was sufficiently high to account for their significantly pinking changes when *Dioscorea spp* is boiled. From these facts, *o*-DPOase-catalysed oxidation of (+)-catechin is useful for measuring the bwrown changes of yams. Other reducing compounds such as ascorbic acid and many polyphenolics will exist as the possible influence factors on the oxidation of yam fleshes. However, it could be thought that the oxidation of (+)-catechin by *o*-DPOase in the presence of oxygen will be the most high major causes of the browing change on their fleshes of the cut yam tubers (Figure 3) [46].

3.2.5. Amylopectin: Supply of Oligosugar, Improvement of Gut

The dry tubers of wild *Dioscorea bulbifera* contains around 83% starch. It is not suitable to eat the raw yams because the wild *Dioscorea bulbifera* contains poisonous alkaloids and bitterness for taste. However, their bitterness might be lost by roasting the tubers for eating as foods. Interestingly, the tubers could be eaten after roasting becasuse the tubers of cultivated *Dioscorea bulbifera* has not their bitterness.

The contents both the starch in wild *Dioscorea bulbifera* and maize starch were quantitatively and qualtatively studied and compared together. First, amylose content and amylopectin content in wild *Dioscorea bulbifera* were 12% and 88%, respectively. On the other hand, amylose content and amylopectin content in maize starch were 23% and 77%, respectively. Apparently, amylopectin content (88%) in wild *Dioscorea bulbifera* showed higher content when compared to amylopectin content in maize starch (77%). Amylopectin is a highly branched, water-insoluble glucan, the more prevalent of the two constituents of starch when compared to amylose. α-Amylase catalyzes the hydrolysis of amylopectin to the limit dextrins. The limit dextrins contain the branched oligoglucosides of a polymer glucoside with around 7 glucoses. It is thought that the branched oligoglucosides will serve to improve the health effect in gut because the branched oligoglucosides could enhance the balance of symbiotic bacteria in gut as the probiotics.

In the study, the period both dextrin formation of wild *Dioscorea bulbifera* and maize starch was 36 min and 40 min, respectively. From these comparisons, *Dioscorea bulbifera* will be very important as a resource of starch [47].

3.2.6. Phytoalexin Stilbenoids: Antibacterial Effect

It is known that phytoalexin could be formed when the bulbs are infected with fungi. Then, it means that phytoalexin is any of a group of compounds which are formed in the plants in response to their fungi injection, physical damage, chemical injury, or a pathogenic process. Their phytoalexin could inhibit or destroy the invading agent from the outside. Originally, phytoalexins might be accumulated for their response against their infection of the plants. Phytoalexins have been established as a mechanism of their plant resistance against the plant diseases. Typically, 6-methoxymellein is essentially a fungal metabolite but also produced as a phytoalexin by carrot of *Daucus carota* to fungal attack [48] and rishitin is accumulated by the tubers of potato of *Solanum tuberosum* infected with the pathogenic fungus *Phytophthora infestans* [48, 49].

Demethylbatatasin IV (**34**) of phytoalexins was found from the tuber ether extraction of *Dioscorea rotundata* infected *Botryodiplodia theobromea*. Demethylbatatasin IV (**34**) showed their antifungal activity against spore germination and gern tube elongation of three pathogens such as *Botryodiplodia theobromea, Aspergillus niger, Penicillium schlerotgenum* and *Cladosporium cladosporioides* at levels with ED_{50} ca 50 μg/mL.

Demethylbatatasin IV (**34**) was less antibacterial activity against four pathogenic bacteria such as *Bacillus cereus*, *Staphylococcus aureus*, *Pseudomonas aeruginosa* and *Escherichia coli* at MIC 10-100 μg/mL (Figure 10) [48, 49].

3.2.7. p-Hydroxy Acetophenones: Hypnotic Action, Perfume (Aromatherapy)

4-Hydroxy-2-[3',7'-dimethyl-octa-2',6'-dienyl]-6-methoxyacetophenone (**35**) and 4,6-dihydroxy-2-*O*-(4'-hydroxybutyl)acetophenone (**36**) of two *p*-hydroxy acetophenones were isolated from the tubers of *Dioscorea bulbifera* (Figure 11) [50].

Acetophenone (**37**) had the hypnotic activity and orange-blossom-like odour. Therefore, these two *p*-hydroxy acetophenones could be also expected to their application such as hypotics and perfumes.

3.2.8. Six Steroids: Antiarthritic and Materials of Hormones

Steroid saponin have been applied in the detergents such as soaps, the foaming agents as the fire extinguishers, or their poisons for the fish catching. It is generally known that these steroid saponins are not toxic for human body and their steroid saponins might not be absorbed by human gut. Because we might not be happened to any toxic problems by steroid saponins though human intakes many steroid saponin foods such as asparagus in daily diets.

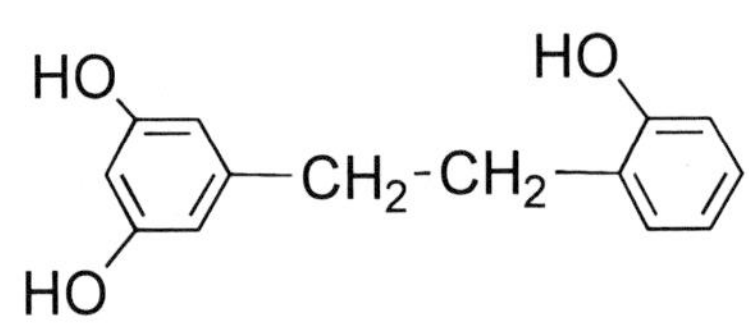

demethylbatatasin IV (**34**)

Fig.10 Demethylbatatasin IV (**34**)

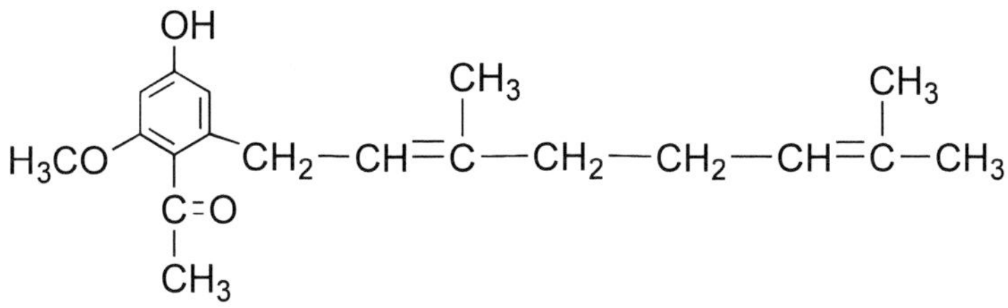

4-hydroxy-2-[3',7'-dimethyl-octa-2',6'-dienyl]-6-methoxyacetophenone (**35**)

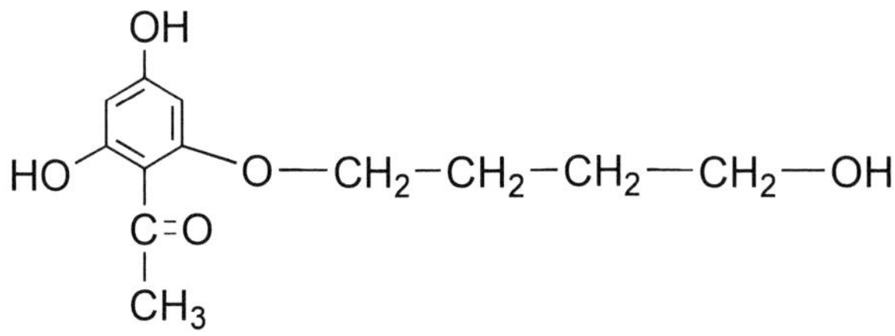

4,6-dihydroxy-2-*O*-(4'-hydroxybutyl)acetophenone (36)

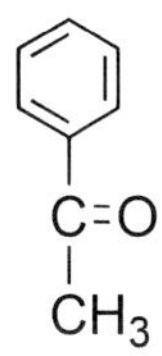

acetophenone (**37**)

Fig.11 4-Hydroxy-2-[3',7'-dimethyl-octa-2',6'-dienyl]-6-methoxyacetophenone (**35**), 4,6-dihydroxy-2-*O*-(4'-hydroxybutyl)acetophenone (36), acetophenone (**37**)

Diosgenin (**1**), β-sitosterol (**38**), stigmasterol (**39**), daucosterol (**40**), diosgenin-3-*O*-α-L-rhamnopyranosyl(1→2)-β-D-glucopyranoside (prosapogenin A of dioscin. **41**), diosgenin-3-*O*-[di-α-L-rhamnopyranosyl(1→2, 1→3)]-β-D-glucopyranoside (taccaoside. **42**) of six steroids were isolated from the tubers of *Dioscorea bulbifera* (Figure 12) [51].

3.2.9. Anthocyanin: Antioxidative, Antitumor

3,7-Dimethoxy-5,4'-dihydroxyflavone (**43**) and 3,7-dimethoxy-5,3',4'-trihydroxyflavone (**44**) of two potential antioxidative flavonoids and antitumor emodin (**45**) of anthraquinones were isolated from the tubers of *Dioscorea bulbifera* (Figure 13) [52].

3.2.10. Diosbulbins: Antitumor Activity

In 1968, diosbulbin A (**4**), diosbulbin B (**5**) and diosbulbin C (**46**) of three diosbulbins in furanoid norditerpenes of scarcely steroid sapogenin were isolated from the root tubers of *Dioscorea bulbifera* (Figure 2) [53, 54].

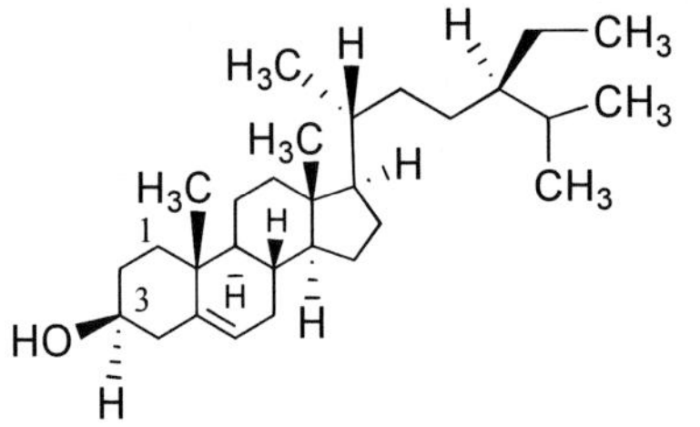

β-sitosterol (38)

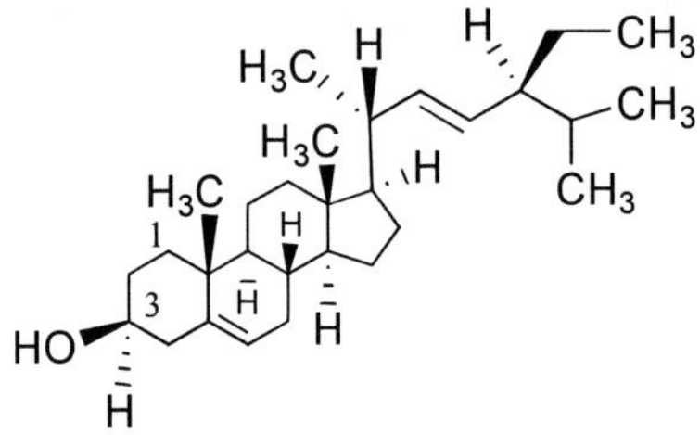

stigmasterol (39)

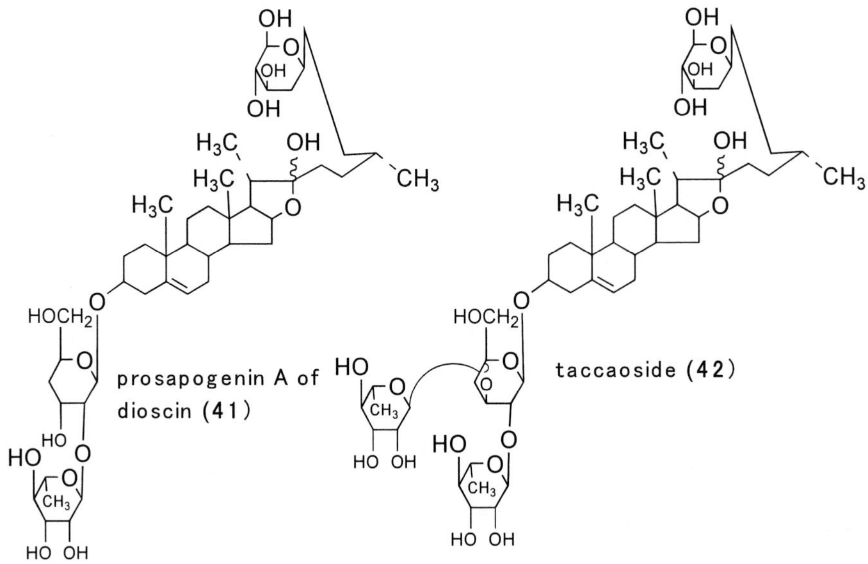

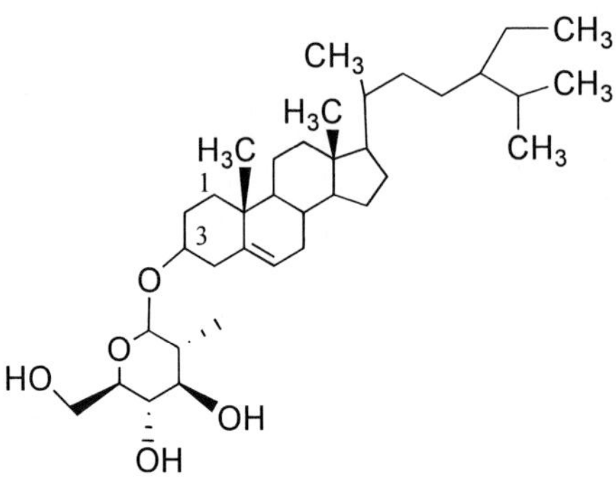

daucosterol (40)

Fig.12 β-Sitosterol (38), stigmasterol (39), daucosterol (40), prosapogenin A of dioscin (41), taccaoside (42)

1984, diosbulbin D (**47**) and 8-epidiosbulbin E (**29**) of a new furanoid norditerpenoid were isolated from the root tubers of *Dioscorea bulbifera* (Figures 7, 14) [55].

Diosbulbin A (**4**), diosbulbin B (**5**) and diosbulbin A 2-*O*-β-D-glucopyranoside (**48**) of three furano-norditerpenes were examined antitumor activity for solid Sarcoma-180 tumor. These three furano-norditerpenes had 1/2 growth inhibition effect when compared to a control (Figure 14) [56].

In 2002, it is confirmed that diosbulbin B (**5**) had their antitumor-promoting effect by the neoplastic transformation assay of mouse epidermal JB6 cell lines [57].

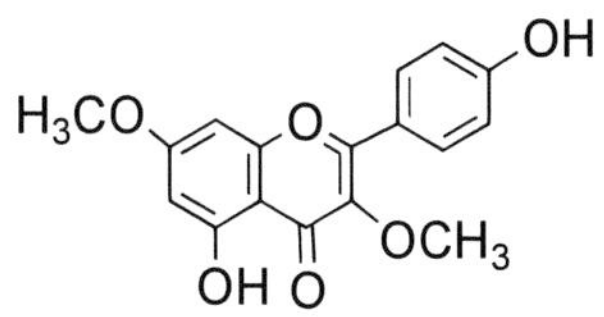

3,7-dimethoxy-5,4'-dihydroxyflavone (43)

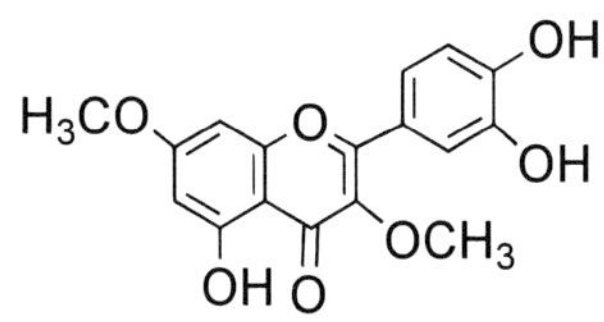

3,7-dimethoxy-5,3',4'-trihydroxyflavone (44)

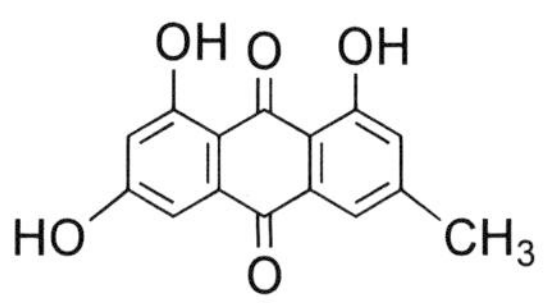

emodin (45)

Fig.13 3,7-Dimethoxy-5,4'-dihydroxyflavone (43)
3,7-dimethoxy-5,3',4'-trihydroxyflavone (44)
emodin (45)

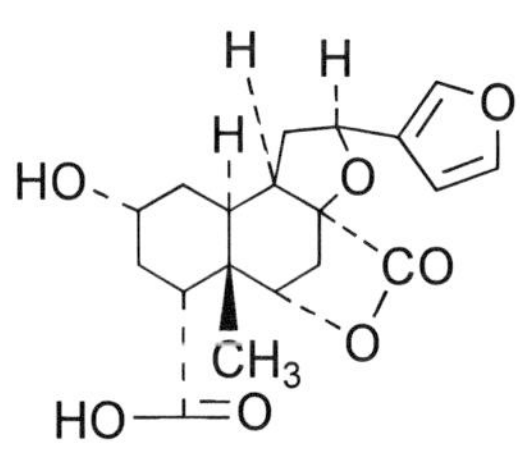

diosbulbin C (**46**)

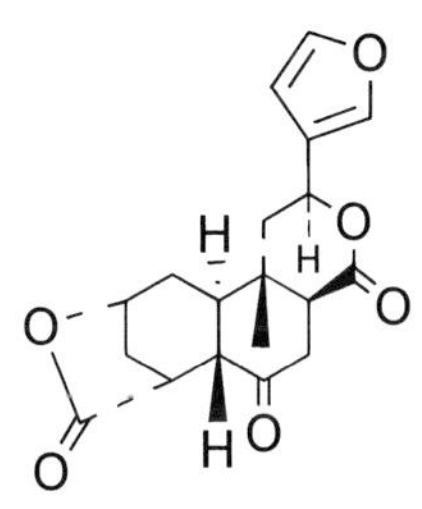

diosbulbin D (**47**)

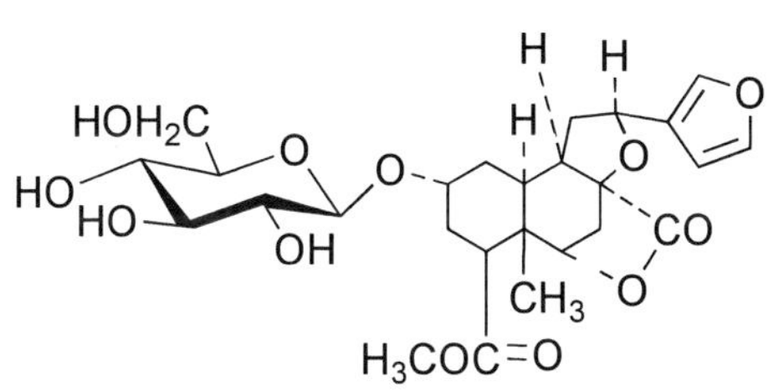

diosbulbin A 2-*O*-β-D-glucopyranoside (**48**)

Fig.14 Diosbulbin C (**46**), diosbulbin D (**47**), diosbulbin A 2-*O*-β-D-glucopyranoside (**48**)

In 2009, diosbulbin B (**5**) and diosbulbin D (**47**) were found from the root tubers of *Dioscorea bulbifera* (Figure 14) [58].

3.2.11. Clerodane Diterpenoids (Bafoudiosbulbin A and Bafoudiosbulbin B): Antibacterial Activity

In 2006, bafoudiosbulbin A (**21**) and bafoudiosbulbin B (**22**) of two novel clerodane diterpenoids were isolated together with tetracosanoic acid (**49**), 1-(tetracosanoyl)-glycerol (**50**), *trans*-tetracosanylferulate (**51**), β-sitosterol (**38**) and 3-*O*-β-D-glucopyranosyl-β-sitosterol (**52**) of five known compounds from the root tubers of *Dioscorea bulbifera* (Figure 15).

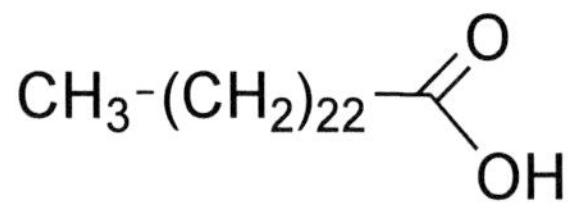

tetracosanoic acid (**49**)

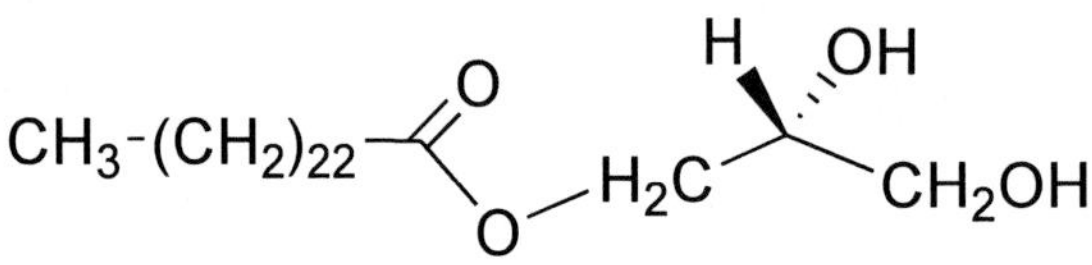

1-(1-(tetracosanoyl)-glycerol (**50**]

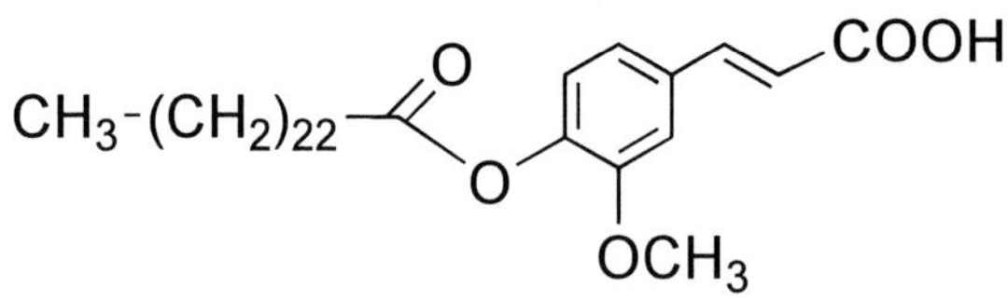

trans-tetracosanylferulate (**51**)

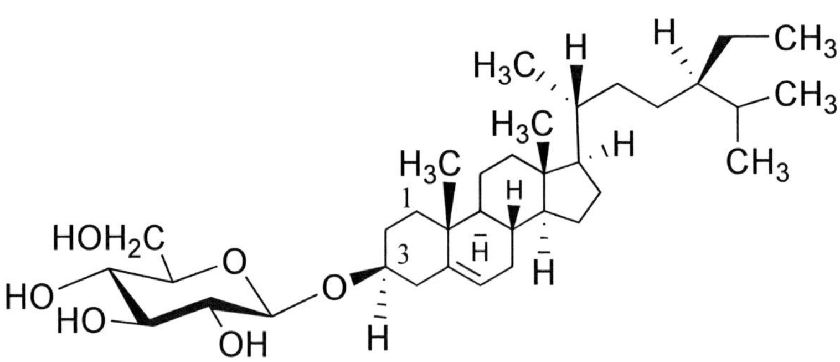

3-*O*-β-D-glucopyranosyl-β-sitosterol (52)

Fig.15 Tetracosanoic acid (**49**), 1-(1-(tetracosanoyl)-glycerol (**50**), *trans*-tetracosanylferulate (**51**), 3-*O*-β-D-glucopyranosyl-β-sitosterol (52)

Bafoudiosbulbin A (**21**) and bafoudiosbulbin B (**22**) have been examined the antibacterial activity agaist seven pathologenic bacteria such as *Pseudomonas aeruginosa*, *Escherichia coli*, *Klebsiella pneumoniae*, *Staphylicoccus aureus*, *Salmonella typhi*, *Salmonella paratyphi* A and *Salmonella paratyphi* B when compared to a control of chloramphenicol of a broad-spectrum antibiotic. Among the inhibition activity against seven bacteria, bafoudiosbulbin A (**21**) and bafoudiosbulbin B (**22**) had significantly their anti-bacterial inhibition against *Pseudomonas aeruginosa*, *Salmonella typhi*, *Salmonella paratyphi* A and *Salmonella paratyphi* B. However, chloramphenicol was no effecive against *Pseudomonas aeruginosa.* Moreover, amoxicillin of a semisynthetic derivative of ampicillin effective against a broad spectrum of Gram-positive and Gram-negative bacteria had the inhibition only against *Pseudomonas aeruginosa* [14].

*3.2.12. Diosbulbin I (**53**) and Diosbulbin J (**54**) of Two Novel Clerodane Diterpenoids: Source Materials of Diosgenin and Related Steroid Saponins*

In 2009, diosbulbin I (**53**) and diosbulbin J (**54**) of two novel clerodane diterpenoids were found of from the root tubers of *Dioscorea bulbifera* (Figure 16) [59].

*3.2.13. Diosbulbisins (**55-58**) of Four Novel Steroidal Saponigenins, Diosbulbiside A (**59**) and Diosbulbiside B (**60**) of Two Novel Spirostane Glycosides, and Diosbulbiside C (**61**) of One Novel Cholestane Glycoside: Antitumor*

In 2009, diosbulbisin A ((25*R*)-17α-hydroxyspirost-4-en-3-one (**55**)), diosbulbisin B ((25*S*)-17α, 25-dihydroxyspirost-4-en-3-one (**56**)) and diosbulbisin C ((25*S*,25*S*)-17α, 26-dihydroxy-22,25-epoxyfurost-4-en-3-one (**57**)), diosbulbisin D ((25*S*,25*S*)-17α, 26-dihydroxy-22,25-epoxyfurost-4,6-dien-3-one (**58**($\Delta^{6,7}$))) of four new steroidal saponigenins, diosbulbiside A ((25*S*)-17α, 25-dihydroxyspirost-5-en-3β-yl-*O*-α-L-rhamnopyranosyl-(1→3)-β-D- -glucopyranoside (**59**)) (Figure 17) and diosbulbiside B ((25*S*)-17α,25-dihydroxyspirost-5-en-3β-yl-*O*-α-L-rhamnopyranosyl-(1→3)-[α-L-rhamnopyranosyl-(1→2)]-β-D-glucopyranoside (**60**)) of two novel spirostane glycosides, and diosbulbiside C ((25*R*)- 26-hydroxycholesta-5,7-dien-16,22-dione-3β-yl-*O*-α-L-rhamnopyranosyl-(1→3)-[α-L-rhamnopyranosyl-(1→2)]-β-D-glucopyranoside (**61**)) of one novel cholestane glycoside were found of from the root tubers of *Dioscorea bulbifera.*

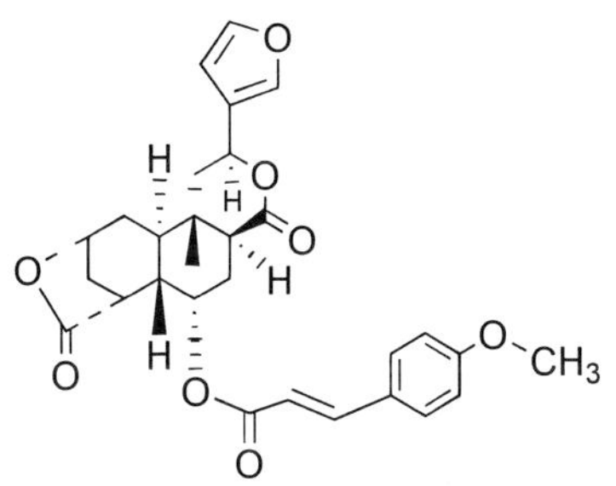

diosbulbin I (53)

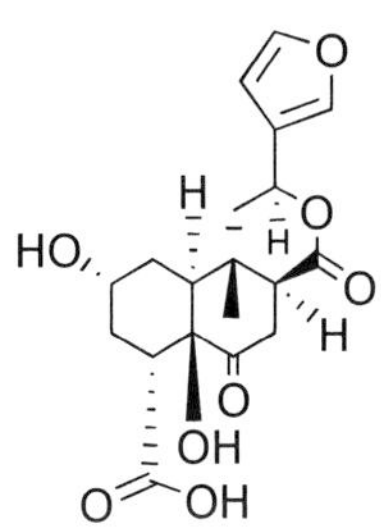

diosbulbin J (**54**)

Fig.16 Diosbulbin I (53), diosbulbin J (**54**)

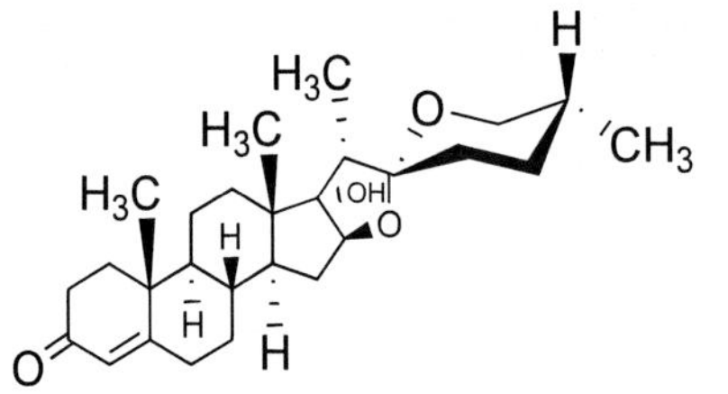

diosbulbisin A [(25*R*)-17α-hydroxyspirost-4-en-3-one (55)]

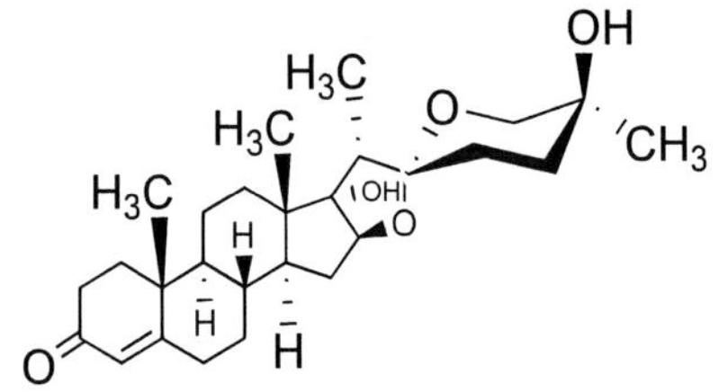

diosbulbisin B [(25*S*)-17α, 25-dihydroxyspirost-4-en-3-one (56)]

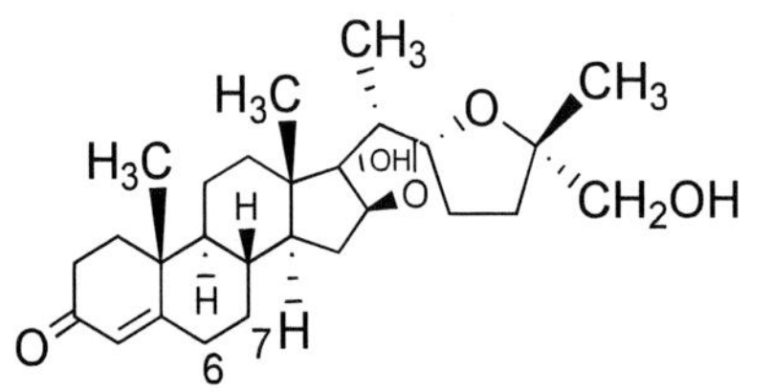

diosbulbisin C [(25*S*,25*S*)-17α, 26-dihydroxy-22,25-epoxyfurost-4-en-3-one (57)]

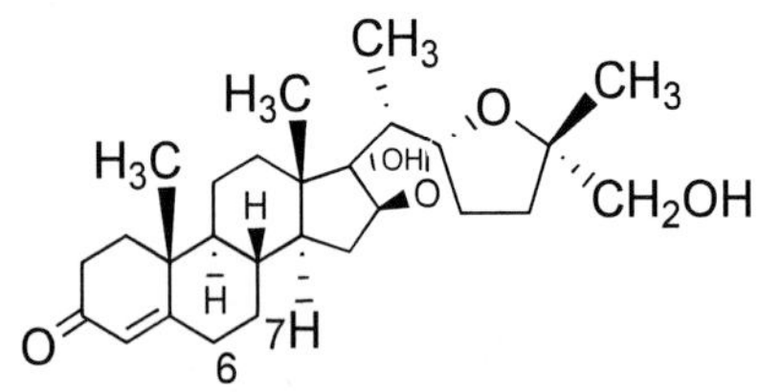

diosbulbisin D [(25*S*,25*S*)-17α, 26-dihydroxy-22,25-epoxyfurost-4,6-dien-3-one (58)($\Delta^{6,7}$)]

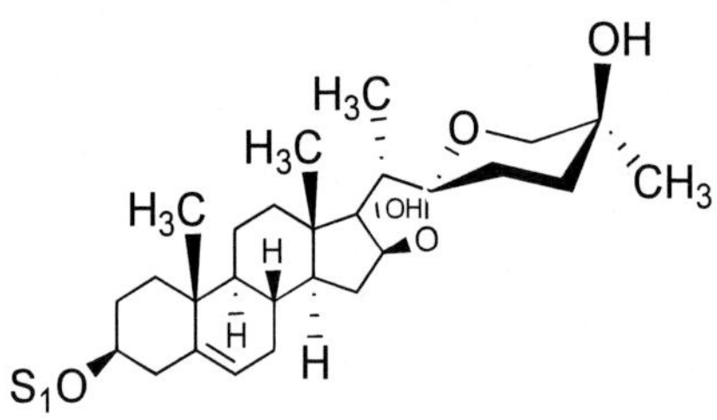

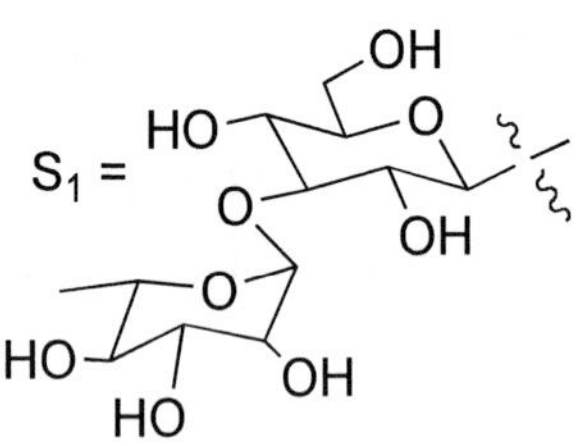

diosbulbiside A [(25*S*)-17α, 25-dihydroxyspirost-5-en-3β-yl-*O*-α-L-rhamnopyranosyl-(1-3)-β-D-glucopyranoside (**59**)]

Fig.17 Diosbulbisin A [(25*R*)-17α-hydroxyspirost-4-en-3-one (55)], diosbulbisin B [(25*S*)-17α, 25-dihydroxyspirost-4-en-3-one (56)],diosbulbisin C [(25*S*,25*S*)-17α, 26-dihydroxy-22,25-epoxyfurost-4-en-3-one (57)], diosbulbisin D [(25*S*,25*S*)-17α, 26-dihydroxy-22,25-epoxyfurost-4,6-dien-3-one (58)($\Delta^{6,7}$)], diosbulbiside A [(25*S*)-17α, 25-dihydroxyspirost-5-en-3β-yl-*O*-α-L-rhamnopyranosyl-(1-3)-β-D-glucopyranoside (**59**)]

At the same time, pennogenin (**62**) (Figure 17(cont.1)), pennogenin-3-*O*-α-rhamnopyranosyl-(1→3)-[α-L-rhamnopyranosyl-(1→2)]- β-D-glucopyranoside (**63**), and pennogenin-3-*O*-α-rhamnopyranosyl-(1→4)-[α-L-rhamnopyranosyl-(1→2)]- β-D-glucopyranoside (**64**) of three known pennogenins were also found of from the root tubers of *Dioscorea bulbifera* (Figure 17(cont.2)).

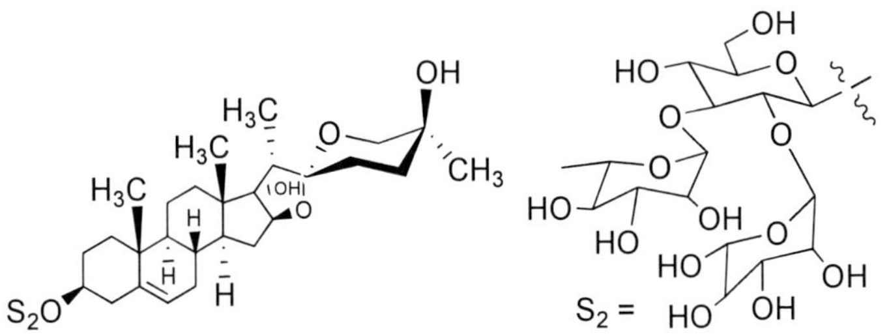

diosbulbiside B [(25*S*)-17α,25-dihydroxyspirost-5-en-3β-yl-*O*-α-L-rhamnopyranosyl-(1-3)-[α-L-rhamnopyranosyl-(1-2)]-β-D-glucopyranoside] (60)

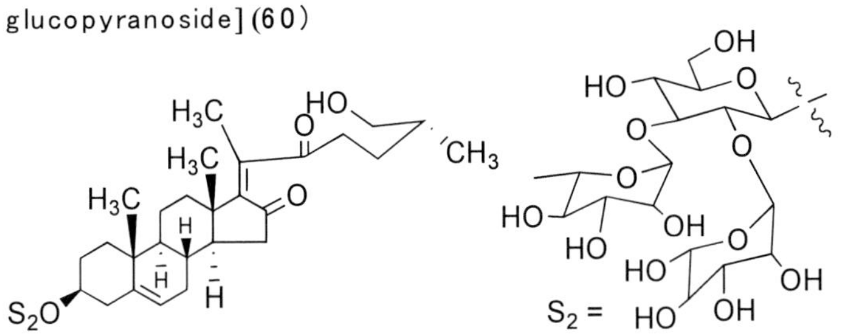

diosbulbiside C [(25*R*)-26-hydroxycholesta-5,7-dien-16,22-dione-3β-yl-*O*-α-L-rhamnopyranosyl-(1-3)-[α-L-rhamnopyranosyl-(1-2)]-β-D-glucopyranoside] (61)

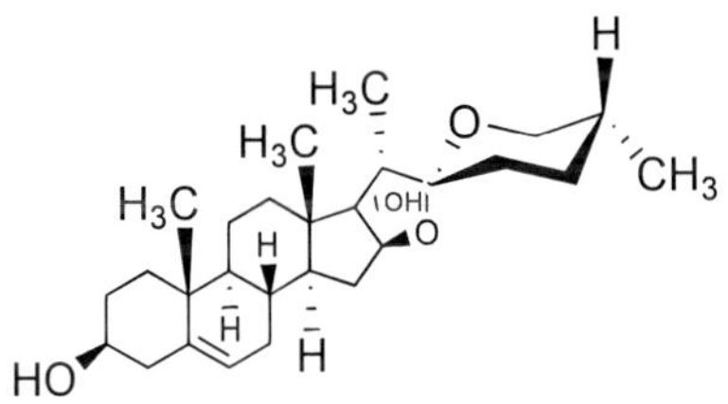

pennogenin (62)

Fig.17 (cont.1) Diosbulbiside B [(25*S*)-17α,25-dihydroxyspirost-5-en-3β-yl-*O*-α-L-rhamnopyranosyl-(1-3)-[α-L-rhamnopyranosyl-(1-2)]-β-D-glucopyranoside] (60), diosbulbiside C [(25*R*)-26-hydroxycholesta-5,7-dien-16,22-dione-3β-yl-*O*-α-L-rhamnopyranosyl-(1-3)-[α-L-rhamnopyranosyl-(1-2)]-β-D-glucopyranoside] (61), pennogenin (62)

These ten compounds (**55-64**) were examined their antitumor activity of the proliferation of against Bel-7402 human hepatocellular carcinoma cells at a concentration of 10 μM by using a standard MTT assay. Only two compounds of pennogenin-3-*O*-α-rhamnopyranosyl-(1→3)-[α-L-rhamnopyranosyl-(1→2)]- β-D-glucopyranoside (**63**), and pennogenin-3-*O*-α-rhamnopyranosyl-(1→4)-[α-L-rhamnopyranosyl-(1→2)]- β-D-glucopyranoside (**64**) showed significantly their cytotoxic activity with 99.1% and 92.6% inhibition, respectively. Furthermore, these two compounds of compound **63** and compound **64** were examined for the cell growth inhibition activity for SMMC7721 human hepatocellular carcinoma cells. And these two compounds of compound **63** and compound **64** showed their IC_{50} values of 4.54 μM compound **63** and 4.85 μM compound **64**, respectively [60].

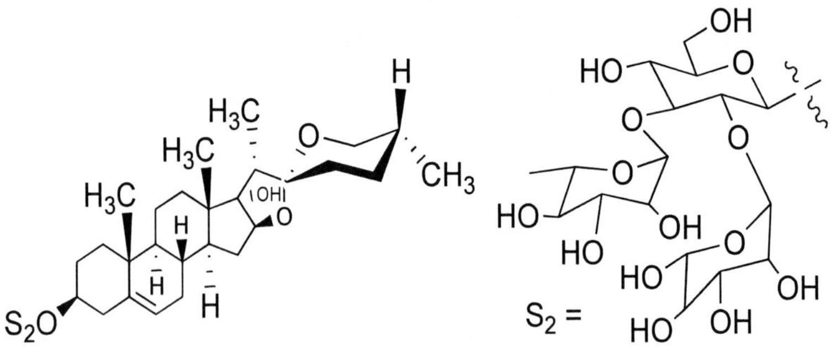

pennogenin-3-*O*-α-L-rhamnopyranosyl-(1–3)-[α-L-rhamnopyranosyl-(1–2)]-β-D-glucopyranoside (63)

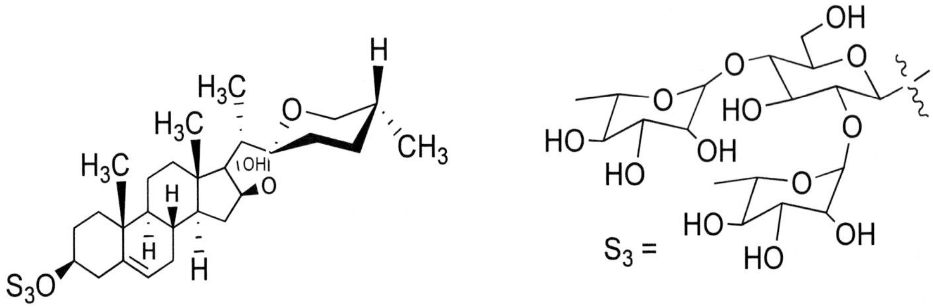

pennogenin-3-*O*-α-L-rhamnopyranosyl-(1–4)-[α-L-rhamnopyranosyl-(1–2)]-β-D-glucopyranoside (64)

Fig.17(cont.2) Pennogenin-3-*O*-α-L-rhamnopyranosyl-(1–3)-[α-L-rhamnopyranosyl-(1–2)]-β-D-glucopyranoside (63), pennogenin-3-*O*-α-L-rhamnopyranosyl-(1–4)-[α-L-rhamnopyranosyl-(1–2)]-β-D-glucopyranoside (64)

3.2.14. Miscellaneous Components

In 2009, stigmasterol (**39**), monoarachidin, (**65**) of glycerol ester of fatty acid, 1,7-*bis*-(4-hydroxyphenyl)-1*E*,4*E*,6*E*-heptatrien-3-one (**66**) of triene derivative, behenic acid (docosanoic acid, **67**) of straight chain saturated fatty acid, demethyl batatasin (**34**), 2,3'-dihydroxy-4',5'-dimethoxybibenzyl (**68**), diosbulbin B (**5**) and diosbulbin D (**47**) of two clerodane diterpenoids, docosyl ferulate (**69**) of a derivative of docosynoic acid (behenolic acid), 1,7-*bis*-(4-hydroxyphenyl)-4*E*,6*E*-heptadien-3-one (**70**), 5,3,4-trihydroxy-3,7-dimethoxyflavone (**71**), tristin (**72**), protocatechuic acid (**73**) of phenolic acid and adenosine (**74**) of a component of nucleic acid were found of from the root tubers of *Dioscorea bulbifera.*

Among these compounds, eight compounds of monoarachidin (**65**), 1,7-*bis*-(4-hydroxyphenyl)-1*E*,4*E*,6*E*-heptatrien-3-one (**66**), behenic acid (**67**), 2,3'-dihydroxy-4',5'-dimethoxybibenzyl (**68**), docosyl ferulate (**69**), 1,7-*bis*-(4-hydroxyphenyl)-4*E*,6*E*-heptadien-3-one (**70**), tristin (**72**) and adenosine (**74**) were isolated from the genus *Dioscore*a for the first time (Figures 18, 18(cont.1)) [58].

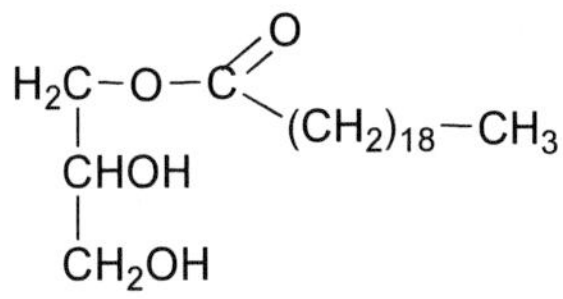

monoarachidin (65)

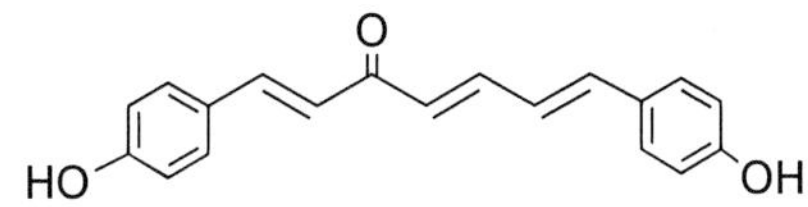

1,7-*bis*-(4-hydroxyphenyl)-1*E*,4*E*,6*E*-heptatrien-3-one (66)

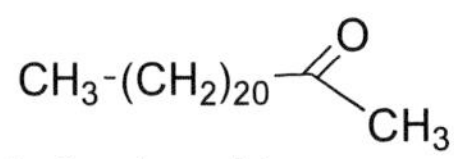

behenic acid, docosanoic acid (67)

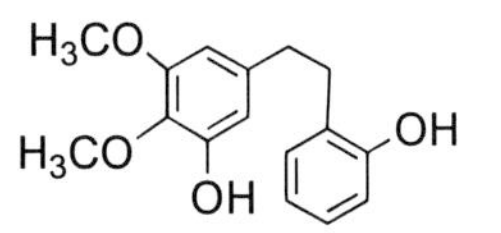

2,3'-dihydroxy-4',5'-dimethoxybibenzyl (68)

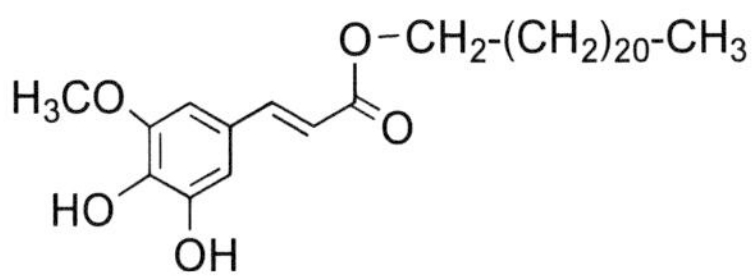

docosyl ferulate (69)

Fig.18 Monoarachidin (65), 1,7-*bis*-(4-hydroxyphenyl)-1*E*,4*E*,6*E*-heptatrien-3-one (66), behenic acid (67), 2,3'-dihydroxy-4',5'-dimethoxybibenzyl (68), docosyl ferulate (69)

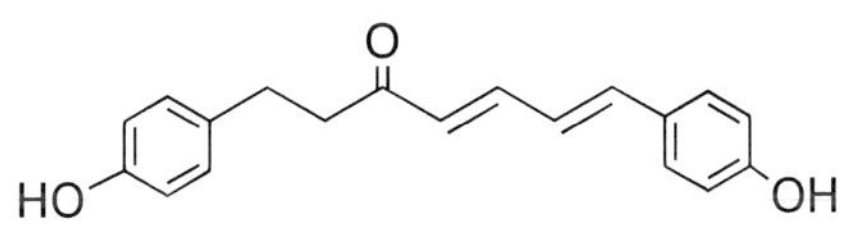

1,7-*bis*-(4-hydroxyphenyl)-4*E*,6*E*-heptadien-3-one (**70**)

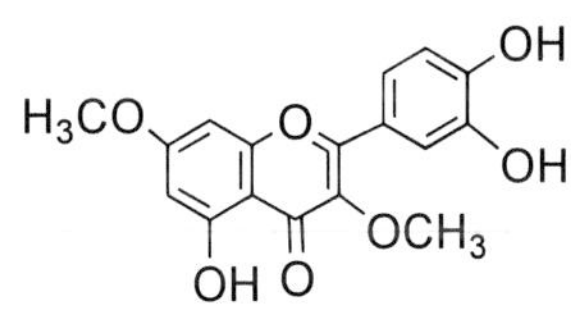

5,3,4-trihydroxy-3,7-dimethoxyflavone (71)

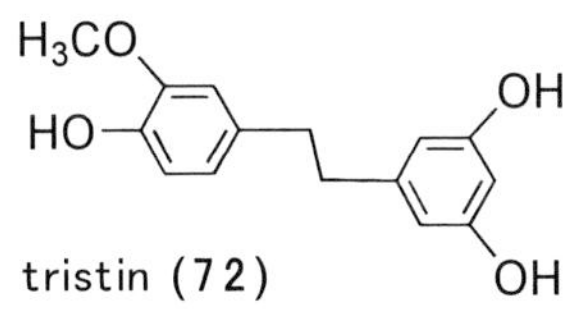

tristin (72)

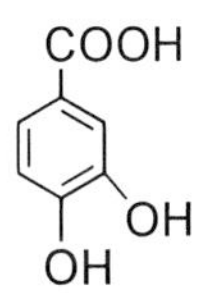

protocatechuic acid (73)

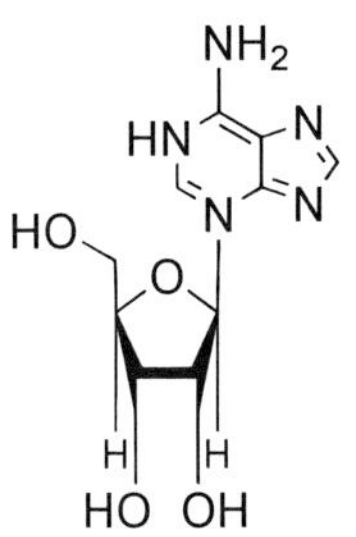

adenosine (74)

Fig.18(cont.1) 1,7-*bis*-(4-Hydroxyphenyl)-4*E*,6*E*-heptadien-3-one (7**0**), 5,3,4-trihydroxy-3,7-dimethoxyflavone (71), tristin (72), protocatechuic acid (3), adenosine (**74**)

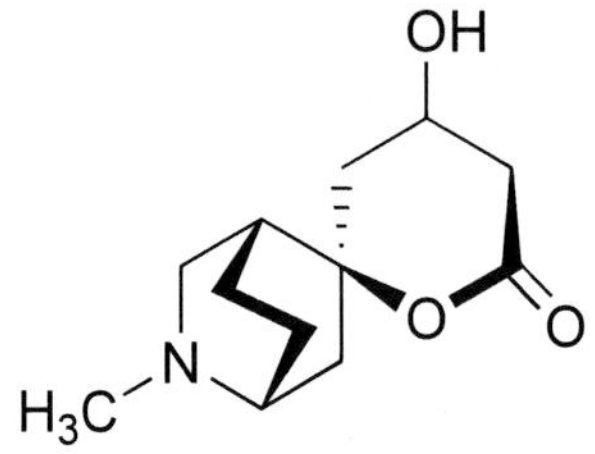

dihydrodioscorine (**75**)

Fig.19 Dihydrodioscorine (**75**)

3.3. Leaves and Stems

3.3.1. Diosbulbin B of Furanoid Diterpenes: Hormone Resource, Improvement of Gut

Diosbulbin B (**5**) of furanoid diterpenes was found of from the leaves and stems of *Dioscorea bulbifera* (Figure 2) [61].

3.4. Aerial Tubers and Underground Tubers

3.4.1. Three Diosbulbins of Bitter Compounds: Antitumor, Source of Diosgenin and Related Steroid Saponins

Dioscorea bulbifera has been used to treat diseases such as thyroid diseases and tumors as traditional folklore medicines in China [59].

In 1974, diosbulbin A (**4**), diosbulbin B (**5**) and diosbulbin C (**46**) of three diosbulbins of furanoid norditerpenes as scarcely steroid sapogenins were found of from the aerial tubers and underground tubers of *Dioscorea bulbifera* (Figures 2, Figure 14) [62].

*3.4.2. Dihydrodioscorine (**77**) of Piperidine Alkaloids: Antifungal Activity*

In 1989, dihydrodioscorine (**75**) of piperidine alkaloids was identified of from the aerial tubers and underground tubers of *Dioscorea bulbifera* (Figure 19).

0.1% dihydrodioscorine (**75**) was found to slow down the rate of their growth against five plant pathogenic fungi such as *Sclerotium rolfsii, Curvularia lunata, Fusarium moniliforme, Macrophomina phaseolina* and *Botryodiplodia theobromae* [63].

References

[1] Ogbuagu, MN. Nutritive and anti-nutritive composition of the wild (Inedible) Species of *Dioscorea bulbifera* (Potato Yam) and *Dioscorea dumentorum* (Bitter Yam). Pacific *J Sci Technol*, 9, 203-207, 2008.

[2] Agbor-Egbe, T; Treche, S. Evaluation of chemical composition of Cameroonian yam germplasm. *J Food Compos Anal*, 8, 274-283, 1995.

[3] Bhandari, MR; Kawabata, J. Bitterness and toxicity in wild yam (*Dioscorea spp.*) tubers of Nepal. *Plant Foods Hum Nutr*, 60, 129-135, 2005.

[4] Pramila, SS; Kumar, A; Raghuvanshi, R. Nutrient composition of some uncommon foods consumed by Kumaon (Uttar Pradesh, India) and Garhwal Hill (Uttar Pradesh, India) subjects. *J Food Sci Technol*, 28, 237-238, 1991.

[5] Rajyalakshmi, P; Geervani, P. Nutritive value of the foods cultivated and consumed by the tribals of South India. *Plant Foods for Human Nutrition*, 46, 53-61, 1994.

[6] Bradbury, JH. Properties and analysis of antinutritional factors in foods. *ASEAN Food J*, 6, 123-128, 1991.

[7] Eka, OU. The chemical composition of yam tubers. In: Osuji G (ed.) *Advances in Yam Research. The Biochemistry and Technology of Yam Tubers.*, 1, 51-75, 1985, Biochemical Society of Nigeria, Enugu, Nigeria.

[8] Benavente-Garcia, O; Castillo, J; Martin, FR; Ortuno, A; Rio, JAD. Uses and properties of *Citrus* flavonoids. *J Agric Food Chem*, 45, 4505-4515, 1997.

[9] Salah, N; Miller, NJ; Paganga, G; Tijburg, L; Bolwell, GP; Rice-Evans, C. Polyphenolic flavonoids as scavenger of aqueous phase radicals and as chain-breaking antioxidants. *Arch Biochem Biophy*, 2, 339-346, 1995.

[10] Martin, FW; Telek, L; Ruberté, RM. Yellow pigments of *Dioscorea bulbifera. J Agric Food Chem*, 22, 335-337, 1974.

[11] Quigley, FR. Diosgenin in West African Discorea plants. *Planta Medica*, 33, 414-415, 1978.

[12] Teponno, RB; Tapondjou, AL; Ju-Jung, H; Nam, JH; Tane, P; Park, HJ. Three new clerodane diterpenoids from the bulbils of *Dioscorea bulbifera* L. var. sativa. *Helv Chim Acta*, 90, 1599-1605, 2007.

[13] Tapondjou, AL; Miyamoto, T; Mirjolet, JF; Guilbaud, N; Lacaille-Dubois, MA. Pursaethosides A-E, triterpene saponins from *Entada pursaetha. J Nat Prod*, 68, 1185-1190, 2005.

[14] Teponno, RB; Tapondjou, AL; Gatsing, D; Djoukeng, JD; Abou-Mansour, E; Tabacchi, R; Tane, P; Stoekli-Evans, H; Lontsi, D. Bafoudiosbulbins A, and B, two anti-salmonellal clerodane diterpenoids from *Dioscorea bulbifera* L. var sativa. Phytochemistry 67, 1957-1963, 2006 and Erratum in: *Phytochemistry*, 68(10), 1480, 2007.

[15] Hanson, JR. Diterpenoids. *Nat Prod Rep*, 19, 125-132, 2002.

[16] Teponno, RB; Tapondjou, AL; Abou-Mansour, E; Stoeckli-Evans, H; Tane, P; Barboni, L. Bafoudiosbulbin F and G, further clerodane diterpenoids from *Dioscorea bulbifera* L. var sativa and revised structure of bafoudiosbulbin B. *Phytochemistry*, 69, 2374-2379, 2008.

[17] Jones, RN; Della-Latta, P; Lee, LV; Biedenbach, DJ. Linezolid-resistant *Enterococcus faecium* isolated from a patient without prior exposure to an oxazolidinone: report from the SENTRY Antimicrobial Surveillance *Program. Diagn Microbiol Infect Dis*, 42, 137-139, 2002.

[18] Wilson, P; Andrews, JA; Charlesworth, R; Walesby, R; Singer, M; Farrell, DJ, Robbins M. Linezolid resistance in clinical isolates of *Staphylococcus aureus*. *J Antimicrob Chemother*, 51, 186-188, 2003.

[19] Schelz, Z; Molnar, J; Hohmann, J. Antimicrobial and antiplasmid activities of essential oils. *Fitoterapia*, 77, 279-285, 2006.

[20] Molnar, A; Amaral, L; Molnar, J. Antiplasmid effect of promethazine in mixed bacterial cultures. *Int J Antimicrob Agents*, 22, 217-222, 2003.

[21] Gunics, G; Motohashi, N; Amaral, L; Farkas, S; Molnar, J. Interaction between antibiotics and non-conventional antibiotics on different bacteria. *Int J Antimicrob Agents*, 14, 239-242, 2000.
[22] Amabile-Cuevas, CF; Heinemann, JA. Shooting the messenger of antibiotic resistance: plasmid elimination as a potential counter-evolutionary tactic. *Drug Discov Today*, 89, 465, 2004.
[23] Shriram, V; Jahagirdar, S; Latha, C; Kumar, V; Puranik, V; Rojatkar, S; Dhakephalkar, PK; Shitole, MG. A potential plasmid-curing agent, 8-epidiosbulbin E acetate, from *Dioscorea bulbifera* L. against multidrug-resistant bacteria. *Int J Antimicrob Agents*, 32, 405-410, 2008.
[24] Newman, DJ; Cragg, GM; Snader, KM. The influence of natural products upon drug discovery. *Nat Prod Rep*, 17, 215-234, 2000.
[25] Lakshmi, VV; Padma, S; Polasa, H. Elimination of multidrug-resistant plasmid in bacteria by plumbagin, a compound derived from a plant. *Curr Microbiol*, 16, 159-161, 1987.
[26] Belofsky, G; Percivill, D; Lebis, K; Tegos, GP; Ekart, J. Phenolic metabolites of *Dalea versicolor* that enhance antibiotic activity against model pathogenic bacteria. *J Nat Prod*, 67, 481-484, 2004.
[27] Beg, AZ; Ahmad, I. Effect of *Plumbago zeylamica* extract and certain curing agents on multidrug resistance bacteria of clinical origin. *World J Microbiol Biotechnol*, 16, 841-844, 2004.
[28] Marquez, B; Neuville, L; Moreau, NJ; Genet, JP; Santos, AF; Andrade, MCC; Sant'sAna, AE. Multidrug resistance reversal agent from *Jatropha elliptica*. Phytochemistry, 66, 1804-1811, 2005.
[29] Oluwatuyi, M; Kaatz, GW; Gibbons, S. Antibacterial and resistance modifying activity of *Rosemarinus officinalis*. Phytochemistry, 65, 3249-3254, 2004.
[30] Michalet, S; Gartier, G; David, B; Mariotte, AM; Dijoux-franca, MG; Kaatz, GW; Stavri, M; Gibbons, A. *N*-caffeoylphenalkylamide derivatives as bacterial efflux pump inhibitors. *Bioorg Med Chem Lett*, 17, 1755-1758, 2007.
[31] Gibbons, S; Oluwatuyi, M; Veitch, NC; Gray, Al. Bacterial resistance modifying agents from *Lycopus europaeus*. Phytochemistry, 62, 83-87, 2003.
[32] Urones, JD; Basabe, P; Lithgow, AM; Marcos, IS; Jimenez, A; Diez, D; Gomez, A; White, AJP; Willians, DJ; Simmonds, MSJ; Blaney, WM. New antifeedant neo-clerodane triol. Semisynthesis and antifeedant activity of neo-clerodane diterpenoids. *Tetrahedron*, 51, 2117-2128, 1995.
[33] Roengsunran, S; Musikui, K; Petson, A; Vilavian, T; Sangvanich, P; Pornpakakul, S; Puthong, S; Chaichantipyuth, C; Jaiboon, N; Chaichit, N. Croblongifolin, a new anticancer clerodane from *Croton oblongifolius*. Plant Medica, 68, 274-277, 2002.
[34] Vaidya, V. *Dioscorea bulbifera*, N.O. *Discoreacea*. Sachitra Ayurveda, 18, 3, 1965.
[35] Jindal, MN; Kelkar, VV; Doctor, RB. The anorexient activity of Kalio-kund (*Dioscorea bulbifera* Linn.), methylphenidate and cocaine in rats: a preliminary study. Indian *J Med Res*, 57, 1075-1080, 1969.
[36] Sollmann, T. A manual of pharmacology and its applications to therapeutics and toxicology, 322-344, 1957.

[37] Harris, SC; Ivy, AC; Searle, LM. The mechanism of amphetamine induced loss of weight: a consideration of therapy of hunger and appetite. *J Amer Med Assoc*, 134, 1468, 1947.
[38] Groliman, A; Groliman, EF. Pharmacology and therapeutics. 239, 1965. 6th ed, edited by Grolman Lea, Febinger A, Philadelphia.
[39] Anderson, B; Larson, S. Water and food intake and inhibitory effect of amphetamine on drinking and eating before and after'prefrontal lobotony'in dog. *Acta Physiol Scand*, 38, 22-30, 1956.
[40] Anderson, B; Larson, S. Physiological and pharmacological aspects of the control of hunger and thirst. *Pharmacol Rev*, 13, 1-16, 1961.
[41] Brobeck, JR; Larsson, S; Reyes, E. A study of electrical activity of the hypothalamic feeding mechanism. *J Physiol*, 132, 328-364, 1956.
[42] Brobeck, JR. Food intake as a mechanism of temperature regulation. *Yale J Biol Med*, 20, 545-552, 1948.
[43] Kennedy, GC. The role of deposit fat in the hypothalamic control of food intake in the rat. *Proc Roy Soc Ser B*, 140, 578-592, 1953.
[44] Mayer, J. The glucostatic therapy of food intake and the problem of obesity. *Bull New Engl Med Centre*, 14, 43-49, 1952.
[45] Datta, K; Datta, SK. Pharmacognostic evaluation of potential 'yams' (*Dioscorea*). *J Econ Tax Bot*, 5, 181-196, 1984.
[46] Ozo, ON; Caygill, JC. *O*-Dihydroxyphenol oxidase action on natural polyhydric phenolics and enzymic browning of edible yams. *J Sci Food Agric*, 37, 283-288, 1986.
[47] Raghunandan, KV; Jolly, CI. Starch from *Dioscorea bulbifera.* Indian *J Pharmaceutical Sciences*, 49, 233-236, 1987.
[48] Fagboun, DE; Ogundana, SK; Adesanya, SA; Robert, MF. Dihydrostilbene phytoalexins from *Dioscorea rotundata*. Phytochemistry 26, 3187-3189, 1987.
[49] Adesanya, SA; Ogundana, SK; Roberts, MF. Dihydrostilbene phytoalexins from *Dioscorea bulbifera* and *Dioscorea dumentorum*. Phytochemistry 28, 773-774, 1989.
[50] Gupta, D; Singh, J. *p*-Hydroxyacetophenone derivatives from *Dioscorea bulbifera.* Phytochemistry, 28, 947-949, 1989.
[51] Li, S; Deng, J; Zhao, S. Steroids from tuber of *Dioscorea bulbifera* L. *J Plant Resources Environment*, 8, 61-62, 1999.
[52] Li, SS; Iliya, IA; Deng, JZ; Zhao, SX. Flavonoids and anthraquinone from *Dioscorea bulbifera* L. *Zhongguo Zhong Yao Za Zhi.*, *25*, 159-160, 2000.
[53] Kawasaki, T; Komori, T; Setoguchi, S. Furanoid norditerpenes from *Dioscoreacea*e plants. I. Diosbulbins A, B and C from *Dioscorea bulbifera* L. *forma spontanea* Makino et Nemoto. *Chem Pharm Bull*, 16, 2430-2435, 1968.
[54] Kamiya, K; Wada, Y; Komori, T; Arita, M; Kawasaki, T. Furanoid-norditerpene from plants of *Dioscoreaceae* family. III. Crystal analysis of 2-keto-3-bromo-tetrahydrodiosbulbin-A. *Tetrahedron Lett*, 19, 1869-1872, 1972.
[55] Robert, DHM; Zacarias, DJ; Nurul, HK; Mohammad, S; Mohammad, Q. Diosbulbin D and 8-epidiosbulbin E acetate, norclerodane diterpenoids from *Dioscorea bulbiffera* tubers. *Phytochemistry*, 23, 623-625, 1984.
[56] Komori, T. Glycosides from *Dioscorea bulbifera*. Toxicon 35, 1531-1535, 1997.

[57] Gao, H; Kuroyanagi, M; Wu, L; Kawahara, N; Yasuno, T; Nakamura, Y. Antitumor-promoting constituents from *Dioscorea bulbifera* L. in JB6 mouse epidermal cells. *Biol Pharm Bull*, 25, 1241-1243, 2002.

[58] Wang, G; Lin, B; Liu, J; Wang, G; Wang, F; Liu, J. Chemical constituents from tubers of *Dioscorea bulbifera*. *Zhongguo Zhong Yao Za Zhi,* 34, 1679-1682, 2009a.

[59] Wang, G; Liu, JS; Lin, BB; Wang, GK; Liu, JK. Two new furanoid norditerpenes from *Dioscorea bulbifera*. *Chem Pharm Bull*, (Tokyo) 57, 625-627, 2009b.

[60] Liu, H; Chou, GX; Wu, T; Guo, YL; Wang, SC; Wang, CH; Wang, ZT. Steroidal sapogenins and glycosides from the rhizomes of *Dioscorea bulbifera*. *J Nat Prod*, 72, 1964-1968, 2009.

[61] Yonemitsu, M; Fukuda, N; Kimura, T; Komori, T. Diosbulbin-B from the Leaves and Stems of *Dioscorea bulbifera*: 1H-1H and 13C-1H COSY NMR Studies. *Planta Med*, 59, 577, 1993.

[62] Telek, L; Martin, FW; Ruberté, RM. Bitter compounds in tubers of *Dioscorea bulbifera* L. *J Agric Food Chem*, 22, 332-334, 1974.

[63] Adeleye, A; Ikotun, T. Antifungal activity of dihydrodioscorine extracted from a wild variety of *Dioscorea bulbifera* L. *J Basic Microbiol*, 29, 265-267, 1989.

Abbreviations

diosgenin (**1**)
pregnenolone (**2**)
progesterone (**3**)
diosbulbin A (**4**)
diosbulbin B (**5**)
dioscorine (**6**)
histamine (**7**)
cyanogens (**8**)
oxalate (**9**)
lutein (**10**)
neoxanthin (**11**)
violaxanthin (**12**)
zeaxanthin (**13**)
auroxanthin (**14**)
cryptoxanthin (**15**)
cyanidin (**16**)
yamogenin (**17**)
bafoudiosbulbin C (**18**)
bafoudiosbulbin D (**19**)
bafoudiosbulbin E (**20**)
bafoudiosbulbin A (**21**)
bafoudiosbulbin B (**22**)
3-*O*-β-D-glucopyranosyl-β-sitosterol (**23**)
6'-stearoyl-3-*O*-β-D-glucopyranosyl-β-sitosterol (**24**)

bafoudiosbulbin F (**25**)
bafoudiosbulbin G (**26**)
3,5,4'-trihydroxy-3'-methoxy-bibenzyl (**27**)
kaempferol (**28**)
8-epidiosbulbin E acetate (**29**)
amphetamine (**30**)
estrone (**31**)
androsterone (**32**)
testosterone (**33**)
demethylbatatasin IV (**34**)
4-hydroxy-[2-*trans*-3',7'-dimethyl-octa-2',6'-dienyl]-6-methoxyacetophenone (**35**)
4,6-dihydroxy-2-*O*-(4'-hydroxybutyl)acetophenone (**36**)
acetophenone (**37**)
β-sitosterol (**38**)
stigmasterol (**39**)
daucosterol (**40**)
diosgenin-3-*O*-α-L-rhamnopyranosyl(1→2)-β-D-glucopyranoside (prosapogenin A of dioscin. **41**)
diosgenin-3-*O*-[di-α-L-rhamnopyranosyl(1→2, 1→3)]-β-D-glucopyranoside (taccaoside. **42**)
3,7-dimethoxy-5,4'-dihydroxyflavone (**43**)
3,7-dimethoxy-5,3',4'-trihydroxyflavone (**44**)
emodin (**45**)
diosbulbin C (**46**)
diosbulbin D (**47**)
diosbulbin A 2-*O*-β-D-glucopyranoside (**48**)
tetracosanoic acid (**49**)
1-(tetracosanoyl)-glycerol (**50**)
trans-tetracosanylferulate (**51**)
3-*O*-β-D-glucopyranosyl-β-sitosterol (**52**)
diosbulbin I (**53**)
diosbulbin J (**54**)
diosbulbisin A (**55**)
diosbulbisin B (**56**)
diosbulbisin C (**57**)
diosbulbisin D (**58**)
diosbulbiside A (**59**)
diosbulbiside B (**60**)
diosbulbiside C (**61**)
pennogenin (**62**)
pennogenin-3-*O*-α-rhamnopyranosyl-(1→3)-[α-L-rhamnopyranosyl-(1→2)-β-D-glucopyranoside (**63**)
pennogenin-3-*O*-α-rhamnopyranosyl-(1→4)-[α-L-rhamnopyranosyl-(1→2)-β-D-glucopyranoside (**64**)
monoarachidin (**65**)

1,7-*bis*-(4-hydroxyphenyl)-1*E*,4*E*,6*E*-heptatrien-3-one (**66**)
behenic acid (**67**)
2,3'-dihydroxy-4',5'-dimethoxybibenzyl, (**68**)
docosyl ferulate (**69**)
7-*bis*-(4-hydroxyphenyl)-4*E*,6*E*-heptadien-3-one (**70**)
5,3,4-trihydroxy-3,7-dimethoxyflavone (**71**)
tristin (**72**)
protocatechuic acid (**73**)
adenosine (**74**)a
dihydrodioscorine (**75**)

In: Isoflavones Biosynthesis, Occurence and Health Effects ISBN: 978-1-61728-113-6
Editor: Melanie J. Thompson, pp. 225-237 © 2010 Nova Science Publishers, Inc.

Chapter VIII

Effects of the Processing Techniques on Isoflavone Profiles

***L. Claudio Aguiar*[1]*, G. Solange Canniatti-Brazaca*[1]*, K. Yong Park*[2] *and M. Severino Alencar*[1]**
[1]Universidade de São Paulo, Escola Superior de Agricultura "Luiz de Queiroz", PO Box 09, Zip code 13418-900, Piracicaba/SP, Brazil
[2]Faculdade de Engenharia de Alimentos, Universidade Estadual de Campinas (UNICAMP), CP 6177, Zip code 13083-970, Campinas, SP, Brazil

Abstract

There is evidence that soy consumption benefits health by contributing to the prevention of chronic diseases due several factors, including isoflavones present in this legume. The isoflavones from soybean are known to have several biological properties such as estrogenic, antimicrobial and antitumoral (breast and prostate) activities, besides inhibiting the activity of enzymes linked to the cellular division. These properties are shown to be more accentuated in the aglycone than glycoside forms. Their presence in soy products have been reported with a number of beneficial health effects. In addition, both USA and Brazil are great soybean producer in the World. The analysis of isoflavone form contents into the soy products take an important role in the understanding of consumer of these products. In this way, with the germination process, the soybean has a positive effect in improving nutritional status including isoflavones. However, the isoflavone content varies with the period of germination. The process of germination showed no significant difference in amount of glycitin, genistein, malonyl daidzin, malonyl genistin, daidzein, glycitein and genistein, although theirs numerical reductions after germination. There were only differences in the amount of daidzein and malonyl glycitin between the samples. Some authors have been reported that the total concentration of isoflavones increased after 1 day of germination and then slowly decrease these concentrations. This increase and decrease observed in the content of isoflavones may be associated with the conversion of isoflavones to other flavonoids. The germination time of 48 hours not improved the amount of isoflavones. Different cultivars present several amount of isoflavones and forms of isoflavones. When the grains were cooked to be soft the malonyl form decreased, and aglycone and glycoside forms increased. It was observed that the heating treatment transformed the malonylglucosides into glucoside isoflavones. After heat treatment at 121°C for 30 min, nearly all malonyl isoflavones were converted into glucoside, but acetylisoflavones were not detected via RP-HPLC analysis. On the other hand, gamma radiation also promoted some reduction on isoflavone profiles, but this effect depends of humidity into soy products. Irradiation from ^{60}Co interfere at amount of isoflavones forms at doses up 2 kGy, glycosides and malonyl forms have tendency to down, but aglycones increase with radiation. For example, when soybean flour was studied, the gamma-radiation had a little effect (loss of 9.9%, w/w) in total isoflavone. However, soy isoflavones are likely to have the beneficial effects when found in aglycone forms in many processes will be proposed to obtain those aglycone isoflavones.

More recently, attention has been paid to the isoflavone analysis of soy-based products and to the behavior of isoflavones during the variety of food processing technologies. During soybean protein processes, the malonylglucoside isoflavones are transformed to glucoside forms, and after the enzyme treatment it may be converted into aglycones (Park et al., 2002; Park et al., 2001a; Park et al., 2001b). There are indications that the aglycone forms might be more bioactive (Grün et al., 2001) than their parent molecules. However, isoflavone profiles should greatly depend on the extent and level of heating during soy-processing.

Similar effects were also observed for Brazilian soybean grown under a large range of environments (Carrão-Panizzi et al., 1998; 1999). Moreover, in a previous study, Aguiar et al. (2007) observed a distinct distribution of isoflavones in hypocotyls and cotyledon from soybean grains. Daidzein, genistein and glycitein are present in soybean mainly as three types of glycosidic conjugate, linked either to glucose, malonyl, or acetyl derivatives of glucose (Liggins et al., 2000; Pandjaitan et al., 2000; Park et al., 2002).

Germination is process during which the seeds broken their latency stage and the reserved substances present in the cotyledons are used for the development and growth of the hypocotyls. This process causes changes in the biochemical, nutritional and sensory characteristics and results in a substantial increase in some biochemical constituents and biologically active components of grain. The changes can vary depending on genotypes, cultivation local, humidity, temperature, light, the stage of germination, and seed physiological metabolism (Kuo et al., 2004; Lin and Lai, 2006; Martinez et al., 2010; Monje et al., 2006; Ribeiro et al., 2006; Sangronis and Machado, 2007; Urbano et al., 2005).

Germinated soybeans improve of nutritional quality by prevention of lipid oxidation, increase of the nutrients such as ascorbic acid and riboflavin contents, raffinose and stachyose, that are responsible for flatulence problems, decrease and trypsin inhibitors decrease too (Ahmad and Patha, 2000). Germination is recommended not only for improving the nutritional quality, but also for lowering the immunoreactive properties of seeds (Troszynska et al., 2007).

The isoflavone content varies with the period of germination. The process of germination showed no significant difference in amount of glycitein, genistein, malonyl daidzin, malonyl genistin, daidzein, glycitein and genistein, although theirs numerical reductions after germination.

The glycosylation is very important for isoflavonoids in vacuoles in legumes cells, because it promotes solubilization, accumulation and mobilization (Noguchi et al., 2007).

Malonylated products are the most abundant forms of isoflavonoids in the grain, followed by the glycosides, the aglycones, and acetylglycosides (Hsieh et al., 2005).

Conjugates can be converted to aglycones for interactions with microorganisms (Suzuki et al., 2006). The malonylation protects glycosides from degradation by enzymes like glycosidases and helps in their intracellular transport (Dhaubhadel et al., 2008).

With the germination the β-glycoside contents decreased and the total aglycone content increase, and malonyl-glycosides were the major fractions in the germinated soybean seeds (Yuan et al., 2009). Between 3 and 7 days of germination, β-glycoside decreased 15 and 60%, respectively. Malonylglycosides increased by 30% after 3 days of germination but decreased again after 7 days. Acetyl-conjugated forms remained at a low and constant level after 7 days, but total aglycone content increased by five times during the same time (Chiarello et al., 2006).

The ratio of aglycones to total isoflavones increased from the mature soybeans to the sprouts (Lin and Lai, 2006).

During the process of germination occur a circadian change in isoflavone distribution profile, but not always, alternating between malonylglycosides and alglycones for day and night. In the hypocotyls after 12, 36, 60, 84 and 108 hours of germination the glycoside contents decreased and the aglycone contents increased. Malonylglycosides were the most abundant forms and only a small amount of β-glycosides and aglycones were detected after 60, 84 or 108 hours of germination. The hypocotyls sampled after 24, 48, 72, 96 and 120 hours of germination the aglycone content markedly increased, and then, aglycones became the most abundant forms and only a very small amount of malonylglycosides and β-glycosides was detected after 72, 96 or 120 hours of germination (Yuan et al., 2009).

In the cotyledons sampled at 12, 36, 60, 84 and 108 hours of germination, malonylglycosides were the most abundant forms, followed by β-glycosides and only a very small amount of aglycones was detects after 60, 84 or 108 hours of germination as an ordinary germination. In the samples after 24, 48, 72, 96 and 120 hours of germination the aglycone contents also increased, malonylgenistin was the main isoflavone being followed by genistein, β-glycosides, malonyldaidzin, daidzein, and β-daidzein after 72, 96, or 120 hours of germination (Yuan et al., 2009).

The presence of aglycones in soybeans is related to β-glucosidase activity, and the germination affect the β-glucosidase activity, and consequently total isoflavone content and their isomeric forms (Ribeiro et al., 2006). The enzyme could be actived by the soaking step prior to germination or by the germination metabolism itself (Chiarello et al., 2006).

The content of total phenolics compounds increased with the germination process at lupinus to produce sprouts. During the germination occur alterations at quality and quantity of phenolics, between them isoflavones, mainly to activation of endogenous enzymes (Dueñas et al., 2009). Lv et al. (2009) found alterations at isoflavones in chickpea seeds raw and sprouts.

Composition and concentrations of isoflavones are different in seed coat, hypocotyls and cotyledons (Yuan et al., 2009). Berger et al. (2008) found that isoflavone synthesis and accumulation resulted from highly differentiated pathways in cotyledons and hypocotyls.

The accumulation in hypocotyls is controlled main by genotypic control and in cotyledons appeared to be more influenced by environment (Berger et al., 2008).

The hypocotyls (12.39 mg/g) have more total isoflavones than cotyledons (1.59 mg/g) and seed coat (0.01 mg/g) at raw grains (Yuan et al., 2009). The hypocotyls contain the higher concentration of isoflavones, approximately 6 to 10 fold higher than the contain in cotyledons

(Kim et al., 2007; Tsangalis et al., 2004; Murphy et al., 2002). But, when occur germination this amount increase, Ribeiro et al. (2006) found hypocotyls content 25.2 fold higher than that in cotyledons in BRS 213 cultivar.

Hypocotyls presented malonyldaidzin, malonyl-glycitin, malonylgenistin, daidzin, and glycitin in the major fractions, small amounts of genistin and daidzein e very small amounts of genistein and glycitein (Yuan et al., 2009).

In the cotyledons were dominant the forms malonylgenistin, malonyldaidzin, genistin, and daidzin and small amount of genistein and daidzein (Yuan et al., 2009).

The content of daidzeins, genisteins and glyciteins in hypocotyls constitute 59.6%, 13.8% and 26.6%, but in cotyledons constitute 38.1%, 61.9% and 0% of the total isoflavone content, respectively (Yuan et al., 2009).

Berger et al. (2008) studied accumulation of total and individual isoflavones in cotyledons and hypocotyls of soybeans and found that early isoflavone accumulation was observed in the hypocotyls and the accumulation in the cotyledons began when a plateau was reached in the hypocotyls. In the hypocotyls glycitein begins its accumulation after daidzein and genistein and daidzein appears to be a precursor of glycitein (Berger et al., 2008; Charron et al. 2005). Glycitein and its conjugates were presented exclusively in the hypocotyls tissues and may represent an additional metabolite of importance to the hypocotyls (Wang and Murphy, 1994; Graham, 1991).

During germination, β-glucosidase activity increased in the hypocotyls and cotyledons, while total isoflavone content increased in the cotyledons and decreased in the hypocotyls (Ribeiro et al., 2006).

During the first day of germination in all part of the soybean the isoflavone increase, but after this time the content decrease (Graham, 1991). This increase and decrease observed in the content of isoflavones may be associated with the conversion of isoflavones to other flavonoids.

Zhu et al. (2005) showed that the isoflavone contents in soybean are affected by soaking. The increases of isoflavone contents at this stage may be induced by metabolic pathways. The maximum amounts of total isoflavone were observed when the hypocotyls length of seeds from soybean variety Hutcheson and Caviness was 0.5 and 2.5 mm respectively and the amount decrease after this stage. β-glucoside conjugates, especially their malonylglucoside were the forms that more increases. The contents of aglycones, genisteins, and daidzeins, reached the highest just after soaking.

The effects of germination vary greatly with respect to plant species, seed variety and germination conditions like temperature, light and time (Troszynska et al., 2007).

With the optimization of light during germination is possible increase the isoflavone content (Phommalth et al., 2008). According Paucar-Menacho et al. (2010) isoflavone aglycone increase its concentration when the soy was germinated during 63 hours at 30°C. Grains non-germinated the total isoflavone concentration of germinated soybean varied from 278.8 to 453.1 mg/100g. The higher isoflavone concentration was obtained with lower germination times and between 18 to 25°C. The total aglycone content varied from 6.7 to 90.3mg/100g of defatted sample for the different treatments (Paucar-Menacho et al., 2010).

The germination time of 48 hours not improved the amount of isoflavones. There were only differences in the amount of daidzein and malonyl glycitin between the samples. Different cultivars present several amount of isoflavones and forms of isoflavones (Martinez et al., 2009).

Considering Park et al. (2001a) reported that the isoflavone contents are affected by genetic aspects or other environmental factors, thus food processing that can change isoflavone concentrations or change its chemical isoforms in those soybean processed foods. For example, Park et al. (2001a) reported that the concentration of isoflavones in five Brazilian soybean was different in each cultivar. The first cultivar, IAC/Fos-31.1 presents about 1400 μg/g soybean and, other cultivar, IAC/15.1 presents about 300 μg/g soybean. According to Park et al. (2001c) the extraction of isoflavones from defatted soybean flour at room temperature gave the highest values for malonyldaidzin and malonylgenistin, with lesser quantities of daidzin and genistin, which are glucoside forms, respectively. During heating treatment, approximately, half of malonylglucoside conjugates were transformed to glucoside isoflavones. In general, flavonoid levels in processed food are approximately 50% lower than in fresh products (Andlauer and Fürst, 1998) and, according to Arditi et al. (2000), not all soybean products contain isoflavones, because these lipophilic compounds are readily removed by the alcohol extraction process commonly used to process soy foods.

Samples irradiated by gamma rays at doses of 4 and 8 kGy and cooked of soybean cultivar E48 decrease significantly the amount of daidzin and genistin when compared with control, but glycitin maintained to irradiated samples at doses 2, 4 and 8 kGy. Aglycones not have difference to daidzein and glycitein, but genistein increased at samples irradiated at dose of 8 kGy. Cultivar BRS 213 presented daidzin and glycitin similar control and increase genistin at dose of 2 kGy, and daidzein, genistein and glycitein maintained the same the control (Tozette et al., 2008). Irradiation make free radical that can alter the aromatic structure of isoflavone molecule and modify the content of isoflavones (Baldasso et al., 2007).

By other hand, the enzyme treatment of methanolic extracts of soybean flours with β-glucosidase, 85.5% of glycoside isoflavones were hydrolyzed by the enzyme to form aglycones (Park et al., 2001b). Therefore, combined application of heat treatment at 121°C and the fungal β-glucosidase could transform efficiently malonylglycoside isoflavones and glycoside isoflavones to aglycones. Furthermore, methanolic extracts of soybean sample treated with β-glucosidase demonstrated higher antioxidant activity than control (without enzyme treatment). According to previous studies at State University of Campinas, College of Food Engineering (Park et al., 2001a; 2001b; 2001c), it was calculated that antioxidant index of methanolic extracts of before and after enzyme reaction (1 h) were 1.48 and 2.53, respectively. All the methanolic extracts of defatted soybean treated with fungal β–glucosidase demonstrated higher antioxidant activity than the other three samples, which had not been treated with the enzyme. Therefore, it is apparent that higher antioxidant activity is due to conversion of glucoside isoflavones to aglycones. The defatted soybean treated at 121°C and enzyme showed the highest antioxidant activity and formed the greatest amounts of isoflavone aglycones (1.365 μg/g) as compared to samples treated at 100°C and at room temperature, respectively (Park et al., 2001a; 2001b; 2001c). The anti–inflammatory activities of the methanolic extracts of soybean treated in different temperatures and with enzyme β–glucosidase. All the samples presented some inhibitory capacities for the hyaluronidase, although the inhibition values have been low. It was found that the samples without treatment were more susceptible to inhibit the hyaluronidase that those samples treated with β-glucosidase. The methanolic extracts of soybean treated at 100°C (with and without enzyme treatment), without any treatment (26°C), and treated at 121°C (without enzyme treatment), showed similar capacities to inhibit hyaluronidase. The results showed none relationship between the content of isoflavone aglycone and the anti-inflammatory activity.

According to Aguiar (2004), the effect of different temperatures were carried out on isoflavone profiles. Isoflavones from soybean flours were treated with three different temperatures such as room temperature (26°C), cooking temperature (100°C), and autoclave temperature (121°C). It was found that the soybean treated at 121°C in autoclave transformed almost all malonylglucoside isoflavones to glucoside isoflavones, whereas, 100°C treatment transformed only 52% of malonylglucoside isoflavones transformed into aglycones. It was showed that the methanolic extract from soybean treated with β-glucosidase after heat at 121°C showed the highest antioxidant activity. This is due to fact that the extract contained highest content of isoflavone aglycones. It was also found that the fungal β-glucosidase hydrolyzed entirely all glucoside isoflavones into aglycones in methanolic extract from soybean treated at 121°C. The enzyme β-glucosidase from *Aspergillus oryzae* efficiently hydrolyzed glucoside isoflavones to aglycones but the enzyme was not efficient for hydrolysis of malonylglucoside isoflavones.

The glucoside isoflavones and the glucosides of flavonoids are very poorly absorbed in the small intestine as compared with their aglycones, because of higher hydrophilicity of the glucosides and their greater molecular weight (Brown, 1988 and Xu et al., 1995). Glucoside isoflavones in soybean foods are also poorly hydrolyzed by mammalian intestinal digestive enzymes (Brown, 1988). β-Glucosidases of intestinal microflora in lower bowel can hydrolyze the glucoside isoflavones to aglycones and promote their absorption (Friend and Chang, 1984). However, human intestinal bacteria can metabolize and degrade isoflavones as well as liberate isoflavone aglycones. Therefore, it is reasonable to infer that the condition of intestinal microflora profoundly influence isoflavone bioavailability (Xu et al, 1995). It was also reported that Lactobacilli, Bacteroides and Bifidobacteria possess β-glucosidase activity and play major roles in the intestinal hydrolysis of numerous plant β-glucosides present in the human diet, such as glucosides of flavonoids and isoflavones to produce aglycones (Hawksworth et al., 1971). In accordance with studies described in this paragraph, isoflavone aglycones are more desirable in foods as compared to other isomeric isoflavones because of possible increasing of bioavailability.

However, the heat treatment of the soybean flour was found to promote the conversion of malonylglucoside to glucoside isoflavones. Increases in the glucoside isoflavone contents during heating were observed in six samples of defatted soybean flour analyzed. Extraction of isoflavones from defatted soybean flour at room temperature gave the highest amounts of malonylglucoside isoflavones, with low quantity of daidzin, glycitin, and genistin (glucoside forms) (Aguiar, 2004; Park et al., 2001b). The malonylglucoside isoflavones were major isoflavone constituents in the defatted soybean flour without any treatment. Nevertheless, the defatted soybean flour treated at 121ºC for 40 min showed higher concentrations of daidzin, glycitin and genistin than their malonylconjugates. At 25ºC, the cv. IAC Foscarin-31 exhibited 1.4 mg g^{-1} as mean concentration of isoflavones, whereas cv. IAC 15-1 showed about 3.0 mg g^{-1} of defatted soy flour. Heating at 121ºC for 40 min promoted a reduction of up to 17.5 times in the malonylcojugate isoflavones and an increase of approximately 2.5 times in the concentration of glucoside isoflavones (Park et al., 2002; 2001a). According to Coward et al. (1993), this reduction is due to the easy decarboxylation of malonylglucoside isoflavones to their corresponding glucoside derivatives, which explains the high content of daidzin, glycitin and genistin (glucoside forms) in soy flour treated by heating. Soybeans and defatted soy flour (which had been minimally heated during their preparation) contained mostly isoflavone malonylglucoside conjugates, with lesser quantities of the β-glucosides and only trace amounts of acetylglucoside conjugates (Barnes et al., 1994).

After the heat treatment, any of the acetyl isoflavone forms were not detected by RP-HPLC analysis. For all cultivars, the extraction after heating showed an increase in the glucoside forms when compared with those samples obtained from extraction at room temperature (Park et al., 2002). According to Coward et al. (1993), malonylconjugates are instable and sensible to heating, and they are converted to glucoside isoflavones.

In general, flavonoids after food processing are about a half contents than in fresh foods (Andlauer and Fürst 1998) and, according to Arditi et al. (2000), some soyfood contain no isoflavones, because they are removed during the alcoholic extraction to produce soy concentrate protein. Aguiar et al. (2001) observed that isoflavone losses occur during industrial processes of soyfoods, such as soymilk, soy concentrate protein or soy isolate protein.

Isoflavone loss was also observed by Mahungu et al. (1999) when investigating the influence of the extrusion processing of corn/soy mixture on the stability of isoflavones. They reported that extrusion barrel temperature influence the most the isoflavone profile, especially the decarboxylation of malonylglucoside, and found that the amount of extractable isoflavones decreased after extrusion. According to Wu (1992), baking degrades isoflavones and cleaves malonyl groups, acetyl groups, and glycosidic bonds due to heating.

Furthermore, boiling, blanching, freezing, and freeze-drying could be responsible for significant reduction in total isoflavone contents. For example, freezing kept 53% of the initial total isoflavones, boiling 46%, and freezing-drying 40%. The authors reported that freeze-drying resulted in the greatest loss (around 60%) of total isoflavones, with the initial loss (56%) caused however by blanching, and that only 4% were due to the freeze-drying process (Simonne et al., 2000).

High-energy ionizing radiation (gamma radiation) is emitted naturally by ^{60}Co and ^{137}Cs and provides high energy to the electrons, which become highly unstable, generating free radicals, which are responsible for several reactions in biological materials (Sokhey and Hanna, 1993). Exposure to ionizing radiation inhibits ripening and senescence, increasing the shelf-life, and reducing spoilage of many fruits and vegetables (Oufedjikh et al., 1998). The commercial use of irradiation technology to extend the postharvest life of fruits and vegetables has been limited due to the adverse effects on the quality of some vegetable, and for the availability of more economical treatments. Many studies have been conducted to identify effects of irradiation on fruit texture, color, and on other quality parameters, such as concentration of the chemical constituents (Fan et al., 2001). Data on the effects of irradiation on the isoflavone contents in soybean and soy products are limited. Qualitative and quantitative differences on phenolic compounds and on other secondary products have been shown in irradiated vegetables (Oufedjikh et al., 1998).

After extracting phenolic compounds with methanol aqueous solution (80%), isoflavones can detect by reverse-phase high-performance liquid chromatography/diode-array detector after treatment with gamma irradiation for doses of 2 and 5 kGy presented a small effect on the isoflavones content of defatted soy flour (Aguiar et al., 2009). Samples irradiated at 50 kGy showed lower isoflavone contents. The observed reduction in the concentration of isoflavones daidzein, glycitein and genistein induced by gamma radiation in soy molasses was not significant in defatted soy flour, thus suggesting that isoflavones in defatted soy flour were not eliminated by gamma radiation at rates up to 50 kGy.

Gamma radiation has been applied for microbial control of food products (FAO, 1981) and according to the rate of the gamma radiation, it can also decrease ripening and senescence of vegetable products, thus increasing their shelf-life. The commercial utilization of gamma irradiation to improve post-harvest quality of fruits and vegetables has been limited due to due to some adverse factors on some quality characteristics (Oufedjikh et al. 1998). Therefore studies about effects of gamma radiation on food products, are necessary.

The isoflavone contents found in defatted soy flour showed no difference after gamma radiation treatment. The radiation doses of 2 and 5 kGy presented a small effect on total isoflavone content in defatted soybean flour, when compared to the values of control (without treatment). The samples irradiated at 50 kGy showed lower total isoflavone content (Aguiar et al., 2009). As reported by Sokhey and Hanna (1993) higher doses of gamma irradiation creates increasing intensities of free radicals, and propagation of free radical reactions is intensified by the reduced moisture content. However, these free radicals could change the isoflavone compounds of defatted soy flour.

Indeed, other compounds of soybean flour, such as cellulose, hemicellulose, and lignin could protect the phenolic compounds from the negative effects of gamma irradiation. According to Vitti et al. (1998) samples of sugarcane bagasse, rice and cotton fibers were submitted to ionizing radiation at different irradiation rates (200, 400, 600, 800, and 1,000 kGy), and the decomposition of cellulosic fibers was observed only at 800 and 1,000 kGy.

Aguiar et al. (2009) still discussed that the defatted soybean flours were treated at rates up to 50 kGy, probably, the applied energy was not enough to cause a significant change in isoflavone contents in defatted soybean flour.

The use of gamma radiation is increasing because of its effectiveness for insect disinfestation (Hallman 1999) and enhanced food safety (Thayer and Rajkowski 1999).

According to Crawford and Ruff (1996), after several years of testing, the Army Surgeon General concluded in 1965 that irradiated foods were safe for consumption at levels up to 56 kGy. Fifteen years later, the Joint Expert Committee on Food Irradiation, convened by the Food and Agricultural Organization (FAO), the International Atomic Energy Agency (IAEA) and the World Health Organization (WHO), concluded that irradiation of any food commodity up to an overall average dose of 10 kGy presents no toxicological hazard; hence toxicological testing of foods so treated with gamma radiation is no longer required.

According to Aguiar et al. (2009), micrographs of the defatted soybean flour at 5,000 times, showed that the surface of each grain of the soybean flour treated with 10 and 25 kGy were more porous, which could explain the increases of total isoflavone contents (1,853.2 and 1,639.5 μg/g, respectively). As it is more porous on the surface, it may have facilitated isoflavones extration from the soybean flour. However, no significant change in the isoflavone profile of the soy flour was observed after the gamma radiation treatment at room temperature, according to the Tukey test ($P<0.05$).

By the other hand, Aguiar et al. (2009) observed that soybean molasses contain more amounts of residual water, which can absorb radiation energy, producing solvated hydroxyl radicals and free electrons. Probably, these free radicals can be attached to structure of isoflavones, decreasing the isoflavone contents in the soybean molasse. The isoflavone contents in soybean molasses were lower than in defatted soybean flour and statistically different, suggesting effects of irradiation on isoflavone contents when the soybean foodstuff has high water content.

Isoflavones were monitored in two soybean derivatives, defatted soybean flour and soybean molasses. Some isoflavones were lost with the gamma irradiation treatment. Because of the increased interest in the composition of soybean isoflavones, alterations on chemistry of isoflavones need to be considered when interpreting data from clinical trials. Alterations observed in this work were related to water content of the soybean samples.

Acknowledgments

The authors are thankful to Brazilian Research Foundations: CNPq, CAPES and FAPESP for the financial support, as well as, those co-workers and students that cooperated with the construction of these valuable data.

References

Aguiar, C. L. (2004). Physical and biochemical transformation of soybean isoflavone (*Glycine max* L.) and the effect on biological activities. PhD thesis. State University of Campinas, UNICAMP: Campinas, Brazil. (in Portuguese).

Aguiar, C. L., Alencar, S. M., Pacheco, T. A. R. C. & Park, Y. K. (2001). Perdas de isoflavonas durante processamento de alimentos de soja. Proceedings of the 4th Simpósio *Latino-Americano de Ciência de Alimentos, 99.*

Aguiar, C. L., Baptista, A. S., Alencar, S. M., Haddad, R. & Eberlin, M. N. (2007). Analysis of isoflavonoids from leguminous plant extracts by RPHPLC/DAD and eletrospray ionization mass spectrometry. Intern. *J. Food Sci. Nutr.*, *58*, 116-124.

Aguiar, C. L., Baptista, A. S., Walder, J. M. M., Tsai, S. M., Carrão-Panizzi, M. C. & Kitajima, E. W. (2009). Changes in isoflavone profiles of soybean treated with gamma irradiation. *Intern J Food Sci Nutr*, *60*, 387-394.

Ahmad, S. & Patha, K. D. K. (2000). Nutritional changes in soybeans during germination. *J. Food Sci. Techn.*, *37*, 665-666.

Andlauer, W. & Fürst, P. (1998). Antioxidative power of phytochemicals with special reference to cereals. *Cereal Foods World*, *43*, 356-360.

Arditi, T., Meredith, T. & Flowerman, P. (2000). Renewed interest in soy isoflavones and saponins. *Cereal Foods World*, *45*, 414-417.

Baldasso, J. G., Araújo, M. M. & Mancini-Filho, J. (2007). Detecção de soja irradiada pelo ensaio do Cometa e Germinação. Conselho Nacional de Energia Nuclear / CNEN. IPEN. Centro de Tecnologia das Radiações. http://www.usp.br/siicusp/11osiicusp/ficha 2054.htm. Acess:13 Jul 2007.

Barnes, S., Kirk, M. & Coward, L. (1994). Isoflavones and their conjugates in soy foods: extraction conditions and analysis by HPLC-MS. *J Agric Food Chem*, *42*, 2466-2474.

Berger, M., Rasolohery, C. A., Cazalis, R. & Daydé, J. (2008). Isoflavone accumulation kinetics in soybeans seed cotyledons and hypocotyls: distinct pathways and genetic controls. *Crop Sci*, *48*, 700-708.

Brown, J. P. (1998). Hydrolysis and esters; In: Role of the gut flora in toxicity and cancer. Eds. Rowland, IR, *Academic Press*, 109-144.

Carrão-Panizzi, M. C., Beléia, A. D. P., Keisuke, K. & Oliveira, M. C. N. (1999). Effects of genetics and environment on isoflavone content of soybean from different regions of Brazil. *Pesq Agropec Bras*, *34*, 1787-1795.

Carrão-Panizzi, M. C., Keisuke, K., Beléia, A. D. P. & Oliveira, M. C. N. (1998). Influence of growth locations on isoflavone contents in Brazilian soybean cultivars. *Breeding Sci*, *48*, 409-413.

Charron, C. S., Allen, F. L., Johnson, R. D., Pantalone, V. R. & Sams, C. E. (2005). Correlations of oil and protein with isoflavone concentration in soybean [*Glycine max* (L.) Merr.] *J Agric Food Chem*, *53*, 7128-7135.

Chiarello, M. D., Le Guerroué, J. L., Chagas, C. M. S., Franco, O. L., Bianchini, E. & João, M. J. (2006). Influence of heat treatment and grain germination on the isoflavone profile of soy milk. *J Food Bioch*, *30*, 234-247.

Coward, L., Barnes, N. C., Setchell, K. D. R. & Barnes, S. (1993). Genistein, daidzein, and their beta-glycoside conjugates: antitumor isoflavones in soybean foods from American and Asian diets. *J Agric Food Chem*, *41*, 1961-1967.

Crawford, L. M. & Ruff, E. H. (1996). A review of the safety of cold pasteurization through irradiation. *Food Cont*, *7*, 87-97.

Dhaubhadel, S., Farhangkhoee, M. & Chapman, R. (2008). Identification and characterization of isoflavonoid specific glycosyltransferase and malonyltransferase from soybean seeds. *J Exper Bot*, *59*, 981-994.

Dueñas, M., Hernández, T., Estrella, I. & Fernández, D. (2009). Germination as a process to increase the polyphenol content and antioxidnt activity of lupin seeds (*Lupines angustifolius* L.) *Food Chem*, *117*, 599-607.

Fan, X., Argenta, L. & Mattheis, J. (2001). Impacts of ionizing radiation on volatile production by ripening Gala apple fruit. *J Agric Food Chem*, *49*, 254-262.

Friend, D. R. & Chang, G. W. (1984). A colon-specific drug-delivery system based on drug glycosides and the glycosidases of colonic bacteria. *J Med Chem*, *27*, 261-266.

Graham, T. L. (1991). Flavonoid and isoflavonoid distribution in developing soybean seedling tissues and in seed and root exudates. *Plant Phys*, *95*, 594-603.

Grün, I. U., Adhikari, K., Li, C., Li, Y., Lin, B., Zhang, J. & Fernando, L. N. (2001). Changes in the profile of genistein, daidzein, and their conjugates during thermal processing of tofu. *J Agric Food Chem*, *49*, 2839-2843.

Hallman, G. J. (1999). Ionizing radiation quarantine treatments against tephritid fruit flies. *Postharvest Biol Technol*, *16*, 93-106.

Hawksworth, G., Drasar, B. S. & Hill, M. J. (1971). Intestinal bacteria and the hydrolysis of glycosidic bonds. *J Med Microbiol*, *4*, 451-459.

Hsieh, H. C., Kao, T. H. & Chen, B. H. (2005). A fast HPLC method for analysis of isoflavones in soybean. *J Liquid Chrom Related Techn*, *27*, 315-324.

FAO/IAEA/WHO Expert Committee. 1981. *Wholesomeness of irradiated foods*. Report of World Health Organization Research Series No. 659. World Health Organization, Geneva, 7-34.

Kim, J. A., Hong, S. B., Jung, W. S., Yu, C. Y., Ma, K. H., Gwag, J. G. & Chung, I. M. (2007). Comparison of isoflavones composition in seed, embryo, cotyledon and seed coat of cooked-with-rice and vegetable soybean (*Glycine max* L.) varieties. *Food Chem., 102*, 738-744.

Kuo, Y. H., Rozan, P., Lambein, F., Frias, J. & Vidal-Valverde, C. (2004). Effects of different germination conditions on the contents of free protein and non-protein amino acids of commercial legumes. *Food Chem., 86*, 537-545.

Liggins, J., Bluck, L. J. C., Runswick, S., Atkinson, C., Coward, W. A. & Bingham, S. A. (2000). Daidzein and genistein content of fruits and nuts. *J Nutr Biochem, 11*, 326-331.

Lin, P. Y. & Lai, H. M. (2006). Bioactive compounds in legumes and their germinated products. *J Agric Food Chem, 54*, 3807-3814.

Lv, Q., Yang, Y., Zhao, Y., Gu, D., He, D., Yili, A., Ma, Q., Cheng, Z., Gao, Y., Aisa, H. A. & Ito, Y. (2009). Comparative study on separation and purification of isoflavones from the seeds and sprouts of chickpea by high-seed countercurrent chromatography. *J Liquid Chrom Related Techn, 32*, 2879-2892.

Mahungu, S. M., Diaz-Mercado, S., Li, J., Schwenk, M., Singletary, K. & Faller, J. (1999). Stability of isoflavones during extrusion processing of corn/soy mixture. *J Agric Food Chem, 47*, 279-284.

Martinez, A. P. C., Martinez, P. C. C., Souza, M. C., Canniatti-Brazaca, S. G., Alencar, S. M., Moreno, I. A. & Odoni, T. L. C. (2008). Isoflavone contents in germinated soybeans. In: *Simpósio Internacional de iniciação científica, 16.* Piracicaba SP. http://www.usp.br/siicusp/Resumos/16Siicusp/4654.pdf. Acess: jan. 12 2010.

Martinez, A. P. C., Martinez, P. C. C., Souza, M. C. & Canniatti-Brazaca, S. G. In press. Chemical change in soybean grains with germination. *Ciên Tecnol Alim.*

Monje, M. C., Berger, M., Farines, V., Reybier, R., Verger, A., Daydé, J., Théodorout, V. & Nepveu, F. (2006). Antioxidant capacity of cotyledons and germs of soybean in relation to their isoflavone content. *Food Techn Biotechn, 44*, 493-498.

Murphy, P. A., Barua, K. & Hauck, C. C. (2002). Solvent extraction selection in the determination of isoflavones in soy foods. *J Chrom B: biomed Sci Appl, 77*, 129-138.

Noguchi, A., Saito, A., Homma, Y., Nakao, M., Sasaki, N., Nishino, T., Takahashi, S. & Nakayama, T. A. (2007). UDP-glucose:isoflavone 7-Oglycosyltransferase from the roots of soybean (*Glycine max*) seedlings: Purification, gene cloning, phylogenetics, and na implication for an alternative strategy of enzyme catalysis. *J Biol Chem, 282*, 23581-23590.

Oufedjikh, H., Mahrouz, M., Lacroix, M., Amiot, M. J. & Taccini, M. (1998). The influence of gamma irradiation on flavonoid content during storage of irradiated Clementina. *Radiat Phys Chem, 52*, 107-112.

Pandjaitan, N., Hettiarachchy, N. & Ju, Z. Y. (2000). Enrichment of genistein in soy protein concentrate with beta-glucosidase. *J Food Sci., 65*, 403-407.

Park, Y. K., Aguiar, C. L., Alencar, S. M., Mascarenhas, H. A. A. & Scamparini, A. R. P. (2002). Conversion of malonyl-beta-glycoside isoflavones into glycoside isoflavones in Brazilian soybeans. *Ciên Tecn Alim, 32*, 130-136.

Park, Y. K., Aguiar, C. L., Alencar, S. M., Mascarenhas, H. A. A. & Scamparini, A. R. P. (2001a). Survey of isoflavone contents in Brazilian soybean. *Cien Tecn Alim 3*, 12-14.

Park, Y. K., Alencar, S. M., Nery, I. A., Aguiar, C. L. & Sato, H. H. (2001b). Enrichment of isoflavone aglycones in extracted soybean isoflavones by heat and fungal beta-glucosidase. *Food Sci Industry, 34*, 14-19.

Park, Y. K., Aguiar, C. L., Alencar, S. M., Nery, I. A. & Sato, H. H. (2001c). Extraction of isoflavones from defatted soy flour by ethanol and conversion of glucoside isoflavones to aglycones by fungal beta-glucosidase. *IFT Annual Meeting, 164.*

Paucar-Menacho, L. M., Berhow, M. A., Mandarino, J. M. G., Mejia, E. G. & de, Chang, Y. K. (2010). Optimisation of germination time and temperature on the concentration of bioactive compounds in Brazilian soybean cultivar BRS 133 using response surface methodology. *Food Chem, 119*, 636-642.

Phommalth, S., Jeong, Y. S., Kim, Y. H., Dhakal, K. H. & Hwang, Y. H. (2008). Effects of light treatment on isoplavone content of germinated soybean seeds. *J Agric Food Chem, 56*, 10123-10128.

Ribeiro, M. L. L., Mandarino, J. M. G., Carrão-Panizzi, M. C., Oliveira, M. C. N. Campo, C. B. H., Nepomuceno, A. L. & Ida, E. I. (2006). β-glycosidase activity and isoflavone content in germinated soybeans radicles and cotyledons. *J Food Bioch, 30*, 453-465.

Sangronis, E. & Machado, C. J. (2007). Influence of germination on the nutritional quality of *Phaseolus vulgaris* and *Cajanus cajan*. LWT- *Food Sci Tech, 40*, 116-120.

Simonne, A. H., Smith, M., Weaver, D. B., Vail, T., Barnes, S. & Wei, C. I. (2000). Retention and changes of soy isoflavones and carotenoids in immature soybean seeds (Edamame) during processing. *J Agric Food Chem, 48*, 6061-6069.

Sokhey, A. S. & Hanna, M. A. (1993). Properties of irradiated starches. *Food Structure, 12*, 397-410.

Suzuki, H., Takahashi, S., Watanabe, R., Fukushima, Y., Fujita, N., Noguchi, A., Yokoyama, R., Nishitani, K., Nishino, T. & Nakayama, T. (2006). An isoflavone conjugate-hydrolyzing beta-glucosidase from the roots of soybean (*Glycine max*) seedlings–Purification, gene cloning, phylogenetics, and cellular localization. *J Biol Chem, 281*, 30251-30259.

Thayer, D. W. & Rajkowski, K. T. (1999). New development in irradiation of fresh fruits and vegetables. *Food Technol*, 53, 62-65.

Tozette, B. N., Toledo, T. C. F., Alencar, S. M., Moreno, I. A. M. & Odoni, T. L. C. (2008). Effects of ionizing radiation from 60Co at isoflavone concentration in cooked soybeans. In: Simpósio Internacional de iniciação científica, 16. Piracicaba SP. http://www.usp.br/siicusp/Resumos/16Siicusp/2585.pdf. *Acess: jan.*, 12, 2010.

Troszynska, A., Szymkiewicz, A. & Wolejszo, A. (2007). The effects of germination on the sensory quality and immunoreactive properties of pea (*Pisum sativum* L.) and soybean (*Glycine max*). *J Food Qual, 30*, 1083-1100.

Tsangalis, D., Ashton, J. F., Stojanovska, L., Wilcox, G. & Shah, N. P. (2004). Development of an isoflavone aglycone-enriched soymilk using soy germ, soy protein isolate and bifidobasteria. *Food Res Intern, 37*, 301-312.

Urbano, G., Aranda, P., Vílchez, A., Aranda, C.., Cabrera, L., Porres, J. M. & López-Jurado, M. (2005). Effects of germination on the composition and nutritive value of proteins in *Pisum sativum, L. Food Chem, 93*, 671-679.

Vitti, D. M. S. S., Del Mastro, N. L., Kikuchi, O. K. & Nogueira, N. L. (1998). Electron irradiation of high fiber by-products. effects on chemical composition and digestibility. *Sci. Agric., 55*, 159-171.

Wang, H. J. & Murphy, P. A. (1994). Isoflavone content in commercial soybean foods. J *Agric Food Chem, 42*, 1666-1673.

Wu, E. S., Loch, J. T., Toder, B. H., Borrelli, A. R., Gawlak, D., Radov, L. A. & Gensmantel, N. P. (1992). Flavones. 3. Synthesis, biological activities, and conformational analysis of isoflavone derivatives and related compounds. *J Med Chem*, *35*, 3519-3525.

Xu, X., Harris, K. S., Wang, H. J., Murphy, P. A. & Hendrich, S. (1995). Bioavailability of soybean isoflavonas depends upon gut microflora in women. *J Nutr.*, *125*, 2307-2315.

Yuan, J. P., Liu, Y. B., Peng, J., Wang, J. H. & Liu, X. (2009). Changes of isoflavone profile in the hypocotyls and colyledons of soybeans during dry heating and germination. *J Agric Food Chem*, *57*, 9002-9010.

Zhu, D., Hettiarachchy, N. S., Horax, R. & Chen, P. (2005). Isoflavone contents in germinated soybean seeds. *Plant Foods Human Nutr*, *60*, 147-151.

In: Isoflavones Biosynthesis, Occurence and Health Effects ISBN: 978-1-61728-113-6
Editor: Melanie J. Thompson, pp. 239-254 © 2010 Nova Science Publishers, Inc.

Chapter IX

Structural Studies and Mechanisms of Isoflavonoid Biosynthesis

Xiaoqiang Wang
Plant Biology Division, Samuel Roberts Noble Foundation,
2510 Sam Noble Parkway, Ardmore, OK 73401, U.S.A.

Abstract

Isoflavonoids play important roles in plant defense, exhibit various health-promoting activities, and have attracted great interest from the pharmaceutical and nutraceutical industries as drug targets and dietary supplements. The isoflavonoids are identified primarily in leguminous plants. They are mainly synthesized through the central phenylpropanoid pathway and the specific isoflavonoid branch pathway. Many key enzymes in the central pathway have been elucidated structurally, providing essential insights into early stages of isoflavonoid biosynthesis. Significant progress has been made on structural studies of enzymes in the isoflavonoid branch pathway. Structures of two isoflavonoid-specific NADPH-dependent reductases revealed how enzymes specifically recognize isoflavonoids and catalyze stereo-specific reductions. Several isoflavonoid methyltransferases and glycosyltransferases were also characterized structurally. Bioactivities and functions of isoflavonoids are largely dictated by their methylation and glycosylation patterns. The structural analysis of these enzymes provides a basis for understanding the complex biosynthetic process and the selective substrate- and regio-specificities. Structure-based enzyme engineering is becoming a new strategy for synthesis of bioactive isoflavonoids.

1. Introduction – Isoflavonoids and Biosynthetic Enzymes

Isoflavonoids are a valuable group of natural products and are primarily synthesized in leguminous plants, where they play important roles in plant defense. They also have significant health benefits for animals and humans and have been ascribed various estrogenic, antiangiogenic, antioxidant, and anticancer activities, leading to their popularity as dietary supplements [1-4]. They may have additional health-promoting activities, including chemoprevention of osteoporosis and other postmenopausal disorders [5-7]. It's been shown that isoflavone genistein inhibits the growth of various cancer cells through the modulation of genes that are related to the control of cell cycle, apoptosis, and cell signaling pathways [8]. Genistein is also a potent inhibitor of oxidative stress, angiogenesis, and metastasis. Epidemiological studies indicate that dietary isoflavones are major factors for reduction of cancer risk in populations consuming a high soy diet [8,9].

Isoflavonoids originate from the amino acid L-phenyalanine, and their biosynthesis has been investigated extensively with nearly all the enzymes now isolated and functionally characterized [10] (Figure 1). Phenylalanine ammonia-lyase (PAL) is the first enzyme in the central phenylpropanoid pathway, catalyzing the non-oxidative conversion of L-phenylalanine into cinnamate, the precursor of a great variety of phenylpropanoids. Cinnamate 4-hydroxylase (C4H), a cytochrome P450 enzyme, introduces a hydroxyl group at the para position of the phenyl ring of cinnamate and produces p-coumarate. Then 4-coumarate CoA-ligase (4CL) catalyses the final step of the general phenylpropanoid pathway, producing the corresponding CoA thiol esters, e.g., *p*-coumaroyl-CoA. Then a ubiquitous plant enzyme, chalcone synthase (CHS), catalyzes the iterative condensation of one *p*-coumaroyl-CoA and three malonyl-CoA molecules to form chalcone scaffold. Chalcone reductase (CHR), a ketoreductase, acts on an intermediate of the CHS reaction, yielding chalcone and deoxychalcone from the coupled catalytic action with CHS. Chalcone isomerase (CHI) carries out cyclization of chalcones to form the flavonoid core flavanone. Through a legume-specific branch of the phenylpropanoid pathway, isoflavone synthase (IFS) converts flavanone to 2-hydroxyisoflavanone by performing a novel aryl-ring migration to transfer the aromatic B-ring from position C-2 to C-3 and hydroxylation in position C-2. Dehydration of 2-hydroxyisoflavanones occurs either spontaneously or through catalysis by 2-hydroxyisoflavanone dehydratase (2-HID) to generate isoflavones, e.g., daidzein and genistein.

2. Phenylpropanoid Pathway – Structural Studies of Upstream Enzymes

Structural study of enzymes is very critical for understanding catalytic mechanisms and structure-function relationships, and also provides a molecular basis for enzyme engineering. Extensive efforts have been made to study the structures of enzymes involved in the biosynthesis of (iso)flavonoids. Several upstream enzymes in the central phenylpropanoid pathway have been elucidated structurally (Figure 2), including L-phenylalanine ammonia-lyase (PAL) [11], chalcone synthase (CHS) [12], chalcone reductase (CHR) [13], and chalcone isomerase (CHI) [14].

Structures of PALs from parsley (*Petroselinum crispum*), yeast (*Rhodosporidium toruloides*) and cyanobacteria have been reported [11,15,16]. They exist as homotetramers, resemble histidine ammonia-lyase (HAL) from the general histidine degradation pathway [17], and possess a conserved polypeptide chain fold with a central core formed by nearly parallel α-helices (Figure 2a). The PAL catalytic site contains an unusual covalent cofactor, electrophilic prosthetic group 4-methylidene-imidazole-5-one (MIO), which is formed autocatalytically by cyclization and dehydration of a conserved active-site Ala-Ser-Gly tripeptide segment. Two mechanisms are proposed, and the PAL catalyzed reaction is initiated by the MIO interacting with the amino group or the aromatic ring of phenylalanine substrate, via carbanion or carbocation intermediates, leading to the elimination of ammonia and production of trans-cinnamate.

Cinnamate 4-hydroxylase (C4H) is the second key enzyme in the central phenylpropanoid pathway, but no structural information is available due to the difficulty in isolating this membrane-associated cytochrome P450 enzyme. For 4-coumarate CoA-ligase (4CL), the next key enzyme in the pathway, also no structure is available. However, several similar ligase structures are available in the Protein Data Bank (PDB), such as benzoate CoA ligases from *Burkholderia xenovorans* LB400 (PDB: 2V7B) and a long-chain-fatty-acid-CoA ligase from Archaeoglobus fulgidus (PDB: 3G7S), which may provide insights for structure-function of 4CL.

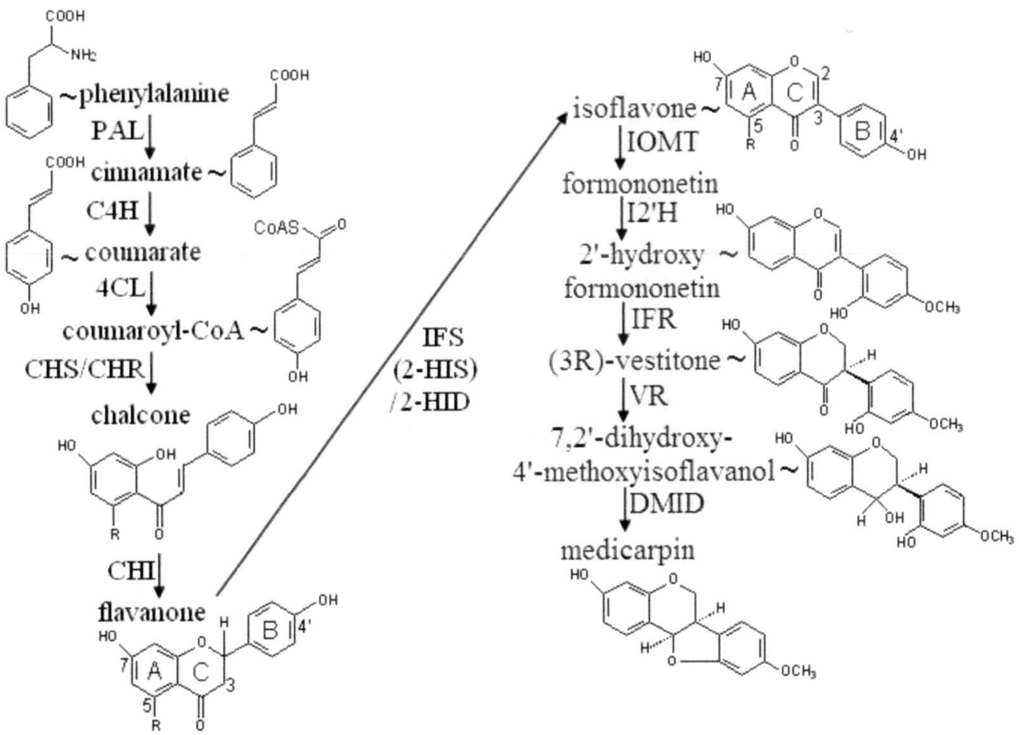

Figure 1. Phenylpropanoid pathway and biosynthesis of isoflavonoids. The enzymes are: PAL, L-phenylalanine ammonia-lyase; C4H, cinnamate 4-hydroxylase; 4CL, 4-coumarate CoA ligase; CHS, chalcone synthase; CHR, chalcone reductase; CHI, chalcone isomerase; IFS, isoflavone synthase (2-HIS:2-hydroxy isoflavanone synthase); 2-HID, 2-hydroxy-isoflavanone dehydratase; IOMT, isoflavone-O-methyltransferase; I2'H, isoflavone 2'-hydroxylase; IFR, isoflavone reductase; VR, vestitone reductase; DMID, 7,2'-dihydroxy,4'-methoxyisoflavanol dehydratase

Chalcone synthase (CHS) is a pivotal enzyme in the phenylpropanoid biosynthetic process. CHS functions as a homodimer, and each CHS monomer consists of two structural domains [12] (Figure 2b). The active site is located at the cleft between the upper and lower domains of each monomer. The crystal structure of a CHS mutant (C164S) acetyl-CoA complex yields insight on the decarboxylation reaction. A CHS complex with hexanoyl-CoA mimics the structure of a linear polyketide-CoA reaction intermediate. The structures of CHS complexes with product analogs naringenin and resveratrol provide information on how the enzyme governs the sequential addition of acetates to the coumaroyl moiety and how CHS controls the stereochemistry of the polyketide cyclization reaction leading to chalcone synthesis [12].

Alfalfa chalcone reductase (CHR) adopts the three-dimensional structure of the aldo/keto reductase superfamily with an $(\alpha/\beta)_8$ TIM-barrel [13] (Figure 2c). The active site is formed at the C-terminal end of the TIM-barrel, and the co-factor NADPH nicotinamide moiety and site of hydride transfer lie at the base of the catalytic cavity. The structure proposes the identity and binding mode of its substrate, namely the non-aromatized coumaryl-trione intermediate of the CHS-catalyzed cyclization of the fully extended coumaryl-tetraketide thioester intermediate [13].

The crystal structure of alfalfa chalcone isomerase (CHI) reveals a novel open-faced β-sandwich fold resembling an upside-down bouquet [14] (Figure 2d). The topology of the active site cleft defines the stereochemistry of the cyclization reaction. The structural and mutational analyses suggest a mechanism in which a tyrosine (Tyr106) activates a water molecule in the active site, which serves as a general acid in the cyclization reaction. Following the nucleophilic attack of the 2'-oxyanion on the α,β-unsaturated double bond, the water molecule stabilized by Tyr 106 acts as a general acid to stabilize the enolate. This results in formation of a flav-3-en-4-ol intermediate that tautomerizes into the reaction product. Together with contributions from general acid-base catalysis, shape complementarity between the CHI binding pocket and chalcone locks the substrate into a constrained conformation and accelerates the cyclization reaction to proceed with a second-order rate constant approaching the diffusion controlled limit [14].

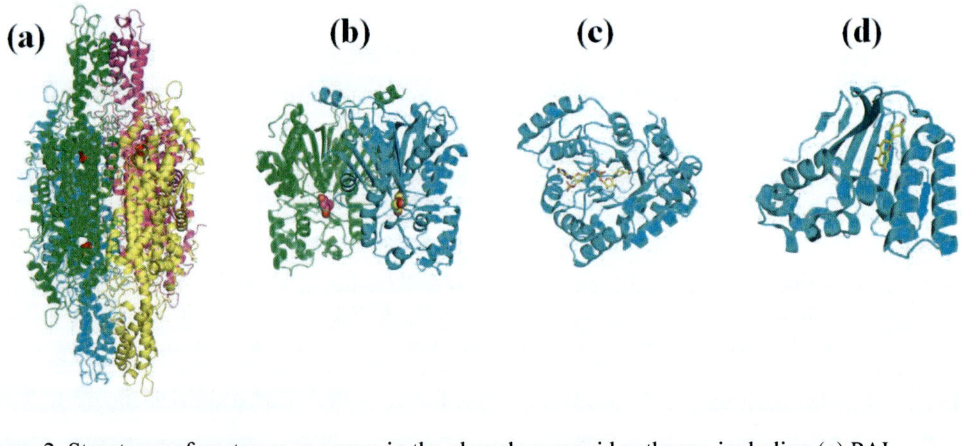

Figure 2. Structures of upstream enzymes in the phenylpropanoid pathway, including (a) PAL homotetramer (PDB ID: 1W27); (b) CHS homodimer (PDB: 1CGK); (c) CHR (PDB: 1ZGD); (d) CHI (PDB: 1EYQ)

3. Isoflavone Synthase and Cytochrome P450s Involved in Isoflavonoid Biosynthesis

Isoflavone synthase (IFS, more correctly named 2-HIS (2-hydroxy-isoflavanone synthase)) is a key enzyme that catalyzes the entry-point reaction into isoflavonoid biosynthesis, converting flavanones to isoflavones by an unusual aryl migration reaction (Figure 1).

IFS is a membrane-associated cytochrome P450 enzyme and belongs to the CYP93C subfamily. Cytochrome P450s are one of the largest superfamilies of enzymes and are ubiquitously distributed in biological organisms. They are hemo-containing enzymes converting a broad variety of substrates and catalyzing a variety of chemistry, including carbon hydroxylation, dehydration, heteroatom oxygenation, heteroatom release (dealkylation), epoxide formation, 1,2-migration, isomerization, reduction, C–C bond cleavage, and so on [18,19].

A very large number of P450s are present in plants, with nearly 2,900 named plant P450s in the P450 database (http://drnelson.utmem.edu/CytochromeP450.html). Many are known to be involved in the biosynthesis of natural products. So far, only two plant P450 structures were reported. They are for guayule allene oxide synthase (AOS), as determined by our group [20], and its ortholog in *Arabidposis* [21]. AOS is a unique P450 enzyme, self-sufficient and requiring no electron donor, and plays a key role in the biosynthesis of oxylipin jasmonates. The AOS structures exhibit a classic P450 fold, but possess a novel heme binding mode with an unusually long heme binding loop and a unique I-helix. The structures also reveal two channels through which substrate and product may access and leave the active site. Asn276 in the substrate binding site may interact with the substrate's hydroperoxy group and play an important role in catalysis.

IFS is a P450 monooxygenase and belongs to the class II P450s which require a NADPH cytochrome P450 reductase for transfer of electrons. IFS is a novel P450, and only presents in legumes for synthesis of isoflavonoids; IFS has been identified from *Glycyrrhiza echinata* [22-24], soybean [25,26]), and *Medicago truncatula* [27]. In *Arabidopsis*, there are 273 putative P450 genes, but none has IFS activity to synthesize the isoflavone genistein. By using transgenic approaches, introducing soybean IFS into *Arabidopsis thaliana* resulted in production of genistein at a high level [28], indicating the importance of IFS and its central role in the biosynthesis of isoflavonoids as well as its extremely high substrate specificity.

Through sequence alignment, homology modeling, multiple mutagenesis and activity assay, several key amino acid residues (Ser310, Lys375 and Leu371) of licorice IFS were identified to be responsible for the unique aryl migration [23,24]. Although IFS specifically catalyzes biosynthesis of isoflavonoids, it's also observed in the *in vitro* IFS reaction in which a byproduct 3-hydroxyflavanone is formed, e.g. 3,7,4'-trihydroxyflavanone from (2S)-liquiritigenin, in addition to the major 2-hydroxyisoflavanone [29,30]. Mutating some residues in the active site may yield dramatically changed ratios of 2-hydroxyisoflavanone to 3-hydroxyflavanone products, and even produces another byproduct, 7,4'-dihydroxyflavone. These observed changes of product distribution may be associated with the switch of the initial hydrogen abstraction from C-3 or C-2 and the presence or absence of aryl migration.

It's reported that engineered artificial IFS enzymes allow very efficient robust *in vivo* synthesis of isoflavones by *E. coli* [31]. IFS is a very valuable tool for modifying/manipulating flavanones and for enzymatic synthesis of novel (iso)flavonone derivatives.

Due to the difficulty in expression and purification, very limited information is available for both biochemical and crystallographic studies of membrane-associated biosynthetic P450s. Besides IFS, other P450s also play important roles in isoflavonoid biosynthesis, such as C4H in the early stage of phenylpropanoid pathway and isoflavone 2'-hydroxylase (I2'H) involved in hydroxylation of isoflavonoids genistein and formononetin. It's critical to obtain structures of this large group of biosynthetic P450s for promoting biochemical and mechanistic studies.

4. Isoflavone O-Methyltransferase and Methylation of Isoflavone

Flavanones are converted to isoflavones by a novel P450 enzyme IFS, then transformed into various isoflavonoid derivatives by other enzymes, including other P450s, methyltransferases, reductases and glycosyltransferases.

Methylation of isoflavonoids modulates their *in vivo* activity by limiting the number of reactive hydroxyl groups, thus altering their solubility and functions. Isoflavone O-methyltransferase (IOMT) was identified in alfalfa and *M. truncatula* [32,33], and is an S-adenosyl-L-methionine (SAM)-dependent methyltransferase, using SAM as a methyl source and yielding S-adenosyl-L-homocysteine (SAH) and the methyl ether derivative.

Structures of alfalfa isoflavone 7-O-methyltransferase (I7OMT) and *M. truncatula* hydroxyisoflavanone 4'-O-methyltransferase (HI4'OMT) have been determined by the Noel group [34,35]. Both crystal structures consist of a large C-terminal catalytic domain with a core α/β Rossmann fold responsible for SAM binding and substrate methylation (Figure 3). Their smaller N-terminal domains are involved in dimerization and formation of the back wall of the substrate binding site, and dimerization appears to be critical for enzyme activity.

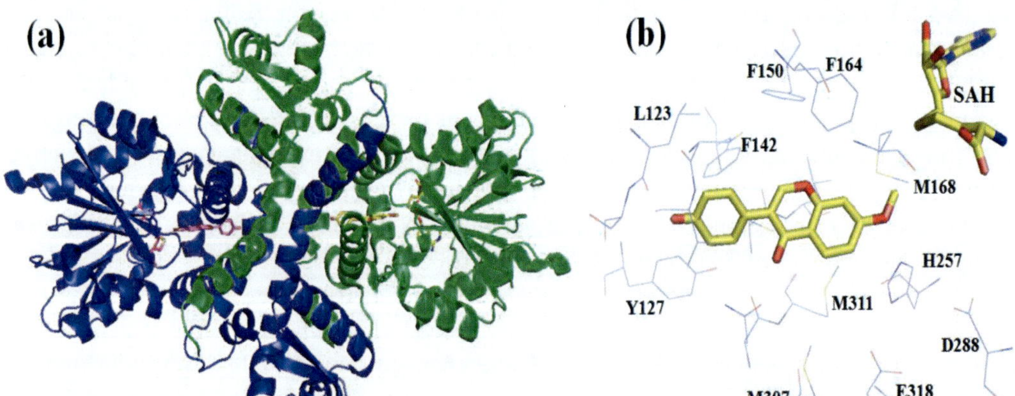

Figure 3. (a) Crystal structures of I7OMT in complex with SAH and isoformononetin (PDB: 1FP2). (b) Substrate binding pocket in I7OMT structure. SAH and isoformononetin (4'-hydroxy-7-methoxyisoflavone) are shown as ball-and-stick models

Catalysis in IOMT proceeds with a histidine (His257 in I7OMT) as the base for deprotonation of the 7-hydroxyl group on the A-ring of isoflavone substrate, then the hydroxyl anion attacks on the methyl group of SAM (Figure 3b). These structural studies provide a molecular basis for understanding the substrate specificity of the isoflavonoid OMTs.

5. Reductases and Their Stereo-Specificity for Isoflavonoids

A large number of isoflavonoid derivatives have been identified with various functions. Some of the isoflavonoid derivatives act as phytoalexins in legumes and play important roles in plant defense with protective activity against microorganisms and herbivores [36-38]. Medicarpin is the major isoflavonoid phytoalexin in alfalfa (*Medicago sativa*) responding to fungal pathogen attack [39]. Isoflavone reductase (IFR) and vestitone reductase (VR) are two key enzymes involved in the biosynthesis of medicarpin [40,41]. They are monomeric, cytosolic, and NADPH-dependent reductases. IFR catalyzes enantiospecific conversions with high specificity for isoflavones, and was identified in alfalfa [40,41], pea (*Pisum sativum*) [42], chickpea (*Cicer arietinum*) [43], and soybean (*Glycine max*) [44]. Alfalfa IFR converts 2'-hydroxyformononetin to (3R)-vestitone.

VR is the immediate downstream enzyme and only recognizes the 3R conformation of vestitone, not (3*S*)-vestitone, and converts (3R)-vestitone into 7,2'-dihydroxy-4'-methoxy-isoflavanol which is further converted to medicarpin by 7,2'-dihydroxy-4'-methoxy-isoflavanol dehydratase (DMID) (Figure 1). VR was identified in alfalfa [41,45], and a VR homolog, sorphorol reductase (SOR), was also identified for the biosynthesis of pisatin in pea [46].

Both IFR and VR structures have been determined by our group [47,48], and they contain a classic Rossmann fold domain in the N-terminus and a small C-terminal domain. They are members of the short-chain dehydrogenase/reductase (SDR) superfamily despite the low sequence identities and the prominent structural differences from other SDR enzymes with known structures. The putative binding sites for the co-factor NADPH and the substrates were defined and located in a large cleft formed between the N- and C-terminal domains of enzymes (Figure 4). The NADPH may bind near the conserved fingerprint region near the N-terminus. Similar to other SDR enzymes, VR possesses a catalytic triad, Ser129-Tyr164-Lys168, for its catalysis and function. Tyr164 acts as the catalytic base residue and attacks the oxygen atom at position 4 of vestitone, and Ser129 may form a hydrogen bond with this oxygen atom. Lys168 may interact with the ribose of NADPH to stabilize the position of the nicotinamide moiety [48].

But for IFR, only Lys144 is present in the equivalent position in its active site without corresponding serine and tyrosine, suggesting that IFR may have a different catalytic mechanism. Lys144 may act as a general base and play an essential role for IFR catalysis. The general base Lys144 acts on the C3 atom of the substrate, and the nicotinamide ring of NADPH may attack the C2 atom of the substrate and transfer its 4-pro-R-hydride (i.e., HA hydrogen) to the pro-S position to form the 3R-conformation of vestitone [47].

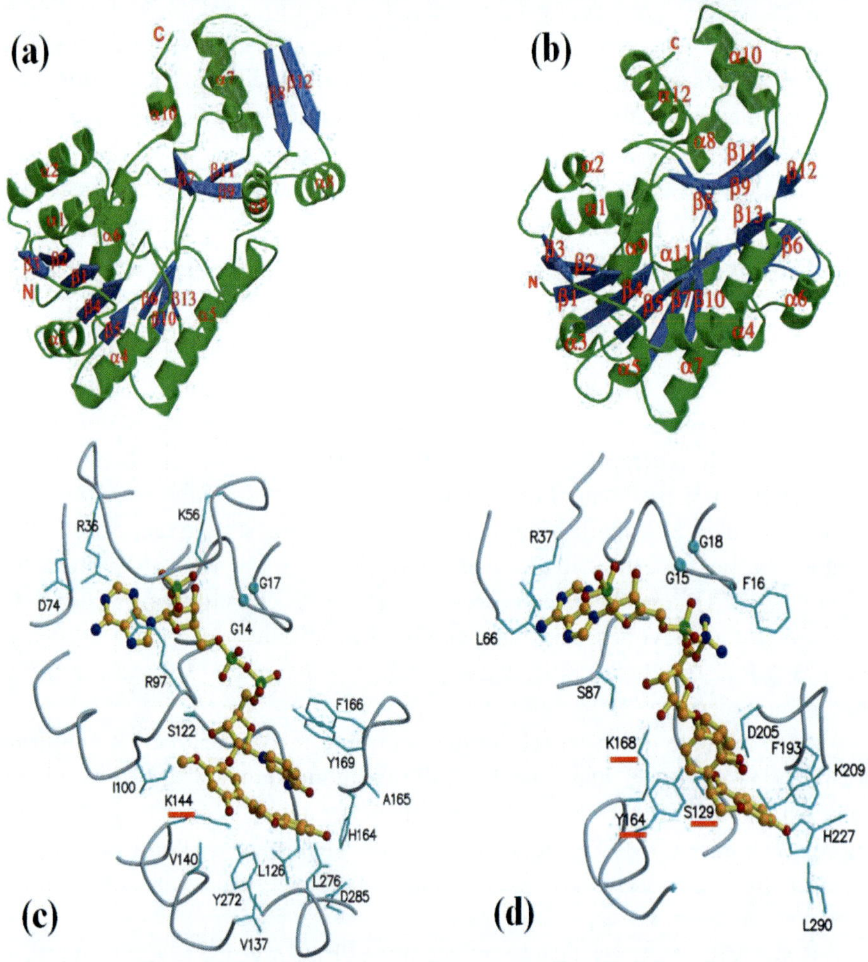

Figure 4. Crystal structures of (a) IFR (PDB: 2GAS) and (b) VR (PDB: 2P4H) and the putative substrate binding pockets of (c) IFR and (d) VR. NADPH and substrate (2'-hydroxyformononetin for IFR, (3R)-vestitone for VR) docked into the pocket are shown as ball-and-stick models

IFR specifically recognizes isoflavones, and 2'-hydroxyl was identified as a required feature for its substrates. IFR structural analysis showed that residue Tyr272 may interact with the 2'-hydroxyl of the substrate and is a possible key residue for substrate recognition and specificity. Mutation of Tyr272 to Phe or other amino acids may alter its specificity to generate novel IFR mutants recognizing isoflavones without a 2'-hydroxyl group or other compounds, as substrates, or produce novel products.

These structures provide a basis for understanding stereo-specific reduction of isoflavonoids and an explanation for these enzymes' strict substrate stereo-specificity. There are a large number of IFR-like proteins and VR homologs for the biosynthesis of isoflavonoid-derivatives in legumes, which are one of the largest families of crop plants with ~20,000 species. The structures of IFR and VR provide a basis for functional characterization of this large protein family, facilitating manipulation of this extensive class of natural product biosynthetic enzymes and their substrate specificity, and the biosynthetic pathway toward producing bioactive isoflavonoids.

6. Glycosylation of Isoflavonoids – Glycosyltransferase Structures and Engineering

Isoflavonoids often exist in glycosylated forms and are decorated with various sugars. Glycosylation is a key modification of plant natural products during their biosynthesis to enhance their solubility and stability and facilitate their storage and accumulation in plant cells [49-51]; it is also one of the major factors determining natural product bioactivity and bioavailability [52,53].

Glycosylation of natural products is catalyzed by uridine diphosphate (UDP) glycosyltransferases (UGTs) [53], members of family 1 glycosyltransferases. Glycosyltransferases are a superfamily of enzymes with over 90 families [54] (www.cazy.org/fam/acc_GT.html). UGTs specifically utilize UDP-sugars as donors and transfer various sugars to plant natural products. Consistent with their variety and complexity, a large number of *UGT* gene sub-families have evolved for the glycosylation of plant natural products. The model legume *Medicago truncatula* possesses over 100 UGTs, and many of them may be involved in the modification of isoflavonoids.

Several plant UGT structures have been determined by the Davies group and our group. *M. truncatula* UGT85H2 is an (iso)flavonoid glycosyltransferase [55]. *M. truncatula* UGT78G1 and grape (*Vitis vinifera*) *Vv*GT1 also recognize some isoflavonoid compounds, although they are able to glycosylate cyanidin to yield cyanidin 3-*O*-glycoside and may be involved in the biosynthesis of anthocyanins *in vivo* [56,57]. *M. truncatula* UGT71G1 also recognizes (iso)flavonoids in spite of its potential role in saponin biosynthesis [58,59].

These plant UGTs all have the GT-B fold, one of two general folds for the GT superfamily of enzymes [60,61], and consist of two N- and C-terminal domains with similar Rossmann-like folds (Figure 5). Each domain of the UGT enzymes contains a central β sheet flanked by α helices on both sides. The N- and C-terminal domains form a cleft which is the substrate binding site. The nucleotide sugar donor is located in this cleft and mainly interacts with the UGT signature PSPG motif in the C-terminal domains of the enzymes, and the acceptor mainly binds to the N-terminal domains.

Plant UGTs are inverting GT enzymes and catalyze glycosyl transfer with a direct displacement S_N2-like mechanism [62]. A highly conserved histidine in the active site (e.g., His22 in UGT71G1 and His20 in *Vv*GT1), close to the glucose moiety of the sugar donor and the acceptors, acts as a general base and catalytic residue for enzyme activity to abstract a proton from the acceptor substrate. A nearby conserved aspartate residue, Asp121 in UGT71G1 and Asp119 in *Vv*GT1, interacts with the histidine by forming a hydrogen bond and may stabilize the histidine and balance its charge after deprotonating the acceptor. The deprotonated acceptor, as a nucleophilic oxyanion, may attack the C1 carbon center of the UDP-sugar with a direct displacement of the UDP moiety and form a β-glucosidic linkage product [59].

UDP-glucose is often used as the favored sugar donor for glycosylation by plant UGTs. UDP-galactose, UDP-glucuronic acid, UDP-xylose and UDP-rhamnose are also used by plant UGTs [52,63]. As revealed in the crystal structures, the UDP-glucose sugar donor mainly interacts with residues in the plant UGT signature PSPG (Putative Secondary Plant Glycosyltransferase) motif [64], which has a highly conserved consensus sequence with 44 amino acids (e.g., residues Trp339-Gln382 in UGT71G1).

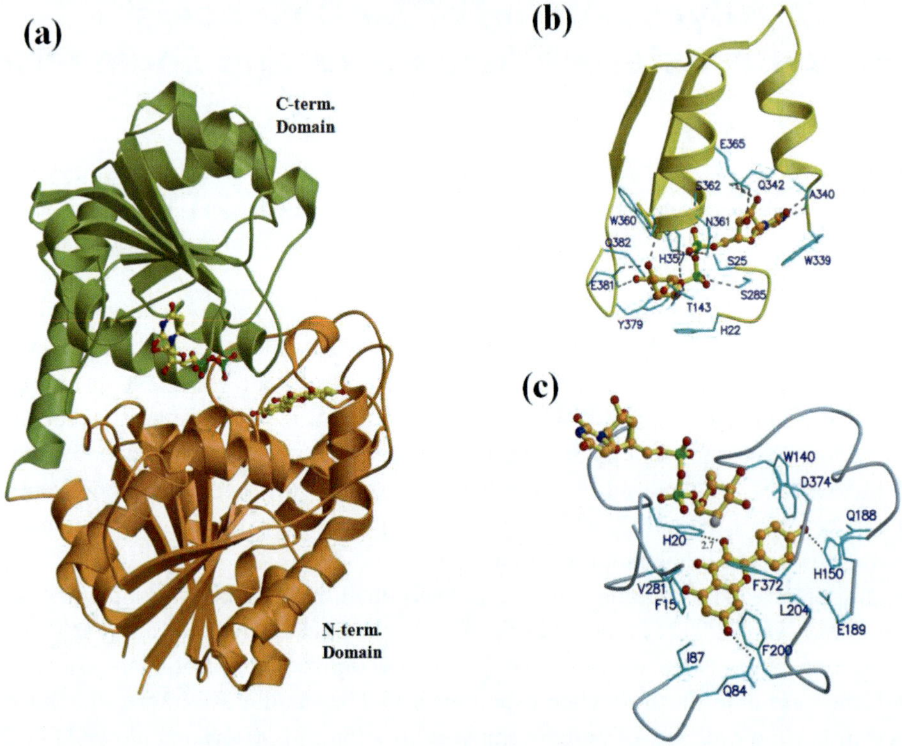

Figure 5. (a) Crystal structure of *M. truncatula* UGT78G1 with bound UDP (upper left) and myricetin (lower right) (PDB: 3HBF). (b) Sugar donor binding site and interaction between the donor molecule UDP-glucose and the enzyme UGT71G1 (PDB: 2ACW). The PSPG motif is shown as a ribbon model in yellow. The structure of UDP-glucose is shown as a ball-and-stick model. (c) Acceptor binding pocket in *Vv*GT1 structure with UDP-2-fluoroglucose (upper left) and acceptor kaempferol (lower right) (PDB: 2C1Z)

The glucose moiety of sugar donor mainly interacts with the last two residues of the PSPG motif (e.g., Glu381 and Gln382 in UGT71G1) and another highly conserved tryptophan (e.g., Trp360 in UGT71G1). Two other regions, around Thr143 and the catalytic residue His22 in UGT71G1, are also close to the glucose moiety and may be the determinants for sugar recognition and binding (Figure 5b).

It has been reported that a group of flavonoid 7-*O*-glucuronosyltransferases from Lamiales plants have an arginine residue (Arg350, corresponding to Trp360 in UGT71G1) which is essential for the glucuronic acid specificity [65]. A UDP-glucuronosyltransferase *Bp*UGT94B1 from red daisy (*Bellis perennis*) has an arginine residue (Arg25) in the N-terminus near the catalytic histidine which was found to be crucial for specificity with UDP-glucuronic acid [66].

The acceptor binding pocket in the UGT structures is adjacent to the donor binding site, mainly consisting of residues in the N-terminal domain and some residues in the C-terminal domain. The acceptor binding regions in UGTs are highly varied, and different UGTs may have quite different topologies of acceptor binding pockets for recognizing different acceptors. Manipulation around these regions affects the UGT enzyme activity and alters the substrate specificity.

Structure-based enzyme engineering on UGT85H2 showed that substitution of Ile305 with threonine enhanced enzyme catalytic efficiency 19-fold and reduced K_m 25-fold with isoflavonoid biochanin A as an acceptor [67].

UGT71G1 is able to glucosylate the A-ring 7-hydroxyl of the isoflavone genistein. Its mutant Y202A converted genistein to two products, the 7-*O*-glucoside and 5-*O*-glucoside [68].

UGT78G1 is a novel enzyme with both glycosylation and deglycosylation activities. It converted biochanin A 7-*O*-glucoside and genistein 7-*O*-glucoside into the corresponding aglycones [69]. Structural comparative analysis showed that an acidic residue Glu192 in UGT78G1, would be a key to its deglycosylation function, and other amino acids, such as another acidic residue Asp376, may also play roles [56].

Engineering work of UGT85H2 showed that a single mutation V200E conferred deglycosylation function and reversibility, and the mutant converted biochanin A 7-*O*-glucoside into the corresponding aglycone, suggesting that the new acidic residue, Glu200, introduced into the enzyme mutant plays a key role in the reverse reaction [67].

The structure-based UGT engineering showed that designing mutants and chimeras can improve catalytic properties and alter substrate specificity and regio-selectivity. The structural and engineering work also revealed key amino acids for UGT reversibility, and a similar approach may be applied to other UGTs for designing effective mutants with reversibility to synthesize activated sugars.

A bacterial whole-cell biocatalysis system has been developed to utilize plant UGTs as biocatalysts for glycoside production. Using endogenous donor sugars in the host cell systems, the bacterial cells expressing plant UGTs may take aglycone substrates added to the culture medium and synthesize glycosides [70,71]. Using *E. coli* expressing glycosyltransferase UGT71G1 and feeding genistein or biochanin A, 10-20 mg/L of either genistein or biochanin A 7-*O*-glucoside was produced [72]. The microbial whole-cell biocatalysis system may be used for efficient production of isoflavonoid glycosides with the UGT biocatalysts. It is also promising to combine other biosynthetic enzymes and include different biocatalysts in one system, such as this bacterial whole-cell biocatalyst system, for synthesis of a diverse range of bioactive isoflavonoid compounds very efficiently.

Isoflavonoids exhibit numerous pharmacological activities and are important drug scaffolds. These isoflavonoid biosynthetic enzymes are excellent templates for designing specific enzyme mutant biocatalysts to produce non-natural isoflavonoid derivatives, facilitating effective isoflavonoid-scaffold drug synthesis and discovery, and impacting the pharmaceutical industry and human health.

Acknowledgments

The work described from the author's laboratory was supported by the National Science Foundation (Grant # 0416833) and by the Samuel Roberts Noble Foundation.

References

[1] Dixon, R. A. (1999). Isoflavonoids: biochemistry, molecular biology and biological functions. In Sankawa, U. (Eds.), *Comprehensive Natural Products Chemistry*, (773-823). Oxford, Elsevier.

[2] Dixon, R. A. & Sumner, L. W. (2003). Legume natural products: Understanding and manipulating complex pathways for human and animal health. *Plant Physiology, 131*, 878-885.

[3] Dixon, R. A. & Ferreira, D. (2002). Molecules of interest: genistein. *Phytochemistry, 60*, 205-211.

[4] Palevitz, B. A. (2000). Soybeans hit main street. *Scientist, 14*, 8-9.

[5] Alekel, D. L., StGermain, A., Pererson, C. T., Hanson, K. B., Stewart, J. W. & Toda, T. (2000). Isoflavone-rich soy protein isolate attenuates bone loss in the lumbar spine of perimenopausal women. *Am J Clin Nutr*, *72,* 844-852.

[6] MerzDemlow, B. E., Duncan, A. M., Wangen, K. E., Xu, X., Carr, T. P., Phipps, W. R. & Kurzer, M. S. (2000). Soy isoflavones improve plasma lipids in normocholesterolemic, premenopausal women. *Am J Clin Nutr*, *71,* 1462-1469.

[7] Uesugi, T., Toda, T., Tsuji, K. & Ishida, H. (2001). Comparative study on reduction of bone loss and lipid metabolism abnormality in ovariectomized rats by soy isoflavones, daidzin, genistin, and glycitin. *Biol Pharm Bull*, *24,* 368-372.

[8] Sarkar, F. H. & Li, Y. (2004). The role of isoflavones in cancer chemoprevention. *Front Biosci*, *9,* 2714-2724.

[9] Banerjee, S., Li, Y., Wang, Z. & Sarkar, F. H. (2008). Multi-targeted therapy of cancer by genistein. *Cancer Lett 269,* 226-42.

[10] Dixon, R. A., Harrison, M. J. & Paiva, N. L. (1995). The isoflavonoid phytoalexin pathway: from enzymes to genes to transcription factors. *Physiol Plant*, *93,* 385-392.

[11] Ritter, H. & Schulz, G. E. (2004). Structural basis for the entrance into the phenylpropanoid metabolism catalyzed by phenylalanine ammonia-lyase. *Plant Cell*, *16,* 3426-3436.

[12] Ferrer, J. L., Jez, J. M., Bowman, M. E., Dixon, R. A. & Noel, J. P. (1999). Structure of chalcone synthase and the molecular basis of plant polyketide biosynthesis. *Nat Struct Biol*, *6,* 775-784.

[13] Bomati, E. K., Austin, M. B., Bowman, M. E., Dixon, R. A. & Noel, J. P. (2005). Structural elucidation of chalcone reductase and implications for deoxychalcone biosynthesis. *J Biol Chem*, *280,* 30496-30503.

[14] Jez, J. M., Bowman, M. E., Dixon, R. A. & Noel, J. P. (2000). Structure and mechanism of the evolutionarily unique plant enzyme chalcone isomerase. *Nat Struct Biol*, 786-791.

[15] Calabrese, J. C., Jordan, D. B., Boodhoo, A., Sariaslani, S. & Vannelli, T. (2004). Crystal structure of phenylalanine ammonia lyase: Multiple helix dipoles implicated in catalysis. *Biochemistry*, *43,* 11403-11416.

[16] Moffitt, M. C., Louie, G. V., Bowman, M. E., Pence, J., Noel, J. P., Moore, B. S. (2007). Discovery of two cyanobacterial phenylalanine ammonia lyases: Kinetic and structural characterization. *Biochemistry*, *46,* 1004-1012.

[17] Schwede, T. F., Retey, J. & Schulz, G. E. (1999). Crystal structure of histidine ammonia-lyase revealing a novel polypeptide modification as the catalytic electrophile. *Biochemistry, 38,* 5355-5361.

[18] Isin, E. M. & Guengerich, F. P. (2007). Complex reactions catalyzed by cytochrome P450 enzymes. *Biochim Biophys Acta*, *1770,* 314-329.

[19] Werck-Reichhart, D. & Feyereisen, R. (2000). Cytochromes P450: a success story. *Genome Biol, 1,* reviews3003.1-9.

[20] Li, L., Chang, Z., Pan, Z., Fu, Z. Q. & Wang, X. (2008). Modes of heme binding and substrate access for cytochrome P450 CYP74A revealed by crystal structures of allene oxide synthase. *Proc Natl Acad Sci*, USA *105,* 13883-13888.

[21] Lee, D. S., Nioche, P., Hamberg, M. & Raman, C. S. (2008). Structural insights into the evolutionary paths of oxylipin biosynthetic enzymes. *Nature*, *455,* 363-368.

[22] Akashi, T., Aoki, T. & Ayabe, S. (1999). Cloning and functional expression of a cytochrome P450 cDNA encoding 2-hydroxyisoflavanone synthase involved in biosynthesis of the isoflavonoid skeleton in licorice. *Plant Physiol*, *121,* 821-828.

[23] Sawada, Y., Kinoshita, K., Akashi, T., Aoki, T. & Ayabe, S. (2002). Key amino acid residues required for aryl migration catalysed by the cytochrome P450 2-hydroxyisoflavanone synthase. *Plant J 31,* 555-564.

[24] Sawada, Y. & Ayabe, S. (2005). Multiple mutagenesis of P450 isoflavonoid synthase reveals a key active-site residue. *Biochem Biophys Res Commun, 330,* 907-913.

[25] Steele, C. L., Gijzen, M., Qutob, D. & Dixon, R. A. (1999). Molecular characterization of the enzyme catalyzing the aryl migration reaction of isoflavonoid biosynthesis in soybean. *Arch Biochem Biophys*, *367,* 146-150.

[26] Jung, W., Yu, O., Lau, S. M., O'Keefe, D. P., Odell, J., Fader, G. & McGonigle, B. (2000). Identification and expression of isoflavone synthase, the key enzyme for biosynthesis of isoflavones in legumes. *Nature Biotechnol*, *18,* 208-212.

[27] Deavours, B. E. & Dixon, R. A. (2005). Metabolic engineering of isoflavonoid biosynthesis in alfalfa (*Medicago sativa* L.). *Plant Physiology*, *138,* 2245-2259.

[28] Liu, C. J., Blount, J. W., Steele, C. L. & Dixon, R. A. (2002). Bottlenecks for the metabolic engineering of isoflavone glycoconjugates in *Arabidopsis*. *Proc Natl Acad Sci* USA *99,* 14578-14583.

[29] Akashi, T., Aoki, T. & Ayabe, S. (1999). Cloning and functional expression of a cytochrome P450 cDNA encoding 2-hydroxyisoflavanone synthase involved in biosynthesis of the isoflavonoid skeleton in licorice. *Plant Physiol*, *121,* 821-828.

[30] Hashim, M. F., Hakamatsuka, T., Ebizuka, Y. & Sankawa, U. (1990). Reaction mechanism of oxidative rearrangement of flavanone in isoflavone biosynthesis. *FEBS Lett*, *271,* 219-222.

[31] Leonard, E. & Koffas, M. A. (2007). Engineering of artificial plant cytochrome P450 enzymes for synthesis of isoflavones by *Escherichia coli*. *Appl Environ Microbiol*, *73,* 7246-7251.

[32] He, X. Z., Reddy, J. T. & Dixon, R. A. (1998). Stress responses in alfalfa (Medicago sativa L). XXII. cDNA cloning and characterization of an elicitor-inducible isoflavone 7-O-methyltransferase. *Plant Mol Biol*, *36,* 43-54.

[33] Deavours, B. E., Liu, C. J., Naoumkina, M. A., Tang, Y., Farag, M. A., Sumner, L. W., Noel, J. P. & Dixon, R. A. (2006). Functional analysis of members of the isoflavone

and isoflavanone O-methyltransferase enzyme families from the model legume *Medicago truncatula*. *Plant Mol Biol*, *62,* 715-733.

[34] Zubieta, C., Dixon, R. A. & Noel, J. P. (2001). Crystal structures of chalcone O-methyltransferase and isoflavone O-methyltransferase reveal the structural basis for substrate specificity in plant O-methyltransferases. *Nature Structural Biology*, *8,* 271-279.

[35] Liu, C. J., Deavours, B. E., Richard, S. B., Ferrer, J. L., Blount, J. W., Huhman, D., Dixon, R. A. & Noel, J. P. (2006). Structural basis for dual functionality of isoflavonoid O-methyltransferases in the evolution of plant defense responses. *Plant Cell*, *18,* 3656-3669.

[36] Dixon, R. A. (2001). Natural products and disease resistance. *Nature*, *411,* 843-847.

[37] Bailey, J. A. & Mansfield, J. W. (1982). *Phytoalexins*, John Wiley, New York.

[38] Blount, J. W., Dixon, R. A. & Paiva, N. L. (1993). Stress responses in alfalfa (*Medicago sativa* L.) XVI. Antifungal activity of medicarpin and its biosynthetic precursors; implications for the genetic manipulation of stress metabolites. *Physiol Mol Plant Pathol*, *41,* 333-349.

[39] Dixon, R. A. & Steele, C. L. (1999). Flavonoids and isoflavonoids - a gold mine for metabolic engineering. *Trends Plant Sci*, *4,* 394-400.

[40] Paiva, N. L., Edwards, R., Sun, Y., Hrazdina, G. & Dixon, R. A. (1991). Stress responses in alfalfa (*Medicago sativa* L.). 11. Molecular cloning and expression of alfalfa isoflavone reductase, a key enzyme of isoflavonoid phytoalexin biosynthesis. *Plant Mol Biol*, *17,* 653-667.

[41] Guo, L., Dixon, R. & Paiva, N. L. (1994). Conversion of vestitone to medicarpin in alfalfa (*Medicago sativa* L.) is catalyzed by two indenpendent enzymes. *J Biol Chem*, *269,* 22372-22378.

[42] Sun, Y., Wu, Q., VanEtten, H. D. & Hrazdina, G. (1991). Stereoisomerism in plant disease resistance: Induction and isolation of the *7*,2'-dihydroxy-4',5'-methylenedioxyisoflavone reductase, an enzyme introducing chirality during synthesis of isoflavanoid phytoalexins in pea (*Pisum sativum* L). *Arch Biochem Biophys*, *284,* 167-173.

[43] Schlieper, D., Tiemann, K. & Barz, W. (1990). Stereospecificity of hydrogen transfer by fungal and plant NADPH: isoflavone oxidoreductases. *Phytochemistry*, *29,* 1519-1524.

[44] Fischer, D., Ebeanau-Jehle, C. & Grisebach, H. (1990). Phytoalexin synthesis in soybean: purification and characterization of NADPH: 2'-hydroxydaidzein oxidoreductase from elicitor-challenged soybean cell cultures. *Arch Biochem Biophys*, *276,* 390-395

[45] Guo, L. & Paiva, N. L. (1995). Molecular cloning and expression of Alfalfa (Medicago sativa L.) vestitone reductase, the penultimate enzyme in medicarpin biosynthesis. *Arch Biochem Biophys*, *320,* 353-360.

[46] DiCenzo, G. L. & VanEtten, H. D. (2006). Studies on the late steps of (+) pisatin biosynthesis: evidence for (-) enantiomeric intermediates. *Phytochemistry*, *67,* 675-683.

[47] Wang, X., He, X., Lin, J., Shao, H., Chang, Z. & Dixon, R. A. (2006). Crystal structure of isoflavone reductase from Alfalfa (*Medicago sativa* L.). *J Mol Biol* , *358,* 1341-1352.

[48] Shao, H., Dixon, R. A. & Wang, X. (2007). Crystal structure of vestitone reductase from Alfalfa (*Medicago sativa* L.). *J Mol Biol, 369,* 265-276.
[49] Jones, P. & Vogt, T. (2001). Glycosyltransferases in secondary plant metabolism: Tranquilizers and stimulant controllers. *Planta, 213,* 164-174.
[50] Gachon, C. M., Langlois-Meurinne, M. & Saindrenan, P. (2005). Plant secondary metabolism glycosyltransferases: The emerging functional analysis. *Trends Plant Sci 10,* 542-549.
[51] Bowles, D., Isayenkova, J., Lim, E. & Poppenberger, B. (2005). Glycosyltransferases: Managers of small molecules. *Current Opinion in Plant Biology, 8,* 254-263.
[52] Bowles, D., Lim, E. K., Poppenberger, B. & Vaistij, F. E. (2006). Glycosyltransferases of lipophilic small molecules. *Annu Rev Plant Biol, 57,* 567-597.
[53] Ross, J., Li, Y., Lim, E. & Bowles, D. J. (2001). Higher plant glycosyltransferases. *Genome Biol, 2,* 3004.1-3004.6.
[54] Campbell, J. A., Davies, G. J., Bulone, V. & Henrissat, B. (1997). A classification of nucleotide-diphospho-sugar glycosyltransferases based on amino acid sequence similarities. *Biochem J, 326,* 929-939.
[55] Li, L., Modolo, L. V., Escamilla-Trevino, L. L., Achnine, L., Dixon, R. A. & Wang, X. (2007). Crystal structure of *Medicago truncatula* UGT85H2 - Insights into the structural basis of a multifunctional (iso)flavonoid glycosyltransferase. *J Mol Biol, 370,* 951-963.
[56] Modolo, L. V., Li, L., Dixon, R. A. & Wang, X. (2009). Crystal structures of glycosyltransferase UGT78G1 reveal the molecular basis for glycosylation and deglycosylation of (iso)flavonoids. *J Mol Biol, 392,* 1292-1302.
[57] Offen, W., Martinez-Fleites, C., Yang, M., Kiat-Lim, E., Davis, B. G., Tarling, C. A., Ford, C. M., Bowles, D. J. & Davies, G. J. (2006). Structure of a flavonoid glucosyltransferase reveals the basis for plant natural product modification. *EMBO J, 25,* 1396-1405.
[58] Achnine, L., Huhman, D. V, Farag, M. A., Sumner, L. W., Blount, J. W. & Dixon, R. A. (2005). Genomics-based selection and functional characterization of triterpene glycosyltransferases from the model legume *Medicago truncatula. Plant J, 41,* 875-87.
[59] Shao, H., He, X., Achnine, L., Blount, J. W., Dixon, R. A. & Wang, X. (2005). Crystal structures of a multifunctional triterpene/flavonoid glycosyltransferase from *Medicago truncatula. Plant Cell, 17,* 3141-3154.
[60] Bourne, Y. & Henrissat, B. (2001). Glycoside hydrolases and glycosyltransferases: families and functional modules. *Curr Opin Struct Biol, 11,* 593-600.
[61] Coutinho, P. M., Deleury, E., Davies, G. J. & Henrissat, B. (2003). An evolving hierarchical family classification for glycosyltransferases. *J Mol Biol, 328,* 307-317.
[62] Lairson, L. L., Henrissat, B., Davies, G. J. & Withers, S. G. (2008). Glycosyltransferases: Structures, functions, and mechanisms. *Annu Rev Biochem, 77,* 521-555.
[63] Lim, E. K. & Bowles, D. J. (2004). A class of plant glycosyltransferases involved in cellular homeostasis. *EMBO J, 23,* 2915-2922.
[64] Hughes, J. & Hughes, M. A. (1994). Multiple secondary plant product UDP-glucose glucosyltransferase genes expressed in cassava (*Manihot esculenta* Crantz) cotyledons. *DNA Seq, 5,* 41-49.

[65] Noguchi, A., Horikawa, M., Fukui, Y., Fukuchi-Mizutani, M., Iuchi-Okada, A., Ishiguro, M., Kiso, Y., Nakayama, T. & Ono, E. (2009). Local differentiation of sugar donor specificity of flavonoid glycosyltransferase in Lamiales. *Plant Cell, 21,* 1556-1572.

[66] Osmani, S. A., Bak, S., Imberty, A., Olsen, C. E. & Moller, B. L. (2008). Catalytic key amino acids and UDP-sugar donor specificity of a plant glucuronosyltransferase, UGT94B1: molecular modeling substantiated by site-specific mutagenesis and biochemical analyses. *Plant Physiology, 148,* 1295-1308.

[67] Modolo, L. V., Escamilla-Trevino L. L., Dixon, R. A. & Wang, X. (2009). Single amino acid mutations of *Medicago* glycosyltransferase UGT85H2 enhance activity and impart reversibility. *FEBS Letters, 583,* 2131-2135.

[68] He, X., Wang, X. & Dixon, R. A. (2006). Mutational analysis of the *Medicago* glycosyltransferase UGT71G1 reveals residues that control regioselectivity for (iso)flavonoid glycosylation. *J Biol Chem, 281*, 34441-34447.

[69] Modolo, L. V., Blount, J. W., Achnine, L., Naoumkina, M. A., Wang, X. & Dixon, R. A. (2007). A functional genomics approach to (iso)flavonoid glycosylation in the model legume *Medicago truncatula. Plant Mol Biol, 64,* 499 -518.

[70] Lim, E. K. (2005). Plant glycosyltransferases: their potential as novel biocatalysts. *Chemistry, 11,* 5486-5494.

[71] Lim, E. K., Ashford, D. A., Hou, B., Jackson, R. G. & Bowles, D. J. (2004). Arabidopsis glycosyltransferases as biocatalysts in fermentation for regioselective synthesis of diverse quercetin glucosides. *Biotech Bioeng, 87,* 623-631.

[72] He, X. Z., Li, W. S., Blount, J. W. & Dixon, R. A. (2008). Regioselective synthesis of plant (iso)flavone glycosides in *Escherichia coli. Appl Microbiol Biotechnol, 80,* 253-260.

In: Isoflavones Biosynthesis, Occurence and Health Effects ISBN: 978-1-61728-113-6
Editor: Melanie J. Thompson, pp. 255-269 © 2010 Nova Science Publishers, Inc.

Chapter X

Photophysics of Genistein and Biochanin a Isoflavones: Solvent Cage and Concentration Effects Studied by UV Visible Spectroscopy

***K. Benthami*[1], *S. Ait Lyazidi*[1*], *M. Haddad*[1], *B. Bennetau*[2] and *S. Shinkaruk*[3]**

[1] Laboratoire de Spectrométrie, des Matériaux et Archéomatériaux (LASMAR), URAC 11, Université Moulay Ismail, Faculté des Sciences, B.P 11201 Zitoune, Meknès, Morocco

[2] Institut des Sciences Moléculaires (ISM), CNRS-U2MR 5255, Université Bordeaux 1, 351 cours de la Libération, F-33405 Talence, France

[3] Ecole Nationale d'Ingénieurs des Travaux Agricoles (ENITA) de Bordeaux 1, cours du Général de Gaulle-CS 40201, 33175 Gradignan, France

Abstract

Genistein and Biochanin (A) isoflavones are shown to exist in two different conformations which are the 90-degrees completely twisted geometry and the 50 degrees less twisted one, with respect to the C_3—$C_{1'}$ single bond. Specific interactions with the solvent cage as well as self association processes seem shifting the isoflavone from the perpendicular conformation towards the less twisted one. From the UV-visible photophysics investigations it's revealed, in both isoflavones, that monomeric species can't exist alone even at very low solute concentration (~10^{-6}M), the self association process occurs already in this concentration range. The theoretical simulation, using analytical atom-atom pair potential, predicts a self dimer in a slipped non sandwich, face to river, perpendicular structure.

[*] Corresponding author: E-mail : aitlyazidisaadia@yahoo.fr, phone: (+212) 666502265, Fax: (+212) 35536808

Keywords: isoflavones, Genistein, Biochanin, geometry, twist, UV-Vis., absorption, simulation, dimer.

1. Introduction

The isoflavonoides Daidzein, Biochanin and Genistein are a subset of phytoestrogens [1] synthesized by higher plants in flowers leafs and roots [2]; soybeans and red clover are the most common source of isoflavones in human food. Biochanin A is the methylated precursor of Genistein, in the gut intestinal bacteria convert conjugated isoflavones to their demethylated forms [3, 4]. Over the past two decades, the isoflavones have been widely studied because of their valuable pharmacological activities. Wang and Col [5] attribute the reason for this rapidly rising interest to the association of many phytoestrogens and other members of polyphenols family with specific diseases [6-10]. Genistein (5, 7-dihydroxy-3-(4-hydroxyphenyl)-4H-1-benzopyran-4-one), the most abundant soy isoflavone, has demonstrated potent biological activity such as anticarcinogenic [11], anti-inflammatory [12], antiviral [13], anti-protozoal [14] and estrogen receptor [15]. This isoflavone has been shown to down-regulate androgen receptor (AR) in androgen-dependent prostate cancer cell lines such as LNCaP [16]. In the same way, using multiple assay techniques as hallmark to detect apoptosis in several cancer cell kinds, Sarkar [17-21] found Genistein induced apoptosis in all cancer cells tested. In these studies it is reported that flow cytometry revealed that the number of apoptotic cells increased 43-57% with longer genistein treatment. This apoptotic effect was further corroborated by findings of Moiseeva et al. [22] who reported that physiological concentrations of dietary phytochemical including genistein results in reduced growth and induction of apoptosis in cancer cells. In recent studies, Raynal et al. [23] concluded that a soy-enriched diet has the potential to produce plasma levels of genistein in the range of the concentration used in vitro that produced an antileukemic activity, also Y. J. Lin et al [24] observed that the serum metabolites of Puerariae radix (Daidein sulphates/glucuronides) inhibited proliferation of the breast cancer cells at a 50% cell growth inhibition (GI_{50}) concentration of 2.35 μM. The genomic effects of phytoestrogens expressing estrogenic or anti-estrogenic effects on ERα or ERβ as well as on the other nuclear receptors are most obvious, but non genomic effects of phytoestrogens may also be very important [3,25]. Earlier A. Andres et al. [26] reported the reducing infectivity character of flavonoids, including Genistein, of a variety of virus affecting humans and animals such as human immunodeficiency virus and porcine respiratory syndrome virus.

Nevertheless it should be noted that the molecular mechanism by which genistein inhibits the growth of cancer cells seams not fully explored. The affinity of phytoestrogens for the ERβ receptor subtype is higher than for the ERα subtype, and this possibility of double binding participates to the complexity recognized for these molecules in terms of mechanism of action, and may explains the confusing balance between their positive versus negative effects on animal and human health [27- 29]. Some years ago, it was reported [30] that the interaction between phytoestrogens and cancerous cells is complicated and would be dependant on several factors, mainly the phytoestrogenic concentration [31, 32]. At high concentration, probably supra physiologic, the genistein would exercise antiestrogenic and antiproliferatif effects. In contrast, at low concentrations the genistein could stimulate the proliferation of tumorous cells with estrogenic receptors [31-37]. So if we take into account the ability of isoflavones to give self associated forms upon concentration, in vivo activity of these flavonoids might be reconsidered because the self association process might impact on the compound subsequent reactivity. This implies that understanding this in vivo activity requires the knowledge of the self association process of the isoflavone in use. The purpose of the present work is the spectroscopic characterization of Genistein and Biochanin monomers and their ground state self dimers. The main questions raised relate to: i) the interaction of the flavonoid with the solvent cage, ii) the determination of the flavonoid concentration under which no self association can occur, and how does the solvent cage influence the self association process, and finally iii) what are the probable configurations of the dimeric species.

It should be mentioned that few works focused on the spectroscopic behaviour of isoflavones [38- 46]. The isoflavones studied in the present work are the Genistein and Biochanin (A); the corresponding structure scheme is shown in figure 1.

2. Experimental

The Genistein and Biochanin samples were synthesized as described in ref [47]. The solvents used are of spectroscopic grade. The UV- visible absorption spectra were recorded on a Cary 50 UV/Visible spectrophotometer.

The geometry optimization of the monomer was obtained using the RHF/ 6-311+G (d) (ground state restricted Hartree Fock method) *ab initio* method [48]. These calculations revealed that the completely twisted geometry, with respect to the C_3—$C_{1'}$ single bond, is the most stable form for both isoflavones. The optimized geometries, considering the isoflavone monomer as in the gas phase, are different from that found in the Genistein crystalline structure [49] in which the phenyl and the flavone rings are rotated by about 50 degrees. This discordance is acceptable because in the crystalline structure two neighboring Genistein monomers lie with their flavone rings parallel and a separation distance of about 3.45 Å as in a dimeric configuration.

From the optimized geometry, energies of the allowed UV–visible absorption transitions and their corresponding oscillator strengths have been calculated for Genistein. The ZINDO/S [50] package was used.

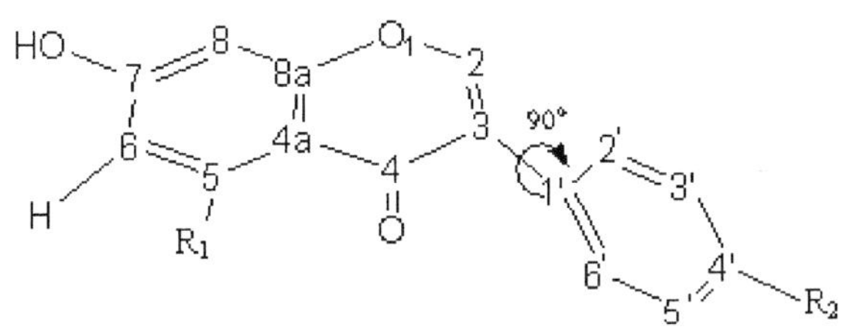

Systematic name	Common name	R1	R2	μ(Debyes)
5, 7, 4'-trihydroxyisoflavone	Genistein	OH	OH	4.09
5, 7-dihydroxy-4'-methoxyisoblavone	Biochanin A	OH	OCH_3	5.40

Figure 1. 90° *ab initio* optimized structure of Genistein and Biochanin A

The potential used for the simulation of the Genistein ground state dimer is the atom-atom pair one proposed at first by Fraga; the method is described below.

3. Results and Discussion

3.1. UV – Visible Absorption Spectra at Low Concentration: Solvent Effect

Figure 2 presents the room temperature UV-visible absorption spectra of Genistein and Biochanin at very low concentration (2.10^{-6}M) in acetonitrile and ethanol solvents. For both compounds, the spectrum shows three parts. The first one consists in a large edge extended between 320 and 420 nm, the second part lies between 270 and 320 nm, and the highest energy band is located around 260 nm. These spectra agree with those reported, for Genistein, by Dunford [38] and Crupi [46]. Table1 shows the corresponding molar extinction coefficients ε_{max} at the bands maxima. These coefficients reveal the $\pi\pi^*$ character of all the bands. Although the lowest energy band, in both isoflavones, shows a low extinction coefficient meaning a forbidden character, the transition could not be of $n\pi^*$ type as reported in ref [38, 39]. This assumption is supported by the observed emitting character of the Genistein and Biochanin [to be published]. The corresponding allowed UV-visible absorption transitions, calculated using ZINDO/S package, are also shown in the case of Genistein. The obtained oscillator strengths values reveal also the $\pi\pi^*$ character of all these transitions. The observed differences between theoretical and experimental wavelengths values are within the magnitude of the error commonly accepted in ZINDO/S calculations [51-55].

Table 1. Experimental spectroscopic parameters of Genistein and Biochanin monomers*

	solvent	$\lambda^{(a)}_{max}$ (nm)	log $\varepsilon^{(b)}_{max}$
Genistein	acetonitrile	260	3.9
		280	4.1
		340	3.4
	ethanol	260	4.3
		280[c]	-
		340	3.4
Biochanin A	acetonitrile	260	4.4
		280	4.3
		340	3.9
	ethanol	260	4.9
		280[c]	-
		340	4.1

* Values taken on Figure 2-a and Figure 2-b.
[a] Wavelength at the band maximum.
[b] Molar extinction coefficient at the band maximum.
[c] Transition completely submerged.

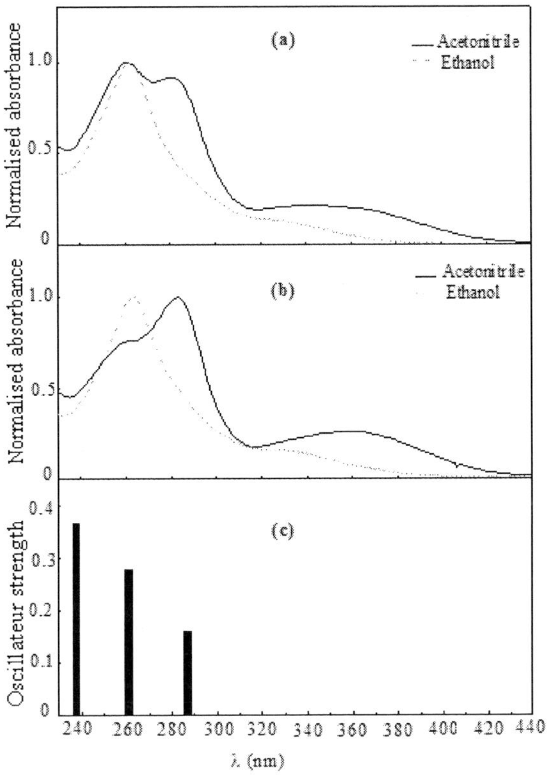

Figure 2. Measured room temperature absorption spectra of Biochanin (a) and Genistein (b) in ethanol and in acetonitrile at 2.10^{-6} M. Calculated allowed absorption transitions of Genistein monomer in its completely twisted 90° geometry (c)

When the solvent polarity is increased from the CH_3CN to the ethanol solvent, the shape of the absorption spectrum changes. In the aprotic CH_3CN solvent the two bands located around 290 and around 260 nm are clearly splitted with comparable extinction coefficients, while in the protic ethanol solvent the band at 260 nm submerges completely the band at 290 nm which appears as a shoulder. Taking into account the high flexibility of the isoflavone with respect to the torsion around the C_3—$C_{1'}$ single bond, this behavior may be explained by the coexistence of different monomeric geometries with different angles of torsion. In the aprotic solvent where no specific interactions occur, the monomeric specie maintains its perpendicular geometry; while in the protic solvent specific interactions with the solvent cage seem to shift the molecular configuration to a less twisted form. Indeed, deviation from the completely twisted to a less twisted configuration has been regarded as the origin of the native fluorescence observed in some isoflavones [45].

3.2. Concentration Effect on the Absorption Spectra

Figure 3 shows the variation of the absorption spectra, of the Genistein and Biochanin with increasing concentration from 10^{-6} (and 2.10^{-6}) to 10^{-4}M in ethanol and in acetonitrile. In both solvents, at the lower concentration, the spectra correspond to those of the monomer species described above in figure 2. In the protic solvent the spectra, of both isofavones, obey to Beer Lambert law; no change in the spectrum shape was observed with concentration increasing. In contrast to this, in the aprotic acetonitrile solvent, as the concentration is rised, the shape of the spectrum changes. Indeed, for both of Genistein and Biochanin, when the product ***c.l*** (where ***c*** is the concentration and ***l*** is the cell length) is maintained constant over the concentration range 10^{-4} – 10^{-6}M, the lowest energy band around 360 nm and the second band around 290 nm decreases whereas the highest energy band around 260 nm increases. Furthermore, the measured spectra at the highest concentration 10^{-4}M are very similar to those observed in the alcoholic ethanol solvent whose shapes don't show any concentration depending. This observed Genistein and Biochanin common behavior reveals that in the aprotic solvent CH_3CN there are at least two absorbing species with different geometries. These ground state species might possibly differ by the angle between the phenyl and the flavone rings induced by the twist, around the C_3—$C_{1'}$ single bond. This is corroborated by the appearance, in both Genistein and Biochanin, of an isosbestic point appears near 270 nm. Hence, in the aprotic solvent, when the solute concentration is increased self association process occurs and seems rotating the molecular skeleton from the completely twisted to a less twisted geometry. With concentration growing, dimeric species are the first self associated forms which can appear at the ground state without perturbing much the absorption spectrum of the monomer specie [56-61]. In the protic ethanol solvent, where no shape change was observed, already at very low solute concentration specific interactions with the solvent cage appear sufficient to induce the twist around de C_3—$C_{1'}$ single bond shifting the monomeric specie from the perpendicular geometry to a less twisted one. So, in the protic medium, the self association process occurring at high concentration seems not inducing any deviation from this non perpendicular form. This behavior leads to the assumption that the dimerization process translates the interacting monomers from the completely twisted configuration to a less twisted one. This is in good agreement with the x-ray crystallographic structure in which neighboring genistein monomers lie as in a sandwich dimeric structure with an inter-planar separation close to 3.45 Å and with a twist angle around de C_3—$C_{1'}$ single bond of about 54 degrees [49]. Similar torsion angles were reported for other genistein derivatives [62].

The reproducible appearance of the isosbestic point, around 270 nm, with concentration rising implies that the completely twisted form absorbing around 280-290 nm disappears to the profit of the less twisted form absorbing at higher energy around 260 nm. This is in accord with the case of the twisted 4-nitroaniline [63]. Furthermore, the apparition of this later band around 260 nm at very low solute concentration, in both Genistein and Biochanin, indicates that van der Waals interactions between two neighboring molecules is so attractive that monomer species cannot exist alone even at very low concentration.

The angle of rotation, induced by the self association process, from the perpendicular geometry towards the less twisted one was estimated using the ratio of the final to the starting molar extinction coefficient at the absorption maximum [63] near 280 nm. This law is:

$$\frac{e_f\,(280nm)}{e_i\,(280nm)} = \cos^2\theta$$

where θ is the angle of rotation around the C_3—$C_{1'}$ single bond from the initial perpendicular geometry, ε_i and ε_f are the extinction coefficients measured at 280 nm respectively at 10^{-6}M (or 2.10^{-6}M for Biochanin) and at 10^{-4} M. Table 2 reports the ε coefficients deduced from Figure 3-b,d as well as the calculated angles of rotation; the obtained values are respectively

40 and 45 degrees for Genistein and Biochanin. So, it seems that the intermediary geometry, in which the flavone and the phenyl rings are not perpendicular, corresponds to an angle of 50 degrees (from 90° to 40°) between the flavone and the phenyl rings planes in the case of Genistein and 45 degrees (from 90° to 45°) in the case of Biochanin. The less twisted configuration is reached when the isoflavone is associated to a neighboring solute molecule.

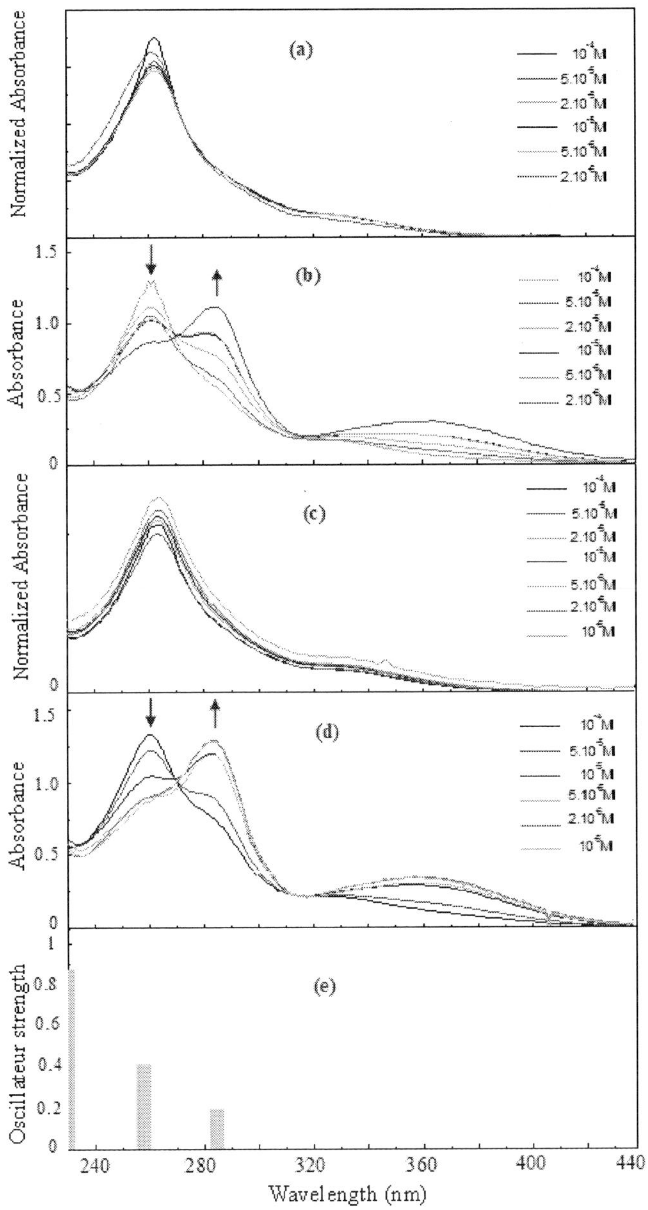

Figure 3. Measured room temperature absorption spectra of: Biochanin in ethanol (a) and in acetonitrile (b), Genistein in ethanol (c) and in acetonitrile (d), (*c.l= conctant*). Calculated allowed absorption transitions of Genistein dimer (e) (*interacting monomers in the 50° less twisted geometry*)

Table 2. Molar extinction coefficients and deduced angle of rotation from 90

λ^*_{max}	Isoflavone in acetonitrile	ε^*_{max}(l.mol^{-1}.cm^{-1}) 10^{-6}M	10^{-4}M	$\varepsilon_f/\varepsilon_i$	θ (degrees)
280 nm	Biochanin A	12330	25140	0.49	45
	Genistein	12750	7590	0.59	40

* Values taken on the spectra measured in acetonitrile (Figure 3-b, d).

3.3. Theoretical Modeling of Genistein Self Dimer

3.3.1. Method

To modelize the isoflavone self associated dimer, we have used the analytical atom- atom pair potential proposed previously by FRAGA [64-68] and improved by Sanchez- Marin [69-71]. In this pairwise potential the interaction energy ΔE between two molecules A and B, assumed to be rigid sets of individual atoms, is obtained as summation of atom i (of A) to atom j (of B) contributions (ΔE_{ij}) which have the form of an expansion in powers of $1/R_{ij}$ such as:

$$\Delta E = \sum_i^A \sum_j^B \Delta E_{ij}^{ab} \tag{1}$$

With

$$\Delta E_{ij}^{ab} = 1389.4168 \frac{q_i^a q_j^b}{R_{ij}} - 694.70838 \frac{f_i^a \alpha_i^a (q_j^b)^2 + f_j^b \alpha_j^b (q_i^a)^2}{R_{ij}^4}$$

$$- 1516.0732 \frac{f_i^a \alpha_i^a f_j^b \alpha_j^b}{\left\{(f_i^a \alpha_i^a / n_i^a)^{1/2} + (f_j^b \alpha_j^b / n_j^b)^{1/2}\right\} R_{ij}^6} + 4.184 \frac{c_i^a c_j^b}{R_{ij}^{12}} \quad (2\ (2)$$

Where q is the net atomic charge, *a* is the isotropic polarizability, n=Z-q is the effective number of electrons, and *f* and *c* are parameters that were adjusted to reproduce Clementi's potentials [72-74]. The superscripts *a* and *b* denote the classes to which belong the atoms *i* and *j*, respectively, taking into account their environments in the respective molecules. Parameters (q, α, *f*, *c*) for H, Li, C, N, O, Na, S, K, and Ca can be assigned to 82 classes reported in refs. [65-68]. R_{ij} is the separation between the two interacting atoms *i* and *j*. The interaction energy is given in kJmol^{-1} when α is in $Å^3$, *c* is in kcal$^{1/2}$Å^6 mol$^{-1/2}$, and R_{ij} is in Angstroms.

To improve the behavior of the potential at short distances, an empirical correction E_{ij}^D of the dispersion energy term is added to the potential like a second R^{-6} term [69-75].

$$\Delta E_{ij}^{D} = -1516.0732 \frac{\alpha_i^a \alpha_j^b}{\left\{(\alpha_i^a / n_i^a)^{1/2} + (\alpha_j^b / n_j^b)^{1/2}\right\} R_{ij}^6} \quad (3)$$

This additional term has the same form and parameters of the original R^{-6} one, but the scaling factors *f* are now set to one. The total pair potential incorporating the new correction is then:

$$\Delta E_{tot} = \sum_i^A \sum_j^B \Delta E_{ij}^{ab} + \sum_i^A \sum_j^B \Delta E_{ij}^{D} \quad (4)$$

The version of the program used in this work is that adopted by Torrens and co-workers [76]. The program allows for the determination of stable conformations of the self associated dimer of two molecules. These conformations are found by minimization of the interaction energy between the two molecules A and B. Both systems are kept rigid throughout the energy minimization process.

The relative position of the molecule B (which is free to translate and rotate) with respect to the fixed molecule A is given by a six fold vector whose three first components are displacements, expressed in Angstroms, along the X-, Y- and Z- axes, and the other three components are rotations expressed in degrees around the three axes taken in the same order. An extensive exploration of the interaction energy 6-D hypersurface has been performed from a standard set of 26 starting points provided as an option in the AMYR program [66-77]. The minimization procedure is stopped when all and each component of the six fold gradient vector has gone under an arbitrary threshold fixed at 10^{-5} units. The components of the 6x6 Hessian matrix are calculated, at each stationary point so reached, as described in ref [76]. The eigenvalues of the Hessian allow the characterization of critical points reached by the minimization. Classification of critical points is made by means of the index of each critical point,λ, which is the number of negative eigenvalues of the Hessian. For saddle point, λ is distinct to zero while λ must be equal to zero for a minimum.

3.3.2. Interacting genistein monomers geometry

Taking into account the previous experimental results, reflecting the high flexibility of the flavonoid relative to the twist around the C_3—$C_{1'}$ single bond, we have assumed that the phenyl ring can rotate easily when the two interacting monomers near each other. Nevertheless, during the minimization process as described above, the two interacting molecules are kept as rigid sets of atoms. So, the phenyl ring in the configuration introduced in the simulating program, for both interacting monomers A and B, was artificially rotated around the C_3—$C_{1'}$ single bond by 40 degrees from the completely twisted initial orientation given by the *ab initio* optimization reported in Figure 1. The 40 degrees torsion value was obtained previously from the evolution of the experimental absorption spectrum upon concentration (Figure 3-d and table 2). Hence this introduced configuration corresponds to the less twisted one in which the phenyl and the flavone rings lie at about 50 degrees (90° - 40°)

from each other. Figure 4 shows this 50 degrees configuration adopted for each interacting Genistein monomer.

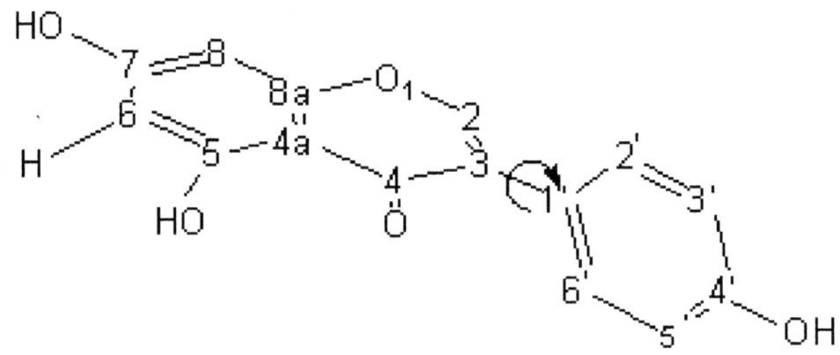

Figure 4. Genistein monomer intermediate, 50 degrees twisted, geometry

3.4. Calculation Results: Genistein Self Dimer

The arrangement of the stacked homo dimer is a slipped non sandwich, face to river, perpendicular structure as shown in Figure 5; the corresponding parameters are presented in table 3. In this rotated structure the two flavone rings are quasi parallel; the angle between their planes is of about 4 degrees and their long in-plane axes, crossing over the carbons C_7 and C_3, lie at about 90 degrees. The mean separation between the two flavones rings planes is 4.8 Å. This value exceeds those commonly reported for sandwich dimers such as furane-dione [78], benzene [68], porphyrin [79] and phthalocyanine [80] which are close to 3.4 - 3.8 Å. This slipped non sandwich structure may be explained by the sterical hindrance due to the phenyl rings; a similar result is reported in the case of diallylquinoxaline self dimer where the interplanar separation was fund between 3 and 5.8 Å [81]. Furthermore the dipole moment corresponding to this Genistein homo dimer is of about 1.42 Debye i.e. lower than that of the monomer which is found equal to 4.1 Debyes by *ab initio* calculation method. The low dipole moment value of the homo dimer is consistent with the observed non dependence of the highest energy absorption band around 260 nm on solvent polarity Figure 3.

The geometrical structure of the simulated dimer seems suitable because it reproduces satisfactory the absorption spectrum measured at high concentration 10^{-4}M Figure 3-d. The corresponding calculated ZINDO/S spectrum is shown in Figure 3-e. As observed experimentally, the main allowed transition of the dimeric specie lies at the high energy side around 260 nm. Indeed, for the monomer the oscillator strengths ratios $\frac{f(240nm)}{f(260nm)}$ and $\frac{f(240nm)}{f(290nm)}$ (deduced from the ZINDO/S spectrum Figure 2-c) are respectively 1.3 and 2.3. For the dimer, these ratios $\frac{f(230nm)}{f(250nm)}$ and $\frac{f(230nm)}{f(285nm)}$ (deduced from the ZINDO/S spectrum Figure 3-e) are respectively 2.1 and 4.6. This is in complete agreement with the variation of the measured extinction coefficients ratio $\frac{e_{\max}(260nm)}{e_{\max}(280nm)}$ (deduced from the spectra shown in Figure 3-d) which is 0.69 at 10^{-6} M and 1.6 at 10^{-4}M.

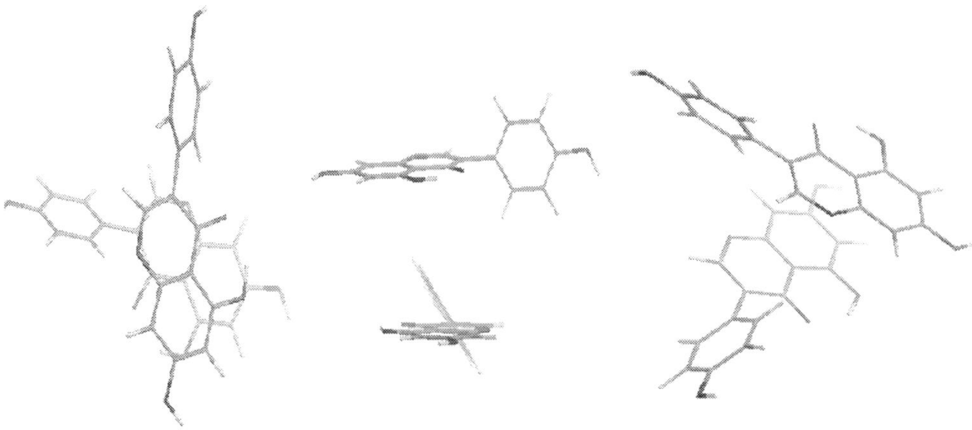

Figure 5. Structure of the most probable Genistein self dimer viewed perpendicularly to the flavone ring planes *(a)*, viewed from the long in-plane axis of the bottom interacting monomer *(b)* and viewed from an intermediate direction *(c)*

Table3. Calculated interaction energy, geometrical parameters and dipole moment of the Genistein simulated dimer

E_{tot} [a]	d [b]	α [c]	β [d]	μ [e]
-21.92	4.8	4	90	1. 41

[a] Total interaction energy (in kJ/mol).
[b] Mean separation distance between the two flavones rings planes in Å.
[c] Angle between the two flavones rings planes in degrees.
[d] Angle between the two C_3—C_7 long in-plane axes in degrees.
[e] Dipole moment in Debye.

4. Conclusion

In this work we have investigated the spectroscopic behavior of Genistein and Biochanin at the ground state. The two isoflavones show similar photophysics behaviors. At very low concentration, the monomer specie exists in the completely twisted conformation where the flavone and the phenyl rings are perpendicular. Nevertheless, in a protic medium, specific interactions with the solvent cage shift the isoflavone towards a less twisted configuration in which the flavone and the phenyl rings are not perpendicular. The experimental as well as the calculated ZINDO/S spectroscopic parameters reveal that all the absorption transitions are of $\pi\pi^*$ character. An eventual $n\pi^*$ character of the first transition is completely ruled out, in this work, because of the observed emitting character of this isoflavone.

With solute concentration rising, the monomeric species disappear to the profit of self dimeric species; the self association process seems also shifting the interacting monomers towards the less twisted form where the angle between the flavone and the phenyl rings planes is estimated to be close to 50 degrees.

The simulation of the ground state self dimer resulted in the stability of a slipped non sandwich, face to river, perpendicular structure, with about 4.8 Å as interplanar separation.

Furthermore, Van der waals interactions between the isoflavone monomers appear so important that free monomers cannot exist alone, in solution at room temperature, even at very low concentration. It is well-timed to underline that in the plasma genistein concentration can reach 4 to 16 microM after soy consummation [82], nevertheless it is difficult to state that in this medium genistein is selfassociated. Further spectroscopic investigations directly on the plasma after soy consummation are required.

Acknowledgments

The authors thank Doctor D. Desmaele, of the University Paris-Sud in France, for his advices and availability during the first submission of these results.

References

[1] Williams, CA; Harborne, JB. in: PM. Dey, Harborne, JB; Eds; Methods in Plant Biochemistry: Plant Phenolics, *Academic Press, London*, 1989, 1, 421-449.
[2] Lin, LZ; He, XG; Lindenmaier, M; Nolan, G; Yang, J; Cleary, M; Qiu, SX; Cordell, GA. *J. Chromatogr. A*., 2000, 876, 87-95.
[3] Beck, V; Rohr, U; Jungbauer, A. *J. Steroid Biochem. Mol. Biol*., 2005, 94, 499-518.
[4] Setchell, KD; Brown, NM; Desai, P; Zimmer-Nechemias, L; Wolfe BE; Brashear, WT; Kirschner, AS; Cassidy, A; Heubi, JE. *J. Nutr*. 2001, 131, 1362S-1375S.
[5] Wang, CC; Prasain, JK; Barnes, S. *J. Chromatogr. B*, 2002, 777, 3-28.
[6] Barnes, S. Proc. Soc. *Exp. Biol. Med*., 1998, 217(3), 386-392.
[7] Cheek, AO; Vonier, PM; Oberdorster, E; Burow, BC; McLachlan, JA. Environ. Health. *Perspect*. 1998, 106, 5-10.
[8] Dixon, RA; Ferreira, D. *Phytochem*., 2002, 60, 205- 211.
[9] Polkowski, K; Skierski, JS; Masurek, AP. Acta. *Pol. Pharm*., 2000, 57(3), 223-232.
[10] Raffoul, JJ; Wang, Y; Kucuk, O; Forman, JD; Sarkar, FH; Hillman, GG. BMC *Cancer*., 2006, 6, 107-117.
[11] Shao, ZM; Wu, J; Shen, ZZ; Barsky, SH. *Cancer. Res*., 58, 1998, 21, 4851-4857.
[12] Sun, L; Tran, N; Tang, F. *J. Med. Chem*., 1998, 41, 2588-2593.
[13] Pandey, VK; Tusi, S; Tusi, Z; Raghubir, R; Dixit, M; Joshi, MN; Bajpai, SK. *Indian J. Chem. B*., 2004, 43(1), 180-183.
[14] Baliani, A; Jimenez-Bueno, G; Stewart, ML; Yardley, V; Brun, R; Barrett, MP; Gilbert, IH. *J. Med. Chem*., 2005, 48, 5570-5579.
[15] Henke, BR; Consler, TG; Go, N; Hale, RL; Hohman, DR; Jones, SA; Lu, AT; Moore, LB; Moore, JT; Orband-Miller, LA; Robinett, RG; Shearin, J; Spearing, PK; Stewart, EL; Turnbull, PS; Weaver, SL; Williams, SP; Wisely, GB; Lambert, MH. *J. Med. Chem*., 2002, 45, 5492-5505.
[16] Basak, S; Pookot, D; Noonan, EJ; Dahiya, R. Mol. *Cancer. Ther*., 2008, 7, 3195-3202.
[17] Banerjee, S; Li, Y; Wang, Z; Sarkar, FH. *Cancer. Lett*., 2008, 269, 226-242.
[18] Davis, JN; Singh, B; Bhuiyan, M; Sarkar, FH. Nutr. *Cancer*., 1998, 32, 123-131.

[19] Lian, F; Bhuiyan, M; Li, YW; Wall, N; Kraut, M; Sarkar, FH. Nutr. *Cancer*., 1998, 31, 184-191.

[20] Li, Y; Upadhyay, S; Bhuiyan, M; Sarkar, FH. *Oncogene*., 1999, 18, 3166-3172.

[21] Alhasan, SA; Pietrasczkiwicz, H; Alonso, MD. Ensley, J; Sarkar, FH. Nutr. Cancer. 1999, 34, 12-19.

[22] Moiseeva, EP; Almeida, GM; Jones, GD; Manson, MM. Mol. *Cancer. Ther*., 2007, 6, 3071-3079.

[23] Raynal, NJM; Momparler, L; Charbonneau, M; Momparler, RL. *J. Nat. Prod*., 2008, 71(1), 3-7.

[24] Lin, YJ; Hou, YC; Lin, CH; Hsu, YA; Sheu, JJC; Lai, CH; Chen, BH; Lee Chao, PD; Wana, L; Tsai, F. *J. Biochem. Biophys. Res. Commun*., 2009, 378, 683-688.

[25] Ho, KJ; Liao, JK. Arterioscler. *Thromb. Vasc. Biol*., 2002, 22, 1952-1961.

[26] Andres, A; Donovanb, SM; Kuhlenschmidt, MS. *J. Nutr. Biochem*., 2009, 20, 563-569.

[27] Antignac, JP; Gaudin-Hirret, I; Naegeli, H; Cariou, R; Elliott, C; Le Bizec, *B. Anal. Chim. Acta*. 2009, 637, 55-63.

[28] Hertrampf, T; Gruca, MJ; Seibel, J; Laudenbach, U; Fritzemeier, KH; Diel, P. Bone. 2007, 40(6), 1529-1535.

[29] Wu, F; Khan, S; Wu, Q; Barhoumi, R; Burghardt, R; Safe, S. *J. Steroid. Biochem. Mol.* Biol. 2008, 110, 104-115.

[30] Arseneault, Y. *Québec pharm*., 2001, 48(6), 452-454.

[31] Adlercreutz, H; Mazur, W. *Ann. Med*., 1997, 29, 95-120.

[32] Tham, DM; Gardner, CD; Haskell, WL. *J. Clin. Endocrinol. Metab*., 1998, 83, 2223-2235.

[33] Brzezinski, A; Debi, A. Eur. *J. Obstet. Gyn. Reprod. Biol*., 1999, 85, 47-51.

[34] Barnes, S; Baillieres. Clin. Endocrinol. *Metab*., 1998, 12, 559-579.

[35] Stoll, BA. *Ann. Oncol*., 1997, 8, 223-225.

[36] Zava, DT; Duwe. G. Nutr. *Cancer*., 1997, 27, 31-40.

[37] Le Bail, JC; Champavier, Y; Chulia, A. *J. et al. Life. Sci*., 2000, 66, 1281-1291.

[38] Dunford, CL; Smith, GJ; Swinny, EE; Markham, KR. *Photochem. Photobiol. Sci*., 2003, 2, 611-615.

[39] Wolfbeis, OS; Leiner, M; Hockmuth, P; Geiger, H. Bunsen-Ges. Ber. Phys. Chem. 1984, 88, 759-767.

[40] Falkovskia, E; Sengupta, PK; Kasha, M. Chem. *Phys. Lett*., 1998, 297, 109-114.

[41] Lai, T; Lim, EC. Chem. *Phys. Lett*., 1979, 62, 507-510.

[42] Lim, EC. *Adv. Photochem*., 1997, 23, 165-211.

[43] Wassam, WA; Lim, EC. *J. Chem. Phys*., 1986, 69, 2175-2180.

[44] Zhao, J; Ren, F. *Spectrochim. Acta*., 2009, 72, 682-685.

[45] de Rijke, E; Joshi, HC; Sanderse, HR; Ariese, F; Brinkman, A. Th; Gooijer, C. Anal. Chim. *Acta*., 2002, 468, 3-11.

[46] Crupi, V; Ficarra, R; Guardo, M; Majolino, D; Stancanelli, R; Venuti, V. *J. Pharm. Biomed. Anal*., 2007, 44, 110-117.

[47] Bennetau-Pelissero, C; Arnal-Schnebelen, B; Lamothe, V; Sauvant, P; Sagne, JL; Verbruggen, MA; Mathey, J; Lavialle, O. *Food. Chem*., 2003, 82(4), 645-658.

[48] RHF/6-311+G (d) (Ground state restricted Hartree Fock method) *ab initio* method were run using the Gaussian 98® 98 w Version 5.2, Copyright ©1995-98 Gaussian, Inc., Pittsburgh, PA, 1998.

[49] Breton, M; Precigoux, G; Courseille, C; M. *Hospital. Acta. Cryst. B.*, 1975, 31, 921-923.
[50] ZINDO/S is a Semi Empirical Quantum Mechanical Program developed by M.C. Zerner at the University of Guelp, Ontario Canada, and at the University of Florida, Gainesville, Florida.
[51] Elber, R; Frank, AJ. *Almog, J. Forensic Sci.*, 2000, 45(4), 757-760.
[52] Clark, T. *Computational Chemistry: A Practical Guide to Chemical Structure and Energy Calculations*, J. Wiley and Sons, New York, 1985.
[53] Sadlej, J; Cooper, IL. Semi-Empirical Methods of Quantum Chemistry, Ellis Horwood Limited. Chichester, 1985.
[54] Moreau, C; Spirau, FS; Létard, JF; Lapouyade, R; Jonusauskas, G; Rullière, C. *J. Phys. Chem*. B. 1998, 102 (9), 1487-1497.
[55] Rasmussen, SC; Sattler, DJ; Mitchell, KA; Maxwell, JJ. *Lumin*., 2004, 109, 111-119.
[56] Rabinowitch, E; Epstein, LF. *J. Am. Chem. Soc*., 1941, 63, 69-78.
[57] Ameloot, M; Boens, N; Andriessen, R; Van Den Bergh, V; De Schryver, FC. *J. Phys. Chem*. 1991, 95, 2041- 2047.
[58] Andriessen, R; Boens, N; Ameloot, M; De Schryver, FC. *J. Phys. Chem*., 1991, 95, 2047-2058.
[59] Declercq, D; Delbeke, P; De Schryver, FC; Van Meervelt, L; Miller. RD. *J. Am. Chem. Soc*. 1993, 115(13), 5702-5708.
[60] Burdett, BC; Jones, EW; Gormally, J. Aggregation Processes in Solution, Elsevier, Amsterdam, 1983.
[61] Valdes Aguilera, O; Neckers, DC. *Acc. Chem. Res*., 1989, 22, 171-177.
[62] Grynkiewicz, G; Zegrocka-Stendel, O; Pucko, W; Ramza, J; Koscielecka, A; Kołodziejski, W; Wozniak, K. *J. Mol. Struct*., 2004, 694, 121-129.
[63] Jaffé, HH; Orchin, M; Theory and Applications of Ultraviolet Spectroscopy, *J. Wiley and Sons*, Inc. New York, 1962, 414.
[64] Fraga, S; Saxena, KMS; Torres, M. Studies in Physical and Theoretical Chemistry, Elsevier, Amesterdam, 1978.
[65] Fraga, S. *Comput. Chem*., 1982, 3, 329-334.
[66] Fraga, S. Comput. *Phys. Commun*., 1983, 29, 351-359.
[67] Nilar, SHM; Fraga, SJ. *J. Comput. Chem*., 1984, 5, 261-262.
[68] Sordo, JA; Fraga, S. *J. Comput. Chem*., 1986, 7, 55-57.
[69] Torrens, F; Sánchez-Marin, J; Orti, E; Nebot-Gil, I. *J. Chem .Soc., Perkin Trans*., 1987, 2, 943-950.
[70] Torrens, F; Sánchez de Meras, AM; Sánchez-Marin, J. *J. Mol. Struct*. (Theochem) 1988, 166, 135-140.
[71] Torrens, F; Sánchez-Marin, J; Tomas, F. *J. Chem. Res*., 1990, 176-177.
[72] Clementi, E. Lecture Notes in Chemistry, *Springer-Verlag, Berlin*, 1980.
[73] Clementi, E; Cavallone, F; Scordamaglia, R. *J. Am. Chem. Soc*., 1977, 99, 5531-5545, 5545-5550.
[74] Scordamaglia, R; Cavallone, F; Clementi, E. *J. Am. Chem. Soc*., 1977, 99(17), 5545-5550.
[75] Torrens, F; Orti, E; Sánchez-Marin, J. *J. Comput. Phys. Commun*., 1991, 66, 341-362.
[76] Rubio, M; Torrens, F; Sánchez-Marin, J. *J. Comput. Chem*., 1993, 14(6), 647-654.
[77] Fraga, S. *J. Comput. Chem*., 1982, 3, 329-334.

[78] Dkaki, M; Ait Lyazidi, S; Haddad, M; Hnach, M; Cazeau-Dubroca, C; Aycard, JP. *J. Phys. Chem. A.*, 1998, 102(27), 5275-5279.

[79] Hunter, CA; Sanders, JKM. *J. Am. Chem. Soc.* 1990, 112, 5525-5534.

[80] Torrens, F; Orti, E; Sanchez-Marin, J. In: RM; Metzger, P; Day, GC. Papavassiliou (Eds.), *Lower-Dimensional Systems and Molecular Electronics*, Plenum Press, New York, 1991; (NATO-ASI Ser. B No. 248), 461.

[81] Azzouzi, F; Ait Lyazidi, S; Haddad, M; Lamotte, M; Essassi, E. M; Hnach, M; Zenkouar, M. *Spectrochim. Acta.*, 2007, 67, 1263-1269.

[82] Takimoto, CH; Glover, K; Huang, X; Hayes, SA; Gallot, L; Quinn, M; Jovanovic, BD; Shapiro, A; Hernandez, L; Goetz, A; Llorens, V; Lieberman, R; Crowell, JA. Poisson, B. A. and Bergan, R. C. Cancer Epidemiol. *Biomarkers Prev.*, 2003, 12, 1213-1221.

In: Isoflavones Biosynthesis, Occurence and Health Effects ISBN: 978-1-61728-113-6
Editor: Melanie J. Thompson, pp. 271-284 © 2010 Nova Science Publishers, Inc.

Chapter XI

Soy Isoflavone Genistein in Cancer Chemoprevention

M.F. Ullah, H. Zubair, H.Y. Khan and S.M. Hadi*
Department of Biochemistry, Faculty of Life Sciences, AMU, Aligarh-202002 (U.P.), India

Abstract

Soy isoflavones have been identified as dietary components having an important role in reducing the incidence of breast and prostate cancers in Asian countries. The isoflavone polyphenol genistein in soybean is considered to be a potent chemopreventive agent against cancer. Several studies, both *in vitro* and *in vivo*, have clearly shown that genistein is a promising therapeutic candidate for cancer chemoprevention as it is able to cause inhibition of human cancer growth via pathways that lead to apoptotic cell death. Genistein has pleiotropic molecular mechanisms of action affecting estrogen mediated signaling pathways, cell cycle regulation, apoptosis and cell growth signal cascades. Of particular interest is the observation that genistein has been found to induce apoptosis in cancer cell lines but not in normal cells. We have proposed a novel mechanism for the cytotoxic action of plant polyphenols including genistein, which has implications for its chemopreventive properties against cancer. We believe that such a mechanism better explains the preferential cytotoxicity of genistein towards cancer cells. In this review we discuss various mechanistic approaches that have been proposed to account for the chemopreventive properties of genistein.

Introduction

Cancer development is a dynamic and long-term process that involves many complex factors through critical steps of initiation, promotion, and progression, leading to an uncontrolled growth

* Corresponding author: E-mail: smhadi@vsnl.com, Tel: +91-571-2700741, Mobile: +919837266761

of cancerous cells throughout the body. Cancer is a growing health problem around the world particularly with the steady rise in life expectancy. It has been estimated that more than two-thirds of human cancers could be prevented through appropriate lifestyle modification including dietary habits [1].Cancer chemoprevention can be defined as the use of specific diets, or natural or synthetic chemicals, to reverse, suppress or prevent carcinogenesis. The concept that cancer can be prevented, or its onset postponed, by certain diet derived substances has in recent years attracted considerable interest. In this regard dietary agents interfering with tumor development are of potential clinical value. Epidemiological studies have indicated that populations with high isoflavone intake through soy consumption have lower incidence of breast, prostate, and colon cancer [2-4]. Soybeans are a rich and relatively unique source of isoflavonoid genistein (4',5,7-Trihydroxyisoflavone) (Figure 1), which is considered to be the major constituent responsible for the chemopreventive efficacy of soy [5]. Glycoside conjugates of genistein account for at least two-third of the total content of isoflavones in soybean, with conjugates of daidzein and other isoflavonoids comprising the remainder [6]. Many mechanisms of action have been proposed for the anticancer activity of genistein including antioxidant effect, induction of cell cycle arrest and apoptosis, interference in signal transduction pathways and induction of prooxidant DNA breakage. Possibly a combination of these mechanisms may be responsible for cancer chemoprevention by genistein. However, precise assessment of the mechanisms by which genistein prevents cancer is necessary before it can be developed as a targeted therapeutic molecule.

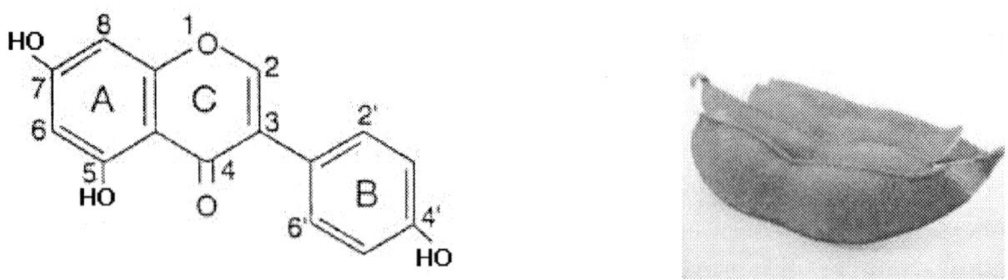

(a) (b)

Figure 1. Chemical structure of genistein (a) and its most common natural source- soybeans (b)

Soy Consumption and Cancer Incidence: An Epidemiological Correlation

Isoflavones are interesting compounds in the cancer prevention domain since epidemiological evidence suggests that high soy diets may reduce the incidence of cancers, primarily breast and prostate cancer [7]. The most compelling evidence for chemoprevention comes from studies of Asian populations that consume traditional diets rich in plant foods, particularly soy products. Compared to western countries, the incidence and mortality from breast and prostate cancer have historically been lower in Asian countries. Although genetic and other environmental factors may also contribute to these observed differences, dietary habits have been thought to account in part for the reduced cancer incidence. A cross-national study involving 50 countries identified soy products as functional foods with substantial protective effects against prostate cancer [8]. It has been reported that Asian women consuming relatively large amounts of soy-derived foods have a low incidence of breast cancer [9,10]. Asian women usually have higher urinary and plasma levels of phytoestrogens such as genistein than women in western populations [7]. Furthermore, documented evidence of increased breast cancer risk among Asian immigrants to the United States and second-generation descendants has been attributed in part to their intake of increasingly western style diets [11,12]. Moreover urinary levels of soy-derived isoflavones including genistein were found to be lower in breast cancer patients compared with case-controls [13,14]. In humans, plasma or serum levels of genistein from soy food ingestion range from less than 1 μM to about 5 μM [15]. Several studies have shown that physiologically relevant concentrations of genistein (2.5-5μM) as present under *in vivo* conditions are associated with cell proliferation in breast cancer cells [16,17]. However, these studies were based on short term exposure of cells to low doses of genistein without resupplimentation. An important consideration of such studies is whether the exposure time exceeds the window of bioavailability of the compound under investigation. Studies by King et al [18] demonstrated the effects of genistein on breast tumor cell growth in culture and observed apparent enhanced growth of breast cancer cells BT-474, MCF- 7 and MDA-MB-468 with the MTS assay after a 72 h treatment with a low, dietarily relevant genistein (3.125 μM) dose. However, this short-term enhancement of growth was not sustained over the long-term, as shown in 7-day growth curve assays with resupplementation of the genistein at 48 h intervals which showed cumulative dose dependent inhibition of MCF-7 and MDA-MB-468 cell growth of 30–50% over the 7 days. Similarly, cell growth inhibition was shown by the low genistein dose of 5 μM over a 72 h growth curve in TRAMP-C2 mouse prostate model cells [19].

Genistein as an Estrogen Mimic: Biphasic Action

The polyphenolic structure of genistein is similar to endogenous estrogen. Genistein shares structural features with the estrogen estradiol-17β, particularly the phenolic ring and the distance (11.5 A) between its 4'- and 7- hydroxyl groups (Figure 2). These features confer ability to bind estrogen receptors and sex hormone binding proteins. Thus genistein can exert both estrogenic and antiestrogenic activity, the latter by competing for receptor binding by estradiol. Several studies have focused on the antiestrogenic activity of genistein as a plausible chemopreventive mechanism of this phytochemical. Interference at the level of estrogen receptor has been suggested to contribute to genistein induced inhibition of tumor-promoting effects exerted by estrogens [20,21]. Genistein was found to inhibit the growth of both hormone-responsive and nonresposive cancer cells in vitro, with an IC50 of 10–50 μmol/l [22-27]. However, at concentrations lower than 10 μmol/l of genistein, the growth of MCF-7 cells, that are estrogen receptor positive, was stimulated [28]. Such a biphasic effect of genistein on cell growth or proliferation has been attributed to the differential response provoked by different doses of the compound: genistein exerts an estrogen-like activity at lower concentrations, but at higher levels, it may act as an estrogen-antagonist. Structural similarities have also been noted between genistein and tamoxifen, a synthetic anti-estrogen that has been clinically tested as a chemopreventive agent in women with high risk of breast cancer. Further, at relatively higher concentrations other mechanisms of genistein might also play an inhibitory role against cancer. For example, an estrogen binding experiment using chimeric proteins of estrogen receptor-alpha and -beta binding domains showed that genistein as an estrogen receptor ligand stimulated the proliferation of alpha receptor-dependent growth of breast cancer cell lines. However the cytotoxicity induced by higher concentrations of genistein was estrogen receptor-independent, suggesting the involvement of alternative mechanisms of action [29].

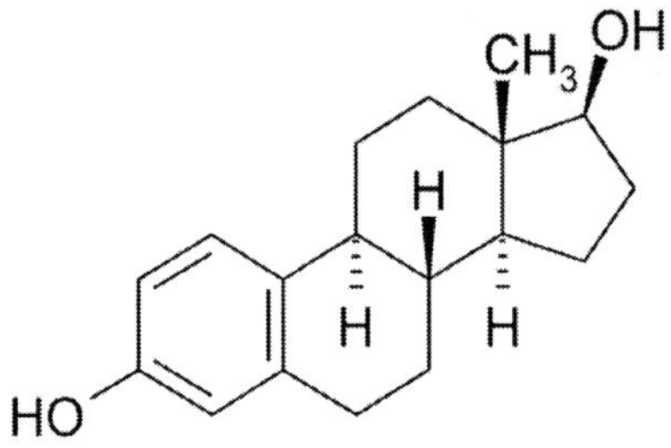

Figure 2. Chemical structure of estrogen estradiol

Induction of Cell Cycle Arrest by Genistein

Cell cycle progression is known to be tightly regulated by different cyclins, cyclin dependent kinases (CDKs) and cyclin dependent kinase inhibitors (CDKIs) in different phases of the cell cycle. Cyclins represent an important family of positive cell cycle regulators that have been reported to be over expressed in cancer cells. The inhibitors of cyclin-dependent kinases such as $p21^{WAF1/CIP1}$, $p27^{KIP1}$, $p16^{INK4A}$, and $p15^{INK4B}$ which are the negative cell cycle regulators are under expressed and are considered to be the target molecules for overexpression by chemopreventive agents. The isoflavone genistein has been shown to exert an inhibitory effect on cell proliferation in both estrogen receptor-positive and -negative human breast cancer cells. The antiproliferative effect of genistein was attributed to G2/M arrest and accompanied by increased $p21^{WAF/CIP1}$ expression [30,31]. Such genistein induced G2/M phase arrest was also observed in prostate [32], gastric [33], lung [34] and melanoma cancer cells [35]. Genistein also decreased the protein level of cyclin B1 with concomitant induction of p21 [36]. In another study genistein induced apoptosis was correlated with G2/M arrest, up-regulation of p21and down-regulation of cyclin B in prostate carcinoma cells [32]. It is generally accepted that genistein can cause G2/M cell cycle arrest. However, reports have also indicated cell cycle arrest by genistein at other transition phases such as G0/G1 [37] and G1/S block [38].

Induction of Programmed Cell Death (Apoptosis) by Genistein

Programmed cell death or apoptosis is a systems approach to get itself rid of defective cells in order to prevent the defect from invading the normal cells. However cancer cells have evolved mechanisms to evade inherent apoptotic signals of the individuals own immune system. Thus chemically synthesized or natural compounds capable of inducing apoptosis in cancer cells are potential chemopreventive agents. The Bcl-2 family is the best characterized component of apoptotic signaling pathways and involves two functional groups including antiapoptotic proteins (Bcl- XL and Bcl-2) and proapoptotic proteins (Bax, Bak and Bad). It is the fine balance of these two functional protein groups that provide the signal for cell survival or cell death which is transmitted to the downstream effector molecules such as intracellular caspases. Moiseeva et al., [39] reported that physiological concentrations of dietary phytochemicals including genistein results in reduced growth and induction of apoptosis in cancer cells. Genistein could induce apoptosis in a number of cancer cells of varied origin including breast cancer cells; prostate cancer cells, non-small cell lung cancer cells,head and neck squamous carcinoma cells, pancreatic cancer cells and leukemia [32,34,39-43]. Sarkar and coworkers [35] have suggested that an up-regulation of Bax and downregulation of Bcl-2 may be one of the molecular mechanism(s) by which genistein induces apoptosis. Studies from their laboratory showed a reduced level in Bcl-2 protein expression in all cancer cells tested after treatment with genistein for 48 h and longer. The expression of Bax was significantly up-regulated in all cancer cells tested after genistein treatment for ≥24 h. The ratio of Bax to Bcl-2 was more significantly increased after 24 h of treatment, corresponding with a significant increase in apoptotic cells after 48 h of genistein treatment. Genistein has also been suggested to induce apoptosis by inhibiting DNA topoisomerase II, thereby preventing ligation of DNA double strand breaks [44,45]. These observations suggest that genistein function as an inducer of apoptotic processes in cancer cells.

Genistein Modulates Cell Growth Signaling Cascades

Protein tyrosine kinases, particularly receptor tyrosine kinases and intracellular tyrosine kinases, participate in a variety of different cellular activities including cell proliferation and differentiation. Increased activity of tyrosine kinases has been linked to malignancy, suggesting the possible potential of tyrosine kinase inhibitors in cancer prevention. Genistein was one of the first tyrosine kinase inhibitors identified [46] and it is thought that this may be one of the primary mechanisms of anticancer effect of genistein. Genistein reduced the incidence of adenocarcinomas in a dose-dependent manner and down-regulated androgen receptor, estrogen receptor-α, progesterone receptor, EGF receptor, insulin-like growth factor-1, and ERK-1 [47]. Such a suppression of carcinogenesis seemed to be mediated through inhibition of phosphorylation of tyrosine protein kinase-dependent growth factor receptors like EGF and PDGF receptor. One of the key signaling cascades involving the regulation of transcription of genes that direct the cells to proliferate is NF-kB pathway which controls the cell growth, differentiation, apoptosis and stress–responses. Under non-stimulating conditions, NF-kB is sequestered in the cytoplasm through tight association with the impeding IkB proteins. Following stimulation, IkB protein is phosphorylated and degraded, allowing the NF-kB to translocate to the nucleus, bind to the NF-kB-specific DNA-binding sites or interact with other transcription factors,and thus regulate gene transcription. The isoflavone genistein has been shown to impede the interation of NF-kB with DNA by binding to it and preventing its translocation to nucleus. It has been demonstrated that genistein treatment could modulate NF-kB DNA binding activity in prostate, breast, head and neck, and pancreatic cancer cells and leukemic cells [40,48-50]. Another signaling pathway that plays a critical role in regulating the balance between cell survival and apoptosis is Akt/PKB signaling cascade [51]. Akt has also been associated with the regulation of NF-kB pathway via phosphorylation and activation of molecules in the NF-kB pathway [52]. Thus, strategies to block the activity of Akt would ideally lead to the inhibition of proliferation and the induction of apoptosis. Indeed it has been shown that genistein inhibits the activation of Akt, which may result in the inhibition of cell survival threshold that lead to the induction of apoptotic signals [42].

A Prooxidant Mechanism of Genistein in Chemoprevention of Cancer

Most of the pharmacological properties of plant polyphenols are considered to reflect their ability to scavenge endogenously generated oxygen radicals or those free radicals formed by various xenobiotics, radiation etc. However, some data in the literature suggests that the antioxidant properties of the polyphenolic compounds may not fully account for their chemopreventive effects [53,54]. Of particular interest is the observation that several plant polyphenols including genistein have been found to induce apoptosis in cancer cell lines but not in normal cells [55-57]. On the basis of the studies carried out in our laboratory we have proposed a novel prooxidant action mechanism for the cytotoxic action of isoflavone genistein, which has implications for its chemopreventive properties against cancer. Using a system of whole and permeabilized lymphocytes we have demonstrated that both genistein and its structural analogue isoflavone biochanin A are able to cause strand breaks in cellular DNA [58]. It is to be noted that permeabilized cells contain only cell organelles attached to the residual cytoskeleton thus allowing a direct interaction of isoflavones with cell nuclei. This direct interaction results in considerably greater DNA degradation in case of permeabilized cells. The molecular basis for such DNA degradation has been suggested to be the mobilization of chromatin bound copper and its redox cycling leading to the generation of reactive oxygen species (ROS) such as hydroxyl radicals which act as proximal DNA cleaving agents. The efficiency of genistein to cause cellular DNA breakage was relatively higher than biochanin A (Figure 3) which also correlated with the greater Cu (II) reduction observed in the case of genistein (Figure 4). Such cellular DNA breakage induced by isoflavones was inhibited by copper chelator (neocuproine) in whole lymphocytes but not by bathocuproine (the water soluble membrane impermeable analogue of neocuproine) and molecules that specifically bind iron (desferrioxamine mesylate) or zinc (histidine). However since nuclear pore complex is permeable to small molecules both copper chelators neocuproine and bathocuproine were able to inhibit the DNA breakage in permeabilized cells. Further, desferrioxamine mesylate and histidine were still found to be ineffective (Table 1). Scavengers of reactive oxygen species (ROS) such as superoxide dismutase, catalase and thiourea were also able to inhibit to significant degrees the isoflavone induced DNA breakage (Table 2).

It is generally understood that antioxidants counteract ROS production and inhibit the latter-induced oxidative DNA damage and therefore reduce the risk of cancer. On the other hand, growing experimental evidence suggests that polyphenolic antioxidant-mediated production of ROS (prooxidant action) may be responsible for their ability to induce apoptosis of cancer cells [59,60]. Compared with normal cells, preneoplastic cells and neoplastic cells have been shown to contain elevated levels of copper [61] and may be more sensitive to electron transfer with polyphenolic antioxidants to generate ROS. Therefore DNA damage induced by polyphenolic antioxidants in the presence of Cu(II) may be an important pathway through which preneoplatic cells and neoplastic cells can be killed while normal cells survived (Figure 5) [59]

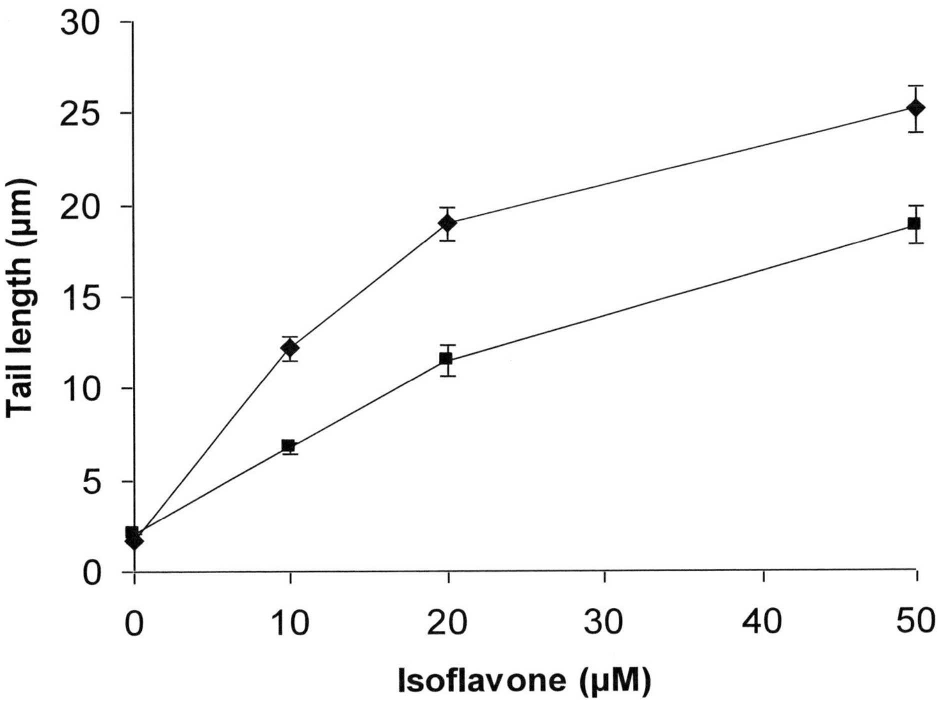

(a)

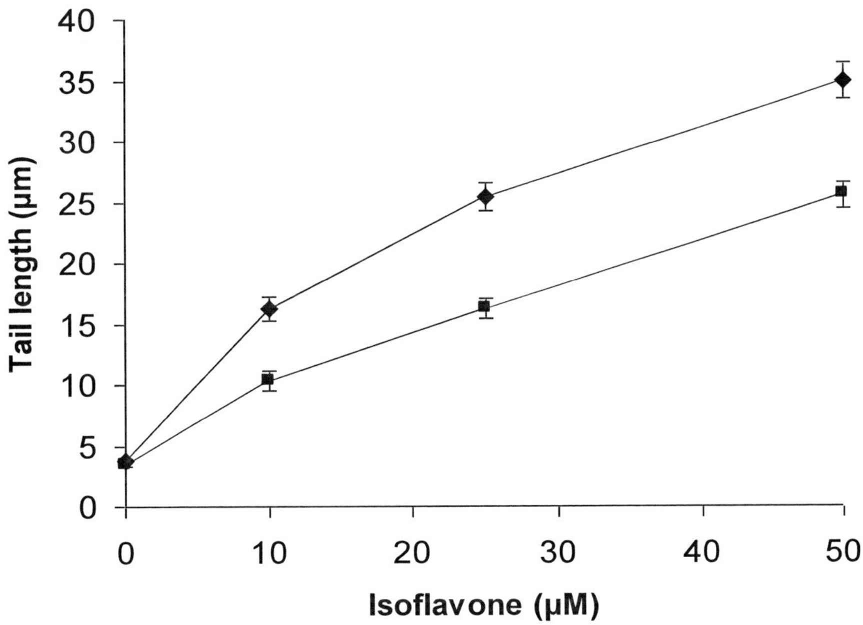

(b)

Figure 3. A comparison of DNA breakage induced by genistein (♦) and biochanin A (■) in (a) Intact lymphocytes and (b) Permeabilized lymphocytes as a function of comet tail lengths. Values reported are ±SEM of three independent experiments. P< 0.01 by comparison with control (in the absence of isoflavone) [58]

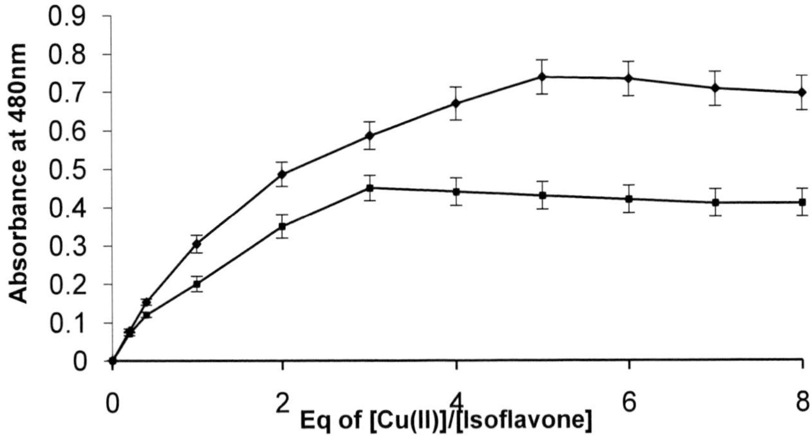

Figure 4. Detection of stoichiomentry of isoflavone-Cu(II) interaction. The concentration of genistein (♦) and biochanin A (■) were 25 μM in the presence of 0.4mM bathocuproine. The absorbance of samples at 480 nm is plotted vs equivalents of Cu(II) per molar equivalent of compounds. All points represent triplicate samples and the mean values are plotted [58]

Table 1. Effect of various metal chelators on DNA degradation in lymphocytes induced by isoflavones. The values shown represent isoflavone induced DNA breakage in whole lymphocytes and permeabilized lymphocytes in the presence of metal chelators measured as a percentage of the control (DNA breakage induced by isoflavone in the absence of any chelator) [58]

	Whole lymphocytes		Permeabilized lymphocytes	
Dose	Tail length (μm)	% of Control	Tail length (μm)	% of Control
Genistein (50μM)	26.32 ±2.13[a]	-	35.76 ±2.67[b]	-
+ Neocuproine (100μM)	17.16 ± 0.78	65.19 ± 4.53	24.84 ± 1.29	69.49 ± 5.21
+ Bathocuproine (100μM)	26.04 ± 1.80	98.93 ± 6.85	20.42 ± 0.69	57.12 ± 3.41
+ Histidine (100μM)	25.59 ± 1.63	97.26 ±6. 37	36.20 ± 2.81	101.25±7.79
+ Desferrioxamine mesylate (100μM)	26.70 ± 2.01	101.48±7.28	35.24 ± 2.30	98.57 ± 6.53
Biochanin A (50μM)	20.76 ±1.82[a]	-	26.94 2.21[b]	-
+ Neocuproine (100μM)	09.34 ± 0.27	44.99 ± 2.93	15.24 ± 0.56	56.57 ± 3.68
+ Bathocuproine (100μM)	20.80 ± 1.52	100.24±7.32	10.58 ± 0.32	39.30 ± 2.98
+ Histidine (100μM)	21.53 ± 2.10	103.75±9.77	26.17 ± 2.00	97.17 ± 7.65
+ Desferrioxamine mesylate (100μM)	20.65 ± 1.72	99.51 ± 8.37	26.46 ± 2.20	98.21 ± 8.32

[a] $p<0.05$ as compared with untreated whole lymphocytes (Tail length = 2.30±0.13)
[b] $p<0.05$ as compared with untreated permeabilized lymphocytes (Tail length =2.72±0.17)
Data represent ± S.E.M of three independent experiments

Table 2. Effect of scavengers of reactive oxygen species on isoflavone induced DNA breakage in permeabilized lymphocytes [58]

Dose	Tail length (μm)	% Inhibition
Untreated	02.53 ± 0.17	
Genistein (50μM)	35.24 ± 2.82#	
+ Catalase (100μg/ml)	24.39 ± 2.14*	30.72%
+ SOD (100μg/ml)	20.62 ± 1.65*	41.48%
+Thiourea (1mM)	14.67 ± 1.18*	58.37%
Biochanin A (50μM)	27.19 ± 2.31#	
+ Catalase (100μg/ml)	17.05 ± 1.23*	37.29%
+ SOD (100μg/ml)	15.28 ± 1.17*	43.80%
+ Thiourea (1mM)	10.12 ± 1.01*	62.78%

*$p<0.05$ as compared with control (#). Data represent ± S.E.M of three independent experiments

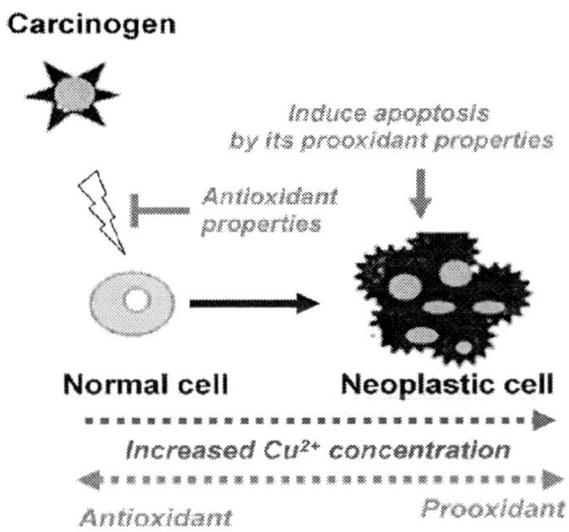

Figure 5. Possible mechanism for chemopreventive effects of phenolic antioxidant [59]

Conclusion

Cancer chemoprevention with dietary constituents has attracted considerable interest in recent years on account of epidemiological data in support of observations that the consumption of fruits and vegetables has an inverse effect on cancer induction. As discussed above, the soy isoflavone genistein has been shown to posses a number of anticancer properties each affecting a different biochemical pathway through alternative mechanisms of action. However an ideal therapeutic strategy would require differential action of drug on cancer cells sparing the normal cells from its effects. This seems to be possible with the identification and targeting of factors specifically altered in cancer cells. For example, polyphenols including isoflavone genistein at physiologically relevant concentrations are cytotoxic to cancer cells but not to normal cells. We propose that it is the elevated levels of copper in cancer cells which make them susceptible to the prooxidant action of polyphenols. Based on their ability to mobilize elevated copper levels in cancer cells plant polyphenols may act as lead compounds for the synthesis and development of novel anticancer drugs.

Further as a chemopreventive strategy a sustained intake of genistein in the form of soy foods may maintain the required physiologically relevant intracellular concentrations cytotoxic to neoplastic cells.

Acknowledgment

The authors acknowledge the financial assistance provided by the University Grants Commission, New Delhi, under the DRS-II program.

References

[1] Doll, R; Peto, R. The causes of cancer: quantitative estimates of avoidable risks of cancer in the United States today. *J. Natl Cancer Inst.,* 1981, 66, 1191-1308.

[2] Adlercreutz, H. Phytoestrogens: epidemiology and possible role in cancer prevention. *Environ. Health Perspect.* 1995, 103, 103-112.

[3] Adlercreutz, H; Goldin, BR; Gorbach, SL; Hockerstedt, K. *et al.* Soybean phytoestrogens intake and cancer risk. *J. Nutr.,* 1995, 125, 757S-770S.

[4] Park, OJ; Sur, YJ. Chemopreventive potential of epigallocatechin gallate and genistein: evidence from epidemiological and laboratory studies. *Toxicol. Lett.,* 2004, 150, 43-56.

[5] Messina, MJ; Persky, V; Detchell, KD; Barnes, S. Soy intake and cancer risk: a review of the *in vitro* and *in vivo* data. *Nutr Cancer,* 1994, 21, 113-131

[6] Coward, L; Barnes, NC; Stechell, KDR; Barnes, S. Genistein and Daidzein and their beta-glycosides conjugates, antitumor isoflavones in soybean foods of American and Asian diets. *J Agric Food Sci.,* 1993, 41, 1961-1967.

[7] Adlercreutz, H. Phytoestrogens and cancer. *Lancet Oncol.*, 2002, 3, 364-373.

[8] Herbert, J; Hurley, T; Olendzki, B; Teas, J; Yunsheng, M; *et al.* Nutritional and socioeconomic factors in relation to prostate cancer mortality: a cross-national study. *J. Nat. Cancer Inst.,* 1998, 90, 1637-1647.

[9] Adlercreutz, H; Honjo, H; Higashi, A; Fotsis, T; Hamalainen, E; Hasegawa, T; Okada, H. Urinary excretion of lignins and isoflavonoid phytoestrogens in Japanese men and women consuming a traditional diet. *Am. J. Clin. Nutr.,* 1991, 54, 1093-1100.

[10] Lee, HP; Gourley, L; Duffy, SW; Esteve, J; Lee, J; Day, NE. Dietary effects on breast cancer risk in Singapore. *Lancet,* 1991, 336, 1197-1200.

[11] Ziegler, RG; Hoover, RN; Pike, MC; Hildesheim, A; Nomura, AM; West, DW; Wu-Williams, AH; Kolonel, LN; Horn-Ross, PL; Rosenthal, JF; Hyer, MB. Migration patterns and breast cancer risk in Asian- American women. *J. Natl. Cancer Inst.,* 1993, 85, 1819-1827.

[12] Wu, HJ; Chan, WH. Genistein protects methylglyoxal-induced oxidative damage and cell injury in human mononuclear cells. *Toxicol. In vitro,* 2007, 21(3), 335-342.

[13] Ingram, D; Sanders, K; Kolybaba, M; Lopez, D. Casecontrol study of phyto-oestrogens and breast cancer. *Lancet,* 1997, 350, 990-994.

[14] Zheng, W; Dai, Q; Custer, L; Shu, XO; Wen, WQ; Jin, F; Franke, AA. Urinary excretion of isoflavonoids and the risk of breast cancer. *Cancer Epidemiol. Biomark. Prev.,* 1999, 8, 35-40.
[15] Wiseman, H; Casey, K; Bowey, EA; Duffy, R; Davies, M; Rowland, IR; Lloyd, AS; Murray, A; Thompson, R; Clarke, DB. Influence of 10 wk soy consumption on plasma concentrations and excretion of isoflavonoids and on gut microflora in healthy adults., *Am. J. Clin. Nutr*., 2004, 8, 692-699.
[16] Young, HJ; Clinton, DA; Kimberly, FA; Kimberly, LK; Daniel, RD; William, GH. Physiological Concentrations of Dietary Genistein Dose-Dependently Stimulate Growth of Estrogen-Dependent Human Breast Cancer (MCF-7) Tumors Implanted in Athymic Nude Mice. *J. Nutr*., 2001, 131, 2957-2962.
[17] Sathyamoorthy, N; Gilsdorf, JS; Wang, TTY. Differential effect of genistein on transforming growth factor beta-1 expression in normal and malignant mammary epithelial cells. *Anticancer Res*., 1998, 18, 2449-54.
[18] King, AA; Leszczynska, J; Hickey, CA; Klein, CB. Prevention of spontaneous and X-ray induced genomic instability in breast cancer cells by the dietary antimutagens genistein and lycopene. *Environ. Mol. Mutagen*., 2006, 47, 45 (Abs.).
[19] Touny, El; Banerjee, LH. Identification of both Myt-1 and Wee-1 as necessary mediators of the p21-independent inactivation of the Cdc-2/ CyclinB1 complex and growth inhibition of TRAMP-C2 cancer cells by genistein. *Prostate,* 2006, 66, 1542-1555.
[20] Martin, PM; Horwitz, KB; Ryan, DS; McGuire, WL. Phytoestrogen interaction with estrogen receptros in human breast cancer cells. *Endocrinology,* 1978, 103, 1860-1867.
[21] Shutt, DA; Cox, RI. Steroid and phytoestrogen binding to sheep uterine receptors in vitro. *J. Endocrinol.,* 1972, 52, 299-310.
[22] Hsieh, CY; Santell, RC; Haslam, SZ; Helferich, WG. Estrogenic effects of genistein on the growth of estrogen receptor-positive human breast cancer (MCF-7) cells in vitro and in vivo. *Cancer Res.,* 1998, 58, 3833-3838.
[23] Peterson, G; Barnes, S. Genistein inhibition of the growth of human breast cancer cells: independence from estrogen receptors and the multi-drug resistance gene. *Biochem. Biophys. Res. Commun.,* 1991, 19, 661-667.
[24] Pagliacci, MC; Smacchia, M; Migliorati, G; Grignani, F; Riccardi, C; Nicoletti, I. Growth-inhibitory effects of the natural phyto-oestrogen genistein in MCF-7 human breast cancer cells. *Eur. J. Cancer,* 1994, 11, 1675-1682.
[25] Peterson, G; Barnes, S. Genistein inhibits both estrogen and growth factor-stimulated proliferation of human breast cancer cells. *Cell Growth Differ.,* 1996, 7, 1345-1351.
[26] Shao, ZM; Shen, ZZ; Fontana, JA; Barsky, SH. Genistine's "ER dependent and independent" actions are mediated through ER pathways in ER-positive breast carcinoma cell lines. *Anticancer Res.,* 2000, 20, 2409-2416.
[27] Twaddle, GM; Turbov, J; Liu, N; Murthy, S. Tyrosine kinase inhibitors as antiproliferative agents against an estrogen-dependent breast cancer cell line in vitro. *J. Surg. Oncol.,* 1999, 70, 83-90.
[28] Zava, DT; Duwe, G. Esrogenic and antiproliferative properties of genistein and other flavonoids in human breast cancer cells in vitro. *Nutr Cancer,* 1997, 27, 31-40.
[29] Maggiolini, M; Bonofiglio, D; Marsico, S; Panno, ML; Cenni, B; Picard, D; Ando, S. Estrogen receptor alpha mediates the proliferative but not the cytotoxic dose-dependent

effects of two major phytoestrogens on human breast cancer cells. *Mol. Pharmacol.,* 2001, 60, 595-602.

[30] Shao, A; Alpaugh, ML; Fontana, JA; Barsky, SH. Geinstein inhibits proliferation similarly in estrogen receptorpositive and negative human breast carcinoma cell lines characterized by p21WAF1/CIP1 induction, G2/M arrest, and apoptosis. *J. Cell. Biochem.,* 1998, 69, 44-54.

[31] Shao, ZM; Wu, J; Shen, ZZ; Barsky, SH. Genistein exerts multiple suppressive effects on human breast carcinoma cell. *Cancer Res.,* 1998, 58, 4851-4857.

[32] Davis, JN; Singh, B; Bhuiyan, M; Sarkar, FH. Genistein-induced upregulation of p21WAF1, downregulation of cyclin B1 and induction of apoptosis in prostate cancer cells. *Nutr. Cancer,* 1998, 32, 123-131.

[33] Matsukawa, Y; Marui, N; Sakai, T; Satomi, Y; Yoshida, M; Matsumoto, K; Nishino, H; Aoike, A. Geinstein arrests cell cycle progression in G2/M. *Cancer Res.,* 1993, 53, 1328-1331

[34] Lian, F; Bhuiyan, M; Li, Y; Wall, N; Kraut, M; Sarkar, FH. Geinstein-induced G2-M arrest, p21WAF1 upregulation, and apoptosis in a non-small-cell lung cancer cell line. *Nutr. Cancer,* 1998, 31, 184-191.

[35] Casagrande, F; Darbon, JM. p21CIP1 is dispensable for the G2 arrest caused by genistein in human melanoma cells. *Exp. Cell Res.,* 2000, 258, 101-108.

[36] Choi, YH; Zhang, L; Lee, WH; Park, K. Genisteininduced G2/M arrest is associated with the inhibition of cyclin B1 and the induction of p21 in human breast carcinoma cells. *Int. J. Oncol.,* 1998, 13, 391-396.

[37] Kuzumaki, T; Kobayashi, T; Ishikawa, K. Genistein induces 21(Cip1/WAF1) expression and blocks the G1 to S phase transition in mouse fibroblast and melanoma cells. *Biochem. Biophys. Res. Commun.,* 1998, 251, 291-295.

[38] Shen, J; Klein, RD; Wei, Q; Guan, Y; Contois, JH; Wang, TTY; Chang, S; Hursting, SD. Low-dose genistein induces cyclin-dependent kinase inhibitors and G1 cell-cycle arrest in human prostate cancer cells. *Mol. Carcinog.,* 2000, 29, 92-102.

[39] Moiseeva, EP; Almeida, GM; Jones, GD; Manson, MM. Extended treatment with physiologic concentrations of dietary phytochemicals results in altered gene expression, reduced growth, and apoptosis of cancer cells. *Mol. Cancer Ther.,* 2007, 6, 3071-3079.

[40] Li, Y; Upadhyay, S; Bhuiyan, M; Sarkar, FH. Induction of apoptosis in breast cancer cells MDA-MB-231 by genistein. *Oncogene,* 1999, 18, 3166-3172.

[41] Alhasan, SA; Pietrasczkiwicz, H; Alonso, MD; Ensley, J; Sarkar, FH. Genistein-induced cell cycle arrest and apoptosis in a head and neck squamous cell carcinoma cell line. *Nutr. Cancer,* 1999, 34, 12-19.

[42] Banerjee, S; Zhang, Y; Wang, Z; Che, M; Chiao, PJ; Abbruzzese, JL; et al., In vitro and in vivo molecular evidence of genistein action in augmenting the efficacy of cisplatin in pancreatic cancer. *Int. J. Cancer,* 2007, 120, 906-917.

[43] Spinozzi, F; Pagliacci, MC; Migliorati, G; Moraca, R; Grignani, F; Riccardi, C; et al. The natural tyrosine kinase inhibitor genistein produces cell cycle arrest and apoptosis in Jurkat T-leukemia cells. *Leuk. Res.,* 1994, 18, 431-439.

[44] Suzuki, K; Koike, H; Matsui, H; Ono, Y; Hasumi, M; Nakazato, H; Okugi, H; Sekine, Y; Oki, K; Ito, K; Yamamoto, T; Fukabori, Y; Kurokawa, K; Yamanaka, H. Genistein, a soy isoflavone, induces glutathione peroxidase in the human prostate cancer cell lines LNCaP and PC-3. *Int. J. Cancer,* 2002, 99, 846-852.

[45] Lynch, A; Harvey, J; Aylott, M; Nicholas, E; Burman, M; Siddiqui, A; Walker, S; Rees, R. Investigations into the concept of a threshold for topoisomerase inhibitor-induced clastogenicity. *Mutagenesis,* 2003, 18, 345-353.

[46] Akiyama, T; Ishida, J; Nakagawa, S; Ogawara, H; Watanabe, S; Itoh, N; Shibuya, M; Fukami, Y. Genistein, a specific inhibitor of tyrosine-specific protein kinases. *J. Biol. Chem.,* 1987, 262, 5592-5595.

[47] Lamatiniere, CA; Cotroneo, MS; Fritz, WA; Wang, J; Mentor-Marcel, R; Elgavish, A. Genistein chemoprevention: timing and mechanisms of action in murine mammary and prostate. *J. Nutr.,* 2002, 132, 552S-558S.

[48] Davis, JN; Kucuk, O; Sarkar, FH. Genistein inhibits NFkappa B activation in prostate cancer cells. *Nutr. Cancer,* 1999, 35, 167-174.

[49] Alhasan, SA; Ensley, JF; Sarkar, FH. Genistein induced molecular changes in a squamous cell carcinoma of the head and neck cell line. *Int. J. Oncol.,* 2000, 16, 333-338.

[50] Natarajan, K; Manna, SK; Chaturvedi, MM; Aggarwal, BB. Protein tyrosine kinase inhibitors block tumor necrosis factor-induced activation of nuclear factor-kappaB, degradation of IkappaBalpha, nuclear translocation of p65, and subsequent gene expression. *Arch. Biochem. Biophys.,* 1998, 352, 59-70.

[51] Franke, TF; Kaplan, DR; Cantley, LC. PI3K: Downstream AKTion blocks apoptosis. *Cell,* 1997, 88, 435-437.

[52] Romashkova, JA; Makarov, SS. NF-kappaB is a target of AKT in anti-apoptotic PDGF signaling. *Nature,* 1999, 401, 86-90.

[53] Ahmad, MS; Fazal, S; Rahman, A; Hadi, SM; Parish, JH. Activities of flavonoids for the cleavage of DNA in the presence of Cu(II): correlation wi±th the generation of active oxygen species. *Carcinogenesis,* 1992, 13, 605-608.

[54] Gali, HU; Perchellet, EM; Klish, DS; Johnson, JM; Perchellet, JP. Hydrolyzable tannins: potent inhibitors of hydroperoxide production and tumor promotion in mouse skin treated with 12-O-tetradecanoyl phorbol-13- acetate in vivo. *Int J Cancer,* 1992, 51, 425-32.

[55] Clement, MV; Hirpara, JL; Chawdhury, SH; Pervaiz, S. Chemopreventive agent resveratrol, a natural product derived from grapes, triggers CD95 signalling-dependent apoptosis in human tumor cells. *Blood,* 1998, 92, 996-1002.

[56] Ahmad, N; Feyes, DK; Nieminen, AL; Agarwal, R; Mukhtar, H. Green tea constituent epigallocatechin-3-gallate and induction of apoptosis and cell cycle arrest in human carcinoma cells. *J. Natl. Cancer Inst.,* 1997, 89, 1881-1886

[57] Chang, KL; Cheng, HL; Huang, LW; Hseih, BS; *et al.,* Combined effects of terazosin and genistein on a metastatic, hormone –independent human prostate cancer cell line. *Cancer Lett.*, 2008, 276, 14-20

[58] Ullah, MF; Shamim, U; Hanif, S; Azmi, AS; Hadi, SM. Cellular DNA breakage by soy isoflavone genistein and its methylated structural analogue biochanin A. *Mol Nutr Food Res.,* 2009, 53, 1376-1385.

[59] Qian, YP; Cai, YJ; Fan, GJ; Wei, QY; Yang, J; Zheng, LF; Li, XL; Fang, JQ; Zhou, B. Antioxidant-Based Lead Discovery for Cancer Chemoprevention: The Case of Resveratrol *J. Med. Chem.*, 2009, 52(7), 1963-1974

[60] Hadi, SM; Bhat, SH; Azmi, AS; Hanif, S; Shamim, U; Ullah, MF. Oxidative breakage of cellular DNA by plant polyphenols: a putative mechanism for anticancer properties. *Semin Cancer Biol.,* 2007, 17, 370-376.

[61] Gupte, A; Mumper, RJ. Elevated copper and oxidative stress in cancer cells as a target for cancer treatment. *Cancer Treat Rev*., 2008, 35, 32-46.

In: Isoflavones Biosynthesis, Occurence and Health Effects ISBN: 978-1-61728-113-6
Editor: Melanie J. Thompson, pp. 285-299 © 2010 Nova Science Publishers, Inc.

Chapter XII

Isoflavones in *Leguminosae*

***S. C. Cunha**[*1]**, T. Visnevschi-Necrasov**[±2]**, M.A. Faria**[£1]**, E. Nunes**[≠2] **and M.B.P.P. Oliveira**[‡1]*

[1]REQUIMTE - Serviço de Bromatologia, Faculty of Pharmacy, University of Porto, Rua Aníbal Cunha, Porto
[2]CIBIO-ICETA, Faculty of Science of University of Porto, Vairão, Portugal,

1. Introduction

Isoflavones are a class of plant secondary metabolites, related to flavonoids which are most commonly known for their antioxidant activity. Flavonoids are a type of polyphenol [Arranz *et al.*, 2009] derived from the shikimate pathway, responsible for plant defense and flower color. Plants use isoflavones and their derivatives as phytoalexin compounds to ward off disease-causing pathogenic fungi and other microbes. Isoflavones are produced via a branch of the general phenylpropanoid pathway that produces flavonoid compounds in higher plants.

Isoflavones are produced almost exclusively by the members of the Fabaceae/Leguminosae (bean) family. Among the members of the Fabaceae family that contain significant quantities of isoflavones, *Glycine max* (soybean) and soy products assume

* REQUIMTE - Serviço de Bromatologia, Faculty of Pharmacy, University of Porto, Rua Aníbal Cunha, 164, 4099-030 Porto, Tel: +351222078910, Fax: +351222003977, Email: sara.cunha@ff.up.pt, Both authors contribute equally for this work

± CIBIO-ICETA, Faculty of Science of University of Porto, R. Padre Armando Quintas 4485-661 Vairão, Portugal, Tel: +351252660402, Fax: +351252661780, Email: tativis74@gmail.com, Both authors contribute equally for this work

£ REQUIMTE - Serviço de Bromatologia, Faculty of Pharmacy, University of Porto, Rua Aníbal Cunha, 164, 4099-030 Porto , Tel: +351222078910, Fax: +351222003977, Email: mfaria@ff.up.pt

≠ CIBIO-ICETA, Faculty of Science of University of Porto, R. Padre Armando Quintas 4485-661 Vairão, Portugal, Tel: +351252660402, Fax: +351252661780, Email: enunes@mail.icav.up.pt

‡ REQUIMTE - Serviço de Bromatologia, Faculty of Pharmacy, University of Porto, Rua Aníbal Cunha, 164, 4099-030 Porto,Tel: +351222078927, Fax: +351222003977, Email: beatoliv@ff.up.pt

particular importance as source of isoflavones in human nutrition. However, the allergenic nature of many soy products for the people and the difficulty in obtaining genetically modified organism (GMO)-free soybeans are causing experts in industry to look for alternative sources of isoflavones [Saviranta *et al.*, 2008]. Thus, other legumes including *Trifolium pratense* (red clover) have recently received considerable attention due to its high isoflavones content in which has been reported to be 2-10 times higher than in soybean seeds, and because its wide adaptaptation to different agro-ecological regions of the world. Red clover contains genistein and daidzein as soybean and, more interestingly, it contains significantly higher concentrations of biochanin A and formononetin, methylderivatives of genistein and daidzein [Delmonte, 2006].

The potential health-protective effects of isoflavones have increased the interest in understanding the chemistry of these phytochemicals in foods and also impulse the development of new products. Products containing extracts or powered red clover for example can be sold as non-prescribed food supplements.

These interests have led to the development of several extraction and analytical methods for the determination of isoflavones in plants. Instrumental techniques based on chromatographic analysis such as high performance liquid chromatography (HPLC) or gas chromatography (GC) are among the most used approaches suggested for monitoring the profile and quantify the contents of isoflavones in plant foods [Rostagno *et al.*, 2009].

In general, chemical pre-treatment of the sample prior to the chromatographic analysis is required. The most widely used technique is based on solvent-extraction followed by solid-phase partition. While allowing an adequate separation of isoflavones from the matrix this kind of procedures are laborious, require a large amount of toxic solvents and are difficult to automate. To overcome these problems recent techniques combining the extraction and cleanup in a single step have been developed namely pressurized liquid extraction, ultrasound-assisted solid-phase extraction, microwave-assisted extraction and matrix solid-phase dispersion [Rostagno *et al.*, 2009].

Isoflavone concentrations and profiles can be affected by genetic and environmental factors. Sivesind and Seguin [2005] examined the effects of the environment, cultivar, plant maturity, plant part, and preservation method on the concentration of the two predominant isoflavones in red clover, formononetin and biochanin A. Different plant parts and the same parts harvested at different growth periods, showed different concentrations of isoflavones. Notwithstanding, some authors have found that sequence variations in the Isoflavone synthase (IFS) gene could have major impacts in the isoflavone levels in soybeen seeds. IFS is the first committed enzyme in the isoflavone pathway that converts flavanones to isoflavones [Jung *et al.*, 2000]. Studies of IFS gene variations and their influence on isoflafones profiles are being conducted in other legume species. Polymorphisms detected in these genes may serve as important molecular markers for breeding.

The aim of the present work is to critically review the isoflavones analysis in several leguminosae species especially from the genera Tr*ifolium,* with an emphasis on the extraction procedures employed before the chromatographic determination, and to discuss the levels found in different legume species of the genus Tr*ifolium, Medicago, Scorpiurus, Lotus, Ornithopus, and Biserrula* cultivated in Portugal. In addition, isoflavone synthase gene (ISF) variations and their influence on isoflafones profiles will be discussed.

2. Isoflavones in Leguminosae: Health Benefits and Function in Plants

Leguminosae (or Fabaceae) is the third largest family of flowering plants, after Orchidaceae and Asteraceae, with over 600 genera and 20 000 species widespread around the world [Doyle, 2004]. Leguminosae include an extensive number of wild and cultivated species.

Among the compounds found in some leguminous family members, the secondary metabolites, in particular isoflavones have been demonstrated to possess a broad spectrum of biological properties, giving them the potential to influence human health [Yamamoto and Gaynor, 2001]. Isoflavones constitute the largest group of natural isoflavonoids, characterized by a common 3-phenyl-chromen-4-one core structure, and differ by substituients such as methoxyl, hydroxyl and glycoside functions on the primary structure [Delmonte, 2006]. According to their chemical functional groups isoflavones are categorized as aglycones (genistein, daidzein, and glycitein), glycosides (genistin, daidzin, and glycitin), malonyl glycosides (malonyl genistein, malonyl daidzein, and malonyl glycitin), and acetyl glycosides (acetyl genistein, acetyl daidzein, and acetyl glycitin) **Figure 1**. In plants, the predominant isoflavones forms are the 6''-*O*-malonyl-7-*O*-*β*-D-glucoside derivatives and 6''-*O*-acetyl-7-*O*-*β*-D-glucosides which are produced from them during food processing or sample preparation [Delmonte, 2006]. Because of their structural similarity to 17*β*-estradiol and weak estrogenic activities, isoflavones can mimic or modulate the actions of endogenous estrogens in vertebrates by binding to estrogens receptors (ER) [Zang *et al.*, 2007].

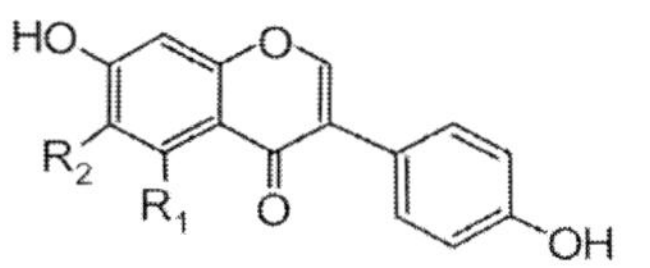

R_1	R_2	aglycones
H	H	daidzein
OH	H	genistein
H	OCH_3	glycitein

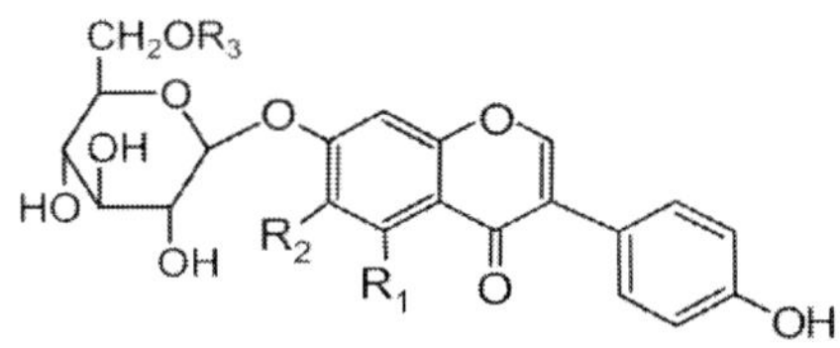

R_1	R_2	R_3	glycoside
H	H	H	daidzin
OH	H	H	genistin
H	OCH_3	H	glycitin
			acetylglycoside
H	H	$COCH_3$	acetyldaidzin
OH	H	$COCH_3$	acetylgenistin
H	OCH_3	$COCH_3$	acetylglycitin
			malonylglycoside
H	H	$COCH_2COOH$	malonyldaidzin
OH	H	$COCH_2COOH$	malonylgenistin
H	OCH_3	$COCH_2COOH$	malonylglycitin

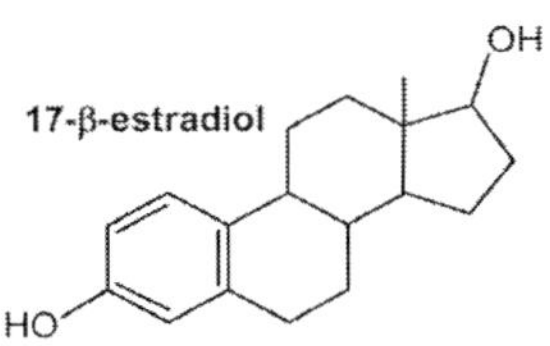

Figure 1. Chemical structure of 17*β*-estradiol and isoflavones aglycones, glycosides, acetyl and malonyl glycosides

Estrogenic activity was first described over 50 years ago following the observation that sheep feeding on pastures containing clover demonstrate hyperestrogenization and infertility [Mazur, 2000].

Genistein, daidzein, biochanin A and formononetin are the compounds most thoroughly investigated and interesting with regard to estrogenic activity. Genistein showed the highest binding affinity to the estrogen receptor contrarily to its methoxy derivative, biochanin A, which does not bind to the estrogenic receptor but have estrogenic activity *in vivo* [Mazur. 2000]. Daidzein has a higher binding affinity for the estrogen receptor than its methoxy derivative, formononetin, but both are weak estrogens *in vivo* [Mazur, 2000].

Several other studies have been devoted in the last two decades to investigate the physiological function of isoflavones in humans and animals as well mentioned in the reviewers of Messina and Bennink [1998], and Wuttke *et al.* [2007]. The health benefits claimed for isoflavones comprise contributions to cancer prevention, reduction postmenopausal symptoms, prevention of coronary heart disease by reducing low-density lipoprotein (LDL) and increasing high-density lipoprotein (HDL) and positive effects on neurobehavioral activities. Although generally accepted, these health benefits have some controversy. As referred by Wuttke, 2007 some results from cell biological and animal experimental investigations on soy/isoflavones and genistein in particular are highly contradictory. Furthermore, there is a considerable lack of studies demonstrating that isoflavones intake is correlated with healthy benefits. On the other hand despite wide-ranging efforts, the exact nature of the function and mode of actions of isoflavones are still not clear.

Concerning to the isoflavones function in plants, they are involved in plant-microbe interactions namely defense and symbiosis. As to defense responses, isoflavones are involved in phytoalexin production. Phytoalexins are a group of chemically diverse, low-molecular mass natural products that posses antimicrobial and/or antiherbivore defence activities. Antimicrobial effects are also showed by isoflavones themselves when tested in vitro [Yu and McGonigle, 2005]. In symbiose response, the colonization of nitrogen-fixing bacteria activity in leguminosae species can be enhanced by the presence of isoflavonoids. Furthermore, isoflavones presumably create positive signal feedback cycles between the pant and the microbe, leading eventually to nodule morphogenesis and nitrogen fixation [Yu and McGonigle, 2005].

3. Manipulation of Plant Isoflavone Production: The Role of Isoflavone Synthase Gene

The isoflavone synthase enzyme, encoded by the isoflavone synthase gene (IFS), catalyzes the first dedicated step of isoflavone biosynthesis, a branch of the general phenyl-propanoid pathway in which flavanones are oxidated into isoflavones (*e.g.* IFS catalyzes the oxidation of 7,4'-dihydroxyflavanone – liquiritigenin – to daidzein) [Jung *et al.*, 2000]. ISF gene was first identified in soybean. Since then ISF genes have been cloned from several other legumes including mung bean, alfalfa, snow pea, red clover, hairy vetch and lupine [Jung *et al*, 2000]. It has been referred that IFSs constitute a multigene family having soybean two IFSs identified and mung bean at least four [Kim *et al.*, 2003]. The recent work developed in the area of identification ISF genes may have several applications in the isoflavone level/profile control in leguminosae and non-leguminosae species. Using gene polymorphism by itself as an isoflavone profile marker to assist plant breeders or by manipulating ISF genes in legumes and non isoflavones-producing species, researchers aim to control the amounts and types of isoflavones present in foodstuffs.

Studies have shown genetic related variation in isoflavone profile and concentration among soybean genotypes [Choi *et al.*, 1996] despite the well known large environmental effect [Hoeck *et al*, 2000]. Several quantitative trait loci (QTLs) have already been identified and related with soybean seed phytoestrogen amounts [Kassem *et al.*, 2004; Primono *et al.*, 2005]. More recently Cheng *et al.* [2008] have identified three single nucleotide polymorphisms (SNPs) in ISF1 gene and two SNPs in IFS2 closely associated with total and individual isoflavone levels in soybean seeds. These polymorphisms could serve as molecular markers for marker assisted selection in future breeding efforts. In early studies Kim and collaborators [2004] have also examined the polymorphism in IFS1 and IFS2 genes in several soybean cultivars and found out that for most of the cultivars there was nucleotide variation resulting in amino acid changes. Notwithstanding the authors did not establish a relation between these polymorphisms in ISF genes and the qualitative/qualitative variability of isoflavones in the studied cultivars.

Studies in *Trifolium pratense* revealed a strong homology to other plant IFS genes at the predicted amino acid sequence level; however the correlation of genetic genetic polymorphisms with isoflavones contents was not evaluated [Kim *et al.*, 2003].

Concerning the manipulation of the ISF genes in legumes the isoflavones levels could be increased and their variability reduced, eliminating the drawbacks of isoflavone concentration changes due to environmental conditions, and their loss during food processing-which can account for 50% in traditional soy foods or protein products [Jung *et al.*, 2000]. In addition, the genetic transformation of other crops to onset isoflavones synthesis would provide food industry with a variety of isoflavone sources for food supplementation. Consequently IFS genes have been used to produce isoflavones in non-legumes crops such as genistein in *Arabidopsis*, tobacco and maize [Kim *et al.*, 2004], lettuce and petunia [Liu *et al.*, 2007]; and genistin in tomato [Shih *et al.*, 2008]. The increase of isoflavone biosynthesis by IFS gene manipulation as well as the modification of isoflavone profiles by the expression of multiple biosynthetic genes in soybean seeds, including IFS2, have been recently described [Yu *et al.*, 2003, Zernova *et al.*, 2009].

Either by identifying molecular markers to assist the development of improved crops or by gene manipulation in legumes and other species researchers are developing tools to take control on the plants isoflavone production.

4. Analytical Methods for Determination of Isoflavones

Determination of isoflavones in plants can be carried out by a variety of methods, from classical chemical methods to more recent chromatographic, enzymatic and electrophoretic methodologies, among others. The most relevant methodologies for separation and quantification of isoflavones in leguminosae in particularly *Trifolium* species are further detailed in this section.

Ideally, the analytical methods developed will be rapid, simple, inexpensive to perform, accurate, precise, sensitive, confirmatory, solvent-sparing (waste minimizing) and safe. In practice, the capability of a method to attain these goals depends on the technology utilized, and it is nearly impossible to meet all the desirable features due the number and complexity of steps required in an analytical method. These steps include: sampling (collection the appropriate sample and it conservation), sample preparation (homogenization, extraction of the analytes with or without hydrolysis, perform cleanup of necessary, etc.), separation and quantification of the analytes and data processing.

An often ignored critical issue in analytical methods is sampling, which involves collection and conservation of samples before the instrumental analysis. The collection of a representative sample of the entire food-stuff is achieved by homogenizing a large sample and takes a representative subsample for analysis. Due to the perishable ability of the isoflavones the plants are normally conversed by drying methods as vacuum drying, freeze-drying or freezing with liquid nitrogen. However, drying methods can affect isoflavones contents in samples as previously reported by Sivesind and Seguin [2005] on leaves of red clover. They found that freeze drying had the minimal impact on glycoside and malony/glycosides contents comparing with vacuum drying that promote the conversion of these compounds into their aglycones, mainly formononetein and biochanin A. Swinny et al., 2005, note that slow thawing vaccum drying samples allowed enzymatic conversion of 95% isoflavones glucosides into their aglycones in red clover. Furthermore, air drying produced a low level of the aglycones, formononetein and biochanin A, and oven drying promoted decarboxylation of malonyl glucosides to acetyl glucosides.

Moreover, isoflavones profiles and total isoflavones contents may be affected during the storage of the samples and during the extraction step. In one of the few published reports dealing specifically with the storage of red clover extracts Rijke *et al.* [2001], evaluated the temperature and stage time on short-term stability of the isoflavones. They concluded that the extract can be stored for 2 weeks at -20°C without significant degradation, but if stored at room temperature the degradation starts immediately.

The simultaneous determination of all chemical groups of isoflavones in plants is usually difficult in part due to the absence of some commercial standards of individual malonyl and acetyl glycosides. Consequently, a basic, acid or enzymatic hydrolysis.are include after, before or during the extraction step in most of the analytical methods Basic hydrolysis act on ester bonds, removing acid groups that are linked to the sugar moiety of isoflavones glycosides. As a result, the malonyl and acetyl glycoside isoflavones are converted to their respective glycosides. Acid hydrolysis breaks the bond between the isoflavones and glycoside moieties transforming all the isoflavones derivatives, into their aglycone forms [Delmonte *et al.*, 2006; Rostagno *et al.*, 2009]. Among the different conditions and agents that can be applied for acid hydrolysis the most common is hydrochloric acid at high temperatures. Basic hydrolysis is usually achieved with solutions of NaOH at room temperature. Enzymatic hydrolysis can be performed with different enzymes such as β-glucosidase, *Helix pomatia* juice containing β -glucuronidase and arylsulfatase activity, cellulase, and a combination of all these three enzymes [Rostagno *et al.*, 2009].

Delmonte and colleagues [2006] reported that acid hydrolysis is useful for the analysis of complex samples of unknown origin, and may be used to identify sugar-isoflavones by comparison of these results with those from basic hydrolysis. In the same work they showed that unhydrolyzed extracts of red clover or soy seems to be more applicable to the quantification of the commercial isoflavones (free or as derivatives) than the hydrolyzed extracts (acid and basic hydrolysis).

An important point to consider during the extraction procedure is the fact that the compartmentation of the cell being disrupted whereupon isoflavone glycosides can be readily converted into their corresponding aglycones due to the native β-glucosidase. In order to reduce this problem Toebes *et al.* [2005] used 350 mM *Tris* in 80% ethanol at pH 7.2 in quantification of malonyl-glycosides from red clover. The same procedure was applied by Sivesind and Seguin [2005] in the evaluation of isoflavones contents in different parts of red clover.

Classical extractions of the isoflavones are based on solvent extraction, employing simple procedures like mixing, shaking, sonicating or Soxhlet extraction. Several solvents (pure or aqueous methanol, ethanol, acetonitrile or acetone with and without the addition of small amounts of acids) are employed in this procedure, generally with a subsequent liquid-liquid partition or column cleanup. The extraction time may range from 30 min. to 24 h and the temperature from room temperature to 80°C. For example, genistein, daidzein, formononetin and biochanin A were extracted from common *Trifolium* species with methanol by shaking at 55°C during 30 min. [Vetter, 1995]; malonyl glycosides were extracted from red clover with 90% methanol using sonication during 1 h at room temperature [Lin *et al.* 2001]. Similar procedure was used by Wu *et al.* [2003] in isoflavones extraction from red clover being the main difference the proportion of methanol (in this case was 80%). Rijke e*t al.* [2001, 2005] reported the use *Tris* solution followed by solvent extraction with 90% methanol in malonyl glycoside determination and other main flavonoids in *Leguminosae*. The extraction of isoflavones from different parts of red clover collected in different growing stages was achieved after an acid hydrolysis with 70% methanol during 16 h at room temperature (22°C) as reported by Tsao *et al.* [2006], whileSaviranta *et al.* [2008] extracted isoflavones from leaves, stems, roots and different coloured flowers from red clover by sonicated 30 min the samples with 80% methanol followed by a vigorous shaking during 10 min. Swinny and Ryan [2005] extract efficiently isoflavones from red clover with 80% methanol followed by 12 h of vortex shaking. Ramos *et al.* [2008] reported the use of 50% methanol in extraction of four isoflavones aglycones from red clover.

Aside from the robustness and adequate separation of isoflavones these procedures are laborious, require a large amount of toxic solvents and laboratorial material and are difficult to automate. To overcome these problems recent techniques have been developed namely solid-phase extraction (SPE), matrix solid-phase dispersion (MSPD), pressurized liquid extraction (PLE), ultrasound-assisted extraction (UAE), and supercritical fluid extraction (SFE).

SPE is an alternative technique to liquid extraction based on the retention of the analyte on a solid sorbent by different mechanisms, like Van der Waals forces or electrostatic interactions. Several sorbents are often used in SPE, such as octadecyl (C18), silica and various other polymers in combination with different solvents like acetonitrile, methanol, water, or mixtures of solvents. SPE is fast and robust with a low solvent consumption. For example, eight isoflavones malonyl glycoside and six acetyl glycosides were extracted from red clover using a two-dimensional SPE (Oasis HLB, RP 105 and abselut sorbents) [Kledjus *et al.*, 2001]. The extraction was performed by loading 3 ml of diluted sample solution (1 ml of a methanolic extract was diluted with 3 ml of water) onto pre-conditioned SPE cartridges. Impurities were washed out with 1 ml of 5% methanol containing 2% acetic acid and the analytes were eluted with 1ml of 80% methanol containing 2% ammonia hydroxide. Sivesind and Seguin [2005] used C18 SPE cartridges for cleanup of extracted of isoflavones of red clover; the elution was performed with 80% methanol. Martin *el al.* [2006] extracted four isoflavones aglycones in lucerne during 1 hour at 80°C using 80% ethanol followed by a cleanup with HLB column being the elution performed with 80% methanol containing 5% of ammonia hydroxide.

MSPD is a sample preparation methodology that combines sample disruption, whilst generating a material with a unique dispersed character for extraction with small amounts of organic solvents. Solid matrices need to be mixed with the solid phase, generally in a mortar, before being transferred to the extraction column, whereas liquid samples can be directly homogenized with the solid phase in the column. Solid phases of different polarities similar to those employed in SPE and more frequently HLB and octadecyl (C18) are employed. Xiao *et al.* [2004] for optimized a MSPD procedure for the extraction of isoflavones in *Radix astragali* which was compared to ultrasonic and Soxhlet extraction methods,using C18 as sorbent and 90% methanol as eluent an acceptable reproducibility, recovery extraction efficiency and consumption was achieved relatively to conventional extraction techniques evaluated. In addition, Rijke *et al.* [2004] isolated isoflavones malonyl glycoside successfully from hydrolyzed samples of red clover using MSDP. In the optimized procedure after grinding with C18 the sample was washed with water and the analytes collected with 90% methanol. More recently the MSPD was used for the determination of 12 isoflavones in leaves of *Trifolium* [Visnevschi-Necrasov *et al.*, 2009]. For the development of the method several extraction parameters were evaluated including type/quantity of the sorbent (HLB, C18) as well as the solvent used (water, methanol, dichloromethane). The best yields were obtained with C18 as sorbent and methanol-dichloromethane mixture (75:25 v/v) as eluent.

PLE is a new technique based on the extraction of samples using small amounts of pressurized solvent at high temperature. The efficiency of extraction increases with temperature and, consequently, the extraction time and volume of solvent can be reduced. The increase in pressure is necessary to keep solvents in liquid state. Moreover, analytes have to be thermally stable under the conditions employed. Zgórka [2009a] evaluated the influence of several parameters, extraction solvents (methanol and acetone or their 75% aqueous solutions), temperatures (75, 100 and 125°C), and the changeable number of static extraction cycles for extraction of isoflavones in *Trifolium* species. Optimum extraction efficiency with PLE, expressed in biochanin A, formononetin , daidzein, and genistein yields from plant material, was obtained using methanol/water as extraction solvent, an oven temperature of 125°C and three 5 min static extraction cycles.

UAE is a simple and rapid method for the extraction of organic compounds from plant materials. It involvestwo types of physical phenomena: diffusion through the cell walls and washing out (rinsing) the cell contents once the cell walls are broken, which are significantly affected by ultrasound irradiation. Ultrasound also exerts a mechanical effect, allowing greater penetration of solvent into the sample matrix, increasing the contact surface area between solid and liquid phase. This coupled with the enhanced mass transfer and significant disruption of cells, via cavitation bubble collapse, increases the release of intracellular product into the bulk medium. The use of higher temperatures in UAE can increase the efficiency of the extraction process due to the increase in the number of cavitation bubbles formed [Rostagno *et al.*, 2003]. Recently, a newly elaborated technique SPE–on BakerBon octadecyl, cyclohexyl and phenyl cartridge was successful used in combination with UAE for isolation of isoflavones aglycones from hydrolyzed samples of red clover [Zgórka 2009b].

SFE uses fluids under conditions where they have similar densities to liquids but higher diffusion coefficients and lower viscosities, which leads to faster extraction of solutes. Moreover, this technique can be automated with a subsequent saving in labour and time-consuming . Carbon dioxide is often used as supercritical fluid, due to its low critical temperature and pressure, although additional clean up may be necessary for some matrices. Bajer *et al.* [2007] used SFE for extraction of isoflavones in *Matricaria recutia, Rosariuns officinalis*, *Foeniculum vulgare* and *Agrimonia eupatoria.* In this work SFE extraction was performed at 35 MPa and 70°C, using carbon dioxide as extractor, which was modified with methanol due the broad polarity range of isoflavones in study. This technique showed very high yields of extraction of apigenin and biochanin A compared with other extraction procedures namely UAE, ultrasonic homogenizer, and Soxhlet.

Quantification of isoflavones is more traditionally achieved using high performance liquid chromatography (HPLC) or gas chromatography (GC). HPLC is the favoured technique for the determination of isoflavones, since many of these analytes lack the thermal stability required for gas chromatographic analysis. In this technique the sample is injected into a flowing solvent and is carried through a column containing a solid phase. Separation is due to differential adsorption on to the solid phase and is determined to some extent by differing polarities of the isoflavones analytes in the sample. Most HPLC methods employ reversed phase chromatography with C8 and C18 and aqueous mobile phases. Isoflavones can efficiently be separated with two different solvent systems, acidified methanol/acidified water and acidified acetonitrile/acidified water.

Detection is often done with ultraviolet (UV) or diode array (DAD) detectors, probably due to its wide applicability and consequent presence in most HPLC systems. The quantification is usually achieved at 254 nm [Krenn *et al.*, 2002; Sivesind and Suguin 2005; Visnevschi-Necrasov *et al.*, 2009], 255 nm [Saviranta *et al.*, 2008] or 260 nm [Xiao *et al.*, 2004; Delmonte *et al.*, 2006; Swinny and Ryan, 2005; Ramos *et al.*, 2008], which corresponding to wavelengths of maximum absorption for most of isoflavones. The use of other detectors as fluorescence detection may seem to be an interesting alternative but, unfortunately, genistein and biochanin A do not display inherent fluorescence. Zgórka [2009b] applied this detector only for quantification of formononetin, being the remaining isoflavones determined by photo diode array. Post-column derivatization with aluminum seems to be the solution, but has been restricted applied to nonflurescence flavones. In recent years, mass spectrometry (MS) coupled to liquid chromatography has been successful applied in quantification and identification of isoflavones. Improved robustness of LC-MS equipments together with good sensitivity and selectivity achieved, has led to more frequent use of this technique in isoflavone analysis [Kledjus *et al.*, 2001; Wu *et al.*, 2003; Lapcík *et al.* 2005, Tsao *et al.*, 2006]. Althoug, multiple MS analyses (MS/MS) may improve sensitivity and selectivity [Rijke *et al.*, 2004], there are still a number of limitations on its application,namely the cost of the equipment and the requirement of trained operatorsLiquid chromatography coupled with nuclear magnetic resonance spectroscopy has also been used in identification of isoflavones [Xiao *et al.*, 2005].

GC is less popular in the analysis of isoflavones. Normally, isoflavones are derivatized and the trimethylsilyl ether derivatives are carried through a column by a flowing inert gas and separated by differential adsorption to a solid phase in the column. Newer instruments use a capillary column where adsorption takes place on the capillary wall rather than packing in the column, which leads to a faster separation and greater sensitivity. Mass spectrometer is probably the most widely used detection in GC, because of their great selectivity and sensitive.

The use of other technique such as capillary electrophoresis (CE) has been reported for determination of isoflavones in *Trifolium* [Peng and Ye, 2006; Zhang *et al.*, 20007]. In CE the separation is based on the differential molibility of compounds with electric charges along with a small diameter capillary. Isoflavones are separate based on the fact that, as weak acids, they achieved negative charged at high pH. Borate buffer is generally applied to the electrophorectic separation because of its interaction with the *cis*-diol moiety of sugars, creating charged complexes [Delmonte and Rader, 2006]. Zhang *et al.*, 2007 developed a micellar electrokinetic capillary chromatography method for determination of four isoflavones in red clover. An electrolyte composed of 30mM borate, 20mM sodium dodecyl sulfate (SDS) and 4mg/ml hydroxypropyl-β-cyclodextrin containing 5% ethanol at pH 10.1 provided a satisfactory separation for all the analytes. The detection of analytes was preformed by UV at 254 nm.

5. Isoflavones Contents in Leguminosae

Several works have been published in the last years reporting the isoflavones content in Leguminosae plants. Isoflavones content in plants is controlled by both genetic and environmental factors. In this section after a brief review of the factors that influence isoflavones concentration in Leguminosae, the results obtained by our group in an experimental field in Portugal, of different legume species of the genus Tr*ifolium* (T), *Medicago, Scorpiurus, Lotus, Ornithopus, and Biserrula* will be presented.

Early studies reported differences in isoflavones concentration among red clover cultivars [Wong, 1963; Deido and Clark 1968]. Deido and Clark [1968] reported 2.7 and 2 fold variation in formononetin and biochanin A, respectively, among 16 field-grown cultivars sampled at a single date. In a study in seven clover species Vetter [1995] revealed that the *T. montamum*, *T. fragiferum*, *T. incarnatum* and *T. repens* contained lower amounts of genistein, daidzein, formononetin and biochanin A compared with the *T alpestre*, *T. pratense* and *T. subterraneum*. In a multi-year, multisite trial study was verified that the total isoflavones concentration in 10 cultivars of red clover ranged from 8.92 mg/g to 127.53 mg/g average across sites, harvests and years; having the cultivar "Star" consistently the lowest isoflavones concentration [Sivesind and Seguin, 2005]. In a recent study Tsao *et al.* [2006] found out that the isoflavone compositions were similar among the 13 red clover cultivars studied, but the total and individual concentration differed significantly, and the predominant isoflavones were formononetin and biochanin A.

Environmental factors have also been reported to affect red clover isoflavones concentration. McMurray *et al.* [1986] in a greenhouse experimental showed that formononetin concentration in expanded leaves was 28% higher in red clover plants grown in a 17/13°C temperature regimen than in 23/15°C. In the same study was also noted that when the first harvest was delayed by 21 days, the formononetin concentration decreased by 39%. For the second harvest, the plots that were allowed the shortest regrowth had the highest formononetin concentration. Differences between harvest dates might be attributable to differences in plant maturity at harvest. A decrease of 57% in the formononetin concentration was reported between the late vegetative and the dying inflorescence stages in greenhouse grown plants. Similar behaviour was observed by Sarelli *et al.* [2003] studding the phytoestrogen content in birdsfoot trefoil and red clover. Formononetin concentration decreased as the plant matured from budding to flowering stage and wilting from 25% to 40% of dry matter content decreased genistein and biochanin A. They also reported that with increasing maturity red clover phytoestrogens content decreased and the effect for daidzein, formononetin biochanin A and total of phytoestrogens was very significant in red clover silage. Maturity stage also affected the isoflavones concentration in different plant parts. Sivesind and Seguin [2005] noted that highest isoflavones concentrations were found in leaves (11970 mg/g) and stems (4896 mg/g) during the vegetative stages, with formononetin concentration declining until plants initiated flowering, especially in stems then stabilized in both parts. Upon initiation, inflorescences containing similar isoflavones concentration than leaves, but concentrations decreased rapidly during flower development to fall even below those observed in stems. Tsao *et al.* [2006] also observed that in leaves, the average total isoflavones concentration significantly increased by 30% from early stage to flowering stage, whereas in the stems and petioles, the total isoflavones concentration decreased significantly as the plants grew to maturity. More recently, Saviranta *et al.* [2008] reported that the highest concentration of daidzein and genistein concentration were found in stems collected in August (0.24 and 0.55 mg/g, respectively), while the highest amounts of formononetin and biochanin A were found in young leaves collected in June (7.47 and 9.69 mg/g, respectively).

Currently only the flowers of red clover have been used for health supplements, while the rest o the plants has been almost completed overlooked. However, other parts of the plants such as roots, leaves and stems contain valuable amounts of isoflavones as verified by Saviranta *et al.* [2008]. This authors noted that the highest amounts of daidzein and genistein were found in petioles (0.11-0.28 and 0.30-0.54 mg/g, respectively), while leaves were rich in formononetin and biochanin A (5.57-9.05 and 10.94-14.59 mg/g, respectively). High formononetin concentration was found in roots, but very little biochanin A. Generally, the leaves contained the highest overall concentration of isoflavones, followed by the stem, petiole and flower [Sivesind and Seguin, 2005; Tsao *et al.* [2006]. Similar behaviour have been reported by Wu *et al.* [2003] who verifiedthat the total isoflavones were greatest in leaves, intermediate in stems and lowest in flowers when average over red clover plants collected in four different environment at one point in time. Notwithstanding, it has also been proved that flowers contain the highest total isoflavones concentration, followed by leaves and stems, when measured at single date [Vetter, 1995].

As described above several factors can influence isoflavone concentration in plants. However, the determination of isoflavones in different plant species is extremely important for the identification of its chemical composition and for the understanding of the benefits or damage of it's consumption by animals and humans. Moreover, the data obtained is useful for the construction of a database reporting isoflavones content in food, taking into account that often products as red clover are used as nutritional supplements.

In our work the content of 12 isoflavones were evaluated in leaves of different species of Leguminosae collected during 2 years in the same experimental fieldconditions. The data presented in **Table 1** were obtained using a previous validated method [Visnevschi-Necrasov *et al.*, 2009].

Table 1 Concentration (mg/g) of isoflavones in different species of leguminosae

Isoflavones	Species					
	Medicago	*Trifolium*	*Ornithop*	*Scorpiurius*	*Lotus*	*Biserrula*
Puerarin	0.004	0.033	0.003	0.010	0.000	0.000
Daidzin	0.000	0.024	0.069	0.020	0.003	0.000
Prunetin	0.010	0.008	0.013	0.002	0.031	0.000
Glycitein	0.007	0.020	0.016	0.047	0.021	0.013
Daidzein	0.012	0.033	0.019	0.022	0.038	0.021
Genistin	0.019	0.019	0.114	0.022	0.022	0.065
Bichanin A	0.019	0.004	0.029	0.014	4.258	0.041
Pratensein	0.015	0.044	0.015	0.086	0.248	0.034
Formononetin	0.843	0.007	0.024	0.068	0.023	0.086
Genistein	0.174	0.094	0.028	0.018	0.094	0.080
Irilone	0.271	0.084	0.020	0.046	0.040	0.014
Total	1.37	0.37	0.35	0.35	4.78	0.35

Isoflavone content varied significantly ($P<0.05$)among the genera studied. Overall, Lotus and *Medicago* had the highest content of isoflavones, 4.78 mg/g and 1.37 mg/g respectively. Overall the most abundant isoflavone was biochanin A (0.73 mg/g). Lotus had the highest concentration of biochanin A (4.258 mg/g), pratensein (0.248 mg/g), and prunetin (0.031 mg/g). Medicago had the highest concentration of formononetin (0.843 mg/g), irilone (0.27 mg/g), and genistein (0.174 mg/g). Ornithopus had the highest concentration of genistin (0.114 mg/g) and of daidzin (0.069 mg/g). . *Scorpirius* had the highest content of glycitein (0.047 mg/g). *Trifolium* and *Lotus* had the highest content of daizein (0.033 and 0.038 mg/g, respectively).

References

Arranz S., Saura-Calixto F., Shaha S., Kroon P.A., High Contents of Nonextractable Polyphenols in Fruits Suggest That Polyphenol Contents of Plant Foods Have Been Underestimated. *J. Agric. Food Chem.,57* 7298–7303 (2009).

Bajer T., Adam M., Galla L., Ventura K., Comparison of various extraction techniques for isolation and determination of isoflavonoids in plants. *J. Sep. Sci.*, 30, 122-127 (2007).

Cheng H., Yu O., Yu D., Polymorphisms of IFS1 and IFS2 gene are associated with isoflavone concentrations in soybean seeds. *Plant Science*, 175, 505-512 (2008).

Choi J.S., Kwon T.W., Kim J.S., Isoflavone contents in some varieties of soybean. *Foods Biotechnol*, 5, 167-169 (1996).

Deido W., Clark K.W., Biochanin A and formononetin content in red clover varieties at several maturity stages. Can. J. Plant. Sci., 48, 175-181 (1968).

Delmonte P., Perry J., Rader J.I., Determination of isoflavones in dietary supplements containing soy, Red Clover and kudzu: Extraction followed by basic or acid hydrolysis. *J. Chromatogr. A*, 1107, 59-69 (2006).

Delmonte P., Rader J.I., Analysis of isoflavones in food and dietary supplements. *J. AOAC Intern.*, 89, 1138-1146 (2006).

Doyle J.J., Legume Phylogeny:content for the family, the major groups, and the relationships of crop and Model species, Chap.3 in Legume Crop Genomics, Stalker H.T., Brummer, E. C., Wison R. F. Editors, AOCS Publishing (2004).

Hoeck J.A., Feher W.R., Murphy P.A., Welke G.A., Influence of genotype and environment on isoflavone contents of soybean. *Crop Sci*, 40, 48-51 (2000).

Jung W., Yu O., Lau S-M. C., O'Keefe D.P., Odell J., Fader G., McGonigle B., Identification and expression of isoflavone synthase, the key enzyme for biosynthesis of isoflavones in legumes.*Nature Biotech*, 18, 208-210 (2000).

Kassem M.A., Meksem K., Iqbal M.J., Njiti V.N., Banz W.J., Winters T.A., Wood A., Lightfoot D.A., Definition of soybean genomic regions that control seed phytoestrogen amounts. *J. Biomed. Biotechnol*, 52-60 (2004).

Kim B.G., Kim S-Y, Song H.E., Lee C, Hur H-G, Kim S.-I, Ahn J.-H., Cloning and expression of the isoflevone synthase gene (IFS-tp) from Trifolium pratense. *Mol Cells*, 15, 301-306 (2003).

Kim H.K, Jang Y.H, Baek I.S, Lee JH, Park M.J., Chung Y.S., Chung J.I., Kim J.K.,Polymorphism and expression of isoflavone synthase genes from soybean cultivars. *Mol Cells*, 19, 67-73 (2004).

Klejdus B., Vitamvásová-Sterbová D., Kubán V., Identification of isoflavone conjugates in red clover (Trifolium pratense) by liquid chromatography–mass spectrometry after two-dimensional solid-phase extraction. *Anal. Chim. Acta*, 450, 81-97 (2001).

Krenn L., Quantification of isoflavones in red clover by high-performance liquid chromatography *J. Chromatogr. B*, 777, 123-128 (2002).

Lapčík O., Klejdus B., Kokoška L., Davidová M., Afandi K., Kubáň V., Hampl R., Identification of isoflavones in Acca sellowiana and two Psidium species (Myrtaceae). *Biochem. Systemat. And Ecolog.*, 33, 983-992 (2005).

Lin, L-Z. He X-G., Lindenmaier M., Yang J., Cleary M., Qiu S.-X., Cordell G. A., LC-ESI-MS Study of the Flavonoid Glycoside Malonates of Red Clover (Trifolium pratense). *J. Agric. Food Chem.*, 448, 354-365 (2000).

Liu R, Hu Y, Li J, Lin Z., Production of soybean isoflavone genistein in non-legume plants via genetically modified secondary metabolism pathway. *Metabol Eng*. 9, 1-7 (2007).

Martin L.M., Castilho M.C., Silveira M.I., Abreu J.M., Liquid Chromatographic Validation of a Quantitation Method for Phytoestrogens, Biochanin-A, Coumestrol, Daidzein, Formononetin, and Genistein, in Lucerne. *J. Liq. Chromatogr. Rel. Tech.*, 29, 2875-2884 (2006).

Mazur W., Phytoestrogens: occurrence in foods, and metabolism of ligans in man and pigs, Academic Dissertation, Medical Faculty of the University of Helsinki, Finland (2000).

McMurray C.H., Laidlaw A.S., McElroy M., The effect of plant development and environment on formononetin concentration in red clover (*Trifolium partense* L.). J. Sci. Food Agric. 37, 333-340 (1986).

Messina M., Bennink M., Soyfoods, isoflavones and risk of colonic cancer: a review of the in vitro and in vivo data. *Baillieres Clin. Endocrinol. Metab.* 12, 707-728 (1998).

Peng Y.Y., Ye J.N., Determination of isoflavones in red clover by capillary electrophoresis with electrochemical detection, *Fitoterapia*, 77, 171-178 (2006).

Primono VS, Poysa V, Ablett GR, Jackson CJ, Gijzen M, Rajcan I., Mapping QTL for individual and total isoflavone content in soybean seeds. *Crop Sci*, 45, 2454-2464 (2005).

Ramos G.P., Dias P.M.B., Morais C.B., Fröehlich P.E., Dall'Agnol M., Zuanazzi J.A.S., LC Determination of Four Isoflavone Aglycones in Red Clover (Trifolium pratense L.). *Chromatographia*, 67, 125-129 (2008).

Rijke E., .Aardenburg, Van Dijk J., Ariese F., Ernst H.O., Gooijer C., Brinkman U.A., Changed isoflavone levels in red clover (Trifolium pratense L.) leaves with disturbed root nodulation in response to waterlogging. *J. Chem. Ecology*, 31, 1285-1298 (2005).

Rijke E., Kanter F., Ariese F., Th.Bronkman U.A., Gooiler C., Liquid chromatography coupled to nuclear magnetic resonance spectroscopy for the identification of isoflavone glucoside malonates in T.pratense L. leaves. *J. Sep. Sci.*, 27, 1061-1070 (2004).

Rijke E., Zafra-Gómez A., Ariese F., Th.Brinkman U.A., Gooijer C.. Determination of isoflavone glucoside malonates in Trifolium pratense L. (red clover) extracts: quantification and stability studies. *J. Chromatogr. A*, 932, 55-64 (2001).

Rostagno M.A., Palma M., Barroso C.G.,Ultrasound-assisted extraction of soy isoflavones. *J. Chromatogr. A*, 1012, 119-128 (2003).

Rostagno M.A., Villares A., Guillamón E., García-Lafuente A., Martínez J.A., Sample preparation for the analysis of isoflavones from soybeans and soy foods. *J. Chromatogr. A*, 1216, 2-29 (2009).

Sarelli L., Tuori M., Syrjälä-Qvist L., Saloniemi H., Phytoestrogen contemt of birdsfoot trefoil and red clover: effects of growth stage and ensiling method. Acta Agric. Scand., Sect. A, *Animal Sci.*, 53, 58-63 (2003).

Saviranta N.M.M., Anttonen M.J., Wright A.von, Kajalainen R.O., Red clover (Trifolium pratense L.) isoflavones: determination of concentrations by plant stage, flower colour, plant part and cultivar. *J. Sci. Food Agric.*, 88, 125-132 (2008).

Shih CH, Chen Y, Wang M, Chu IK, Lo C., Accumulation of isoflavone genistin in transgenic tomato plants overexpressing a soybean isoflavone synthase gene. *J Agric Food Chem*. 56, 5655-5661 (2008).

Sivesind E., Seguin Ph., Effects of the environm*ent, cultivar, maturity, and preservation method on red clover isoflavone concentration. J. Agric Food Chem.*, 10, 53, 6397-402 (2005).

Swinny E.E., Ryan K.G.. Red Clover Trifolium pratense L. Phytoestrogens: UV-B Radiation Increases Isoflavone Yield, and Postharvest Drying Methods Change the Glucoside Conjugate Profiles. *J. Agric. Food Chem.*, 53, 8273-8278 (2005).

Toebes A.H., Boer V.de, Verkleij J.A., Lingeman H., Ernst W.H., Extraction of isoflavone malonylglucosides from Trifolium pratense L. *J Agric Food Chem.*, 53, 4660-6 (2005).

Tsao R., Papadopoulos Y., Yang R., Young J.Ch., McRae K., Isoflavone profiles of red clovers and their distribution in different parts harvested at different growing stages. *J. Agric. Food Chem.*, 54, 5797-5805 (2006).

Vetter J., Isoflavones in different parts of common tirfolium species. *J. Agric. Food Chem.*, 43, 106-108 (1995).

Visnevschi-Necrasov T., Cunha S.C., Nunes E., Oliveira M.B.P.P., Optimization of matrix solid-phase dispersion extraction method for the analysis of isoflavones in Trifolium pratense. *J. Chromatogr. A*, 1216, 3720-3724 (2009).

Wong E., Isoflavone contents of red and subterranean clovers. J. Sci. Food. Agric. 14, 376-379 (1963).

Wu Q., Wang M., Simon J.E., Determination of isoflavones in red clover and related species by high-performance chromatography combined with ultraviolet and mass spectrometric detection. *J. Chromatogr. A*, 1016, 195-209 (2003).

Wuttke W., Jarry H., Seidlová-Wuttke D., Isoflavones—Safe food additives or dangerous drugs? *Ageing Research Reviews*, 6, 150-188 (2007).

Xiao H.B., Krucker M., Albert K., Liang X.M., Determination and identification of isoflavones in Radix astragali by matrix solid-phase dispersion extraction and high-performance liquid chromatography with photodiode array and mass spectrometric detection. *J. Chromatogr. A,* 1032, 117-124 (2004).

Xiao H.B., Krucker M., Putzbach K., Albert K., Capillary liquid chromatography-microcoil H nuclear magnetic resonance spectroscopy and liquid chromatography-ion trap mass spectrometry fot on-line structure elucidation of isoflavones in Radix astragali. *J. Chromatogr. A*, 1067, 135-143 (2005).

Yamamoto Y., Gaynor R.B., Therapeutic potential of inhibition of the NF-κB pathway in the treatment of inflammation and cancer. *J. Clin. Invest.* 107, 135-142 (2001).

Yu O., McGonigle B., Metabolic engineering of isoflavone biosynthesis, Advances in Agronomy, 86, pp.148-181, Elsevier Academic Press, USA (2005).

Yu O., Shi J., Hession AO, Maxwell CA, McGonigle B., Metabolic engineering to increase the isoflavone biosynthesis in soybean seed. *Phytochemistry*, 63, 753-763 (2003).

Zernova OV, Lygin AV, Widholm JM, Lozovaya VV., Modification of isoflavones in soybean seeds via expression of multiple phenolic biosynthetic genes. Plant Physiol Biochem, 47, 769-777 (2009).

Zgórka G., Pressurized liquid extraction versus other extraction techniques in micropreparative isolation of pharmacologically active isoflavones from Trifolium L. species. *Talanta*, 79, 46-53 (2009a).

Zgórka G., Ultrasound-assisted solid-phase extraction coupled with photodiode-array and fluorescence detection for chemotaxonomy of isoflavone phytoestrogens in Trifolium L. (Clover) species. *J. Sep. Sci.*, 32, 965-972 (2009b).

Zhang Y., Chen J., Zhao L., Shi Y-P., Separation and determination of isoflavones in red clover by micellar electrokinetic capillary chromatography. *Biomed. Chromatogr.* 21, 987-992 (2007).

Index

B

C

D

E

F

G

H

I

K

L

M

N

O

P

Q

R

S

T

U

V

W

X

Y

Z